ELECTROCHEMICAL SCIENCE AND TECHNOLOGY
Fundamentals and Applications

ELECTROCHEMICAL SCIENCE AND TECHNOLOGY
Fundamentals and Applications

Keith B. Oldham

Trent University
Peterborough, Ontario, Canada
KOldham@trentu.ca

Jan C. Myland

Trent University
Peterborough, Ontario, Canada
JMyland@trentu.ca

Alan M. Bond

Monash University
Melbourne, Victoria, Australia
Alan.Bond@monash.edu

A John Wiley & Sons, Ltd., Publication

Registered office

John Wiley & Sons Ltd, The Atrium, Southern Gate, Chichester, West Sussex, PO19 8SQ, United Kingdom

For details of our global editorial offices, for customer services and for information about how to apply for permission to reuse the copyright material in this book please see our website at www.wiley.com.

Library of Congress Cataloging-in-Publication Data

Oldham, Keith B.
  Electrochemical science and technology: fundamentals and applications / Keith B. Oldham, Jan C. Myland, Alan M. Bond.
     p. cm.
  Includes bibliographical references and index.
  ISBN 978-0-470-71085-2 (cloth) -- ISBN 978-0-470-71084-5 (pbk.) -- ISBN 978-1-119-96588-6 (ePDF) -- ISBN 978-1-119-96599-2 (oBook) -- ISBN 978-1-119-96684-5 (ePub.) -- ISBN 978-1-119-96685-2 (Mobi)
  1. Electrochemistry. I. Myland, Jan C. II. Bond, A. M. (Alan Maxwell), 1946- III. Title.
  QD553.O42 2011
  541'.37--dc23

                              2011037230

A catalogue record for this book is available from the British Library.

Print ISBN: 9780470710852 (HB)

Print ISBN: 9780470710845 (PB)

Set in 11pt Times New Roman by the authors using WordPerfect

Contents

Index 393

Preface

This book is addressed to all who have a need to come to grips with the fundamentals of electrochemistry and to learn about some of its applications. It could serve as a text for a graduate, or senior undergraduate, course in electrochemistry at a university or college, but this is not the book's sole purpose.

The text treats electrochemistry as a scientific discipline in its own right, not as an offshoot of physical or analytical chemistry. Though the majority of its readers will probably be chemists, the book has been carefully written to serve the needs of scientists and technologists whose background is in a discipline other than chemistry. Electrochemistry is a quantitative science with a strong reliance on mathematics, and this text does not shy away from the mathematical underpinnings of the subject.

To keep the size and cost of the book within reasonable bounds, much of the more tangential material has been relegated to "Webs" – internet documents devoted to a single topic – that are freely accessible from the publisher's website at www.wiley.com/go/EST. By this device, we have managed largely to avoid the "it can be shown that" statements that frustrate readers of many textbooks. Other Webs house worked solutions to the many problems that you will find as footnotes scattered throughout the pages of *Electrochemical Science and Technology*. Another innovation is the provision of Excel® spreadsheets to enable the reader to construct accurate cyclic (and other) voltammograms; see Web#1604 and Web#1635 for details.

It was in 1960 that IUPAC (the International Union of Pure and Applied Chemistry) officially adopted the *SI* system of units, but electrochemists have been reluctant to abandon centimeters, grams and liters. Here, with some concessions to the familiar units of concentration, density and molar mass, we adopt the *SI* system almost exclusively. IUPAC's recommendations for symbols are not always adhered to, but (on pages 195 and 196) we explain how our symbols differ from those that you may encounter elsewhere. On the same pages, we also address the thorny issue of signs.

Few references to the original literature will be found in this book, but we frequently refer to monographs and reviews, in which literature citations are given. We recommend Chapter IV of F. Scholz (Ed.), *Electroanalytical Methods: guide to experiments and applications 2E*, Springer, 2010, for a comprehensive listing of the major textbooks, monographs and journals that serve electrochemistry.

The manuscript has been carefully proofread but, nevertheless, errors and obscurities doubtless remain. If you discover any such anomalies, we would appreciate your bringing it to our attention by emailing Alan.Bond@monash.edu. A list of errata will be maintained on the book's website, www.wiley.com/go/EST.

Electrochemical Science and Technology: fundamentals and applications has many shortcomings of which we are aware, and doubtless others of which we are ignorant, and for which we apologize. We are pleased to acknowledge the help and support that we have received from Tunde Bond, Steve Feldberg, Hubert Girault, Bob de Levie, Florian Mansfeld, David Rand, members of the Electrochemistry Group at Monash University, the Natural Sciences and Engineering Research Council of Canada, the Australian Research Council, and the staff at Wiley's Chichester office.

July 2011   Keith B.Oldham
                                                              Jan C. Myland
                                                              Alan M. Bond

1

Electricity

At the heart of electrochemistry lies the coupling of chemical changes to the passage of electricity. The science of electricity is a branch of physics, but here we start our study of electrochemistry by reviewing the principles of electricity from a more chemical perspective.

Electric Charge: the basis of electricity

Charge is a property possessed by matter. It comes in two varieties that we call **positive charge** and **negative charge**. The salient property of electric charges is that those of opposite sign attract each other, while charges of like sign repel, as illustrated in Figure 1-1.

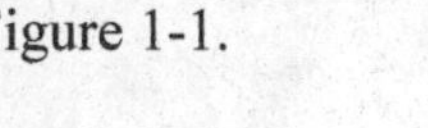

Figure 1-1 Charges of unlike sign attact each other, those of like sign repel.

Charge is measured in **coulombs**, C, and it occurs as multiples of the **elementary charge**

$$1{:}1 \qquad\qquad Q_0 = 1.6022 \times 10^{-19}\,\text{C} \qquad \text{elementary charge}$$

Charge is not found in isolation, it always accompanies matter. Such fundamental particles as the proton H^+ and the electron e^- possess single charges, that is $\pm Q_0$, as do many **ions**[101] such as the sodium Na^+, chloride Cl^-, and hydronium H_3O^+ ions. Other ions, such as the magnesium Mg^{2+} cation and phosphate PO_4^{3-} anion are multiply charged. Even neutral molecules, which have no net charge, are held together electrically and frequently have charges on their surfaces. For example, one side of the water molecule pictured in Figure 1-2 has a negative region, the other side being positively charged. Such structures,

---

[101] Ions are charged atoms or groups of atoms; if positively charged, they are called **cations**, whereas **anion** is the name given to a negative ion.

Electrochemical Science and Technology: Fundamentals and Applications, First Edition. Keith B. Oldham, Jan C. Myland, Alan M. Bond.
© 2012 John Wiley & Sons, Ltd. Published 2012 by John Wiley & Sons, Ltd.

called **dipoles**[102], behave as if they contain small (generally less than Q_0) localized positive and negative charges separated by a small distance.

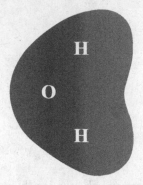

Figure 1-2 The architecture of the **dipolar** water molecule. The red and blue surface regions are charged positively and negatively respectively.

Ions and electrons are the actors in the drama of electrochemistry, as are molecules. Most often these charged particles share the stage and interact with each other, but in this chapter we mostly consider them in isolation. The electrical **force**, f, between two charges Q_1 and Q_2 is independent of the nature of the particles on which the charges reside. With r_{12} as the distance between the two charges, the force[103] obeys a law

1:2
$$f = \frac{Q_1 Q_2}{4\pi\varepsilon r_{12}^2} \qquad \text{Coulomb's law}$$

attributed to Coulomb[104]. The *SI* unit of force is the **newton**[105], N. Here ε is the **permittivity** of the medium, a quantity that will be discussed further on page 13 and which takes the value

1:3
$$\varepsilon_0 = 8.8542 \times 10^{-12} \, C^2 \, N^{-1} \, m^{-2} \qquad \text{permittivity of free space}$$

when the medium is free space[106]. The force is repulsive if the charges have the same sign, attractive otherwise. To give you an idea of the strong forces involved, imagine that all the Na^+ cations from 100 grams of sodium chloride were sent to the moon, then their attractive force towards the earthbound chloride anions probably exceeds your weight[107].

A consequence of the mutual repulsion of two or more similar charges is that they try to get as far from each other as possible. For this reason, the interior of a phase[108] is usually free of net charge. Any excess charge present will be found on the surface of the phase, or very close to it. This is one expression of the **principle of electroneutrality**.

---

[102] Read more at Web#102 about the water dipole and **dipole moments**.

[103] Calculate the repulsive force (in newtons) between two protons separated by 74.14 pm, the internuclear distance in the H_2 molecule. See Web#103 to check your result.

[104] Frenchman Charles Augustin de Coulomb, 1736–1806, first confirmed the law experimentally.

[105] Sir Isaac Newton, 1643–1727, renowned English scientist.

[106] The permittivity of free space, ε_0, is also known as the **electric constant**.

[107] Justify this statement. See Web#107.

[108] A **phase** is a region of uniform chemical composition and uniform physical properties.

Charges at Rest: electric field and electrical potential [109]

Coulomb's law tells us that an electric charge can make its presence felt at points remote from its site. An **electric field** is said to exist around each charge. The electric field is a vector; that is, it has both direction and strength. Figure 1-3 shows that the field around an isolated positive charge points away from the charge, at all solid angles.

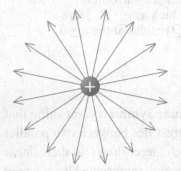

Figure 1-3 The field created by a positive charge is directed away from the charge in all three-dimensional directions, the converse being true for negative charges.

The strength of an electric field at a point can, in principle, be assessed by placing a very small positive "test charge" Q_{test} at the point. The choice of a sufficiently *small* test charge ensures that the preexisting field is not disturbed. The test charge will experience a small **coulombic force**. The **electric field strength**[110], or more simply the **field**, X, is then defined as the quotient of the force by the test charge:

1:4
$$X = \frac{f}{Q_{test}}$$
definition of field

and therefore it has the unit[111] of newtons per coulomb, $N\,C^{-1}$. Thus, for any static charge distribution, it is possible to calculate field strengths using Coulomb's law[112].

Force, and therefore also electric field, is a vector quantity. In this book, however, we shall avoid the need to use vector algebra by addressing only the two geometries that are of paramount importance in electrochemistry. These two geometries are illustrated in Figures 1-4 and 1-5. The first has **spherical symmetry**, which means that all properties are uniform on any sphere centered at the point $r = 0$. Thus, there is only one spatial coordinate to consider; any property depends only on the distance r, where $0 \le r < \infty$. The

---

[109] "Electric" and "electrical" are adjectives of identical meaning. A quirk of usage is that we usually speak of electric field and electric charge but electrical potential and electrical conductivity.

[110] Physicists use E for field strength, but traditionally electrochemists reserve that symbol for potential, a holdover from the antiquated term "electromotive force".

[111] or, equivalently and more commonly, volt per meter, $V\,m^{-1}$. See equation 1:9 for the reason.

[112] Consider two protons separated by 74.14 pm, the internuclear distance in the H_2 molecule. Find the field strength and direction at points 25%, 50% and 75% along the line connecting the protons. For a greater challenge, find the field at some point not on the line of centers. See Web#112.

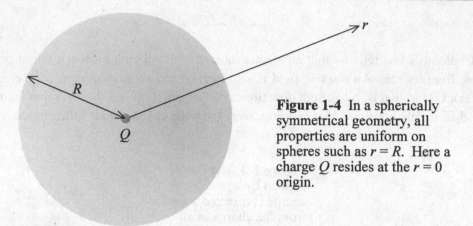

Figure 1-4 In a spherically symmetrical geometry, all properties are uniform on spheres such as $r = R$. Here a charge Q resides at the $r = 0$ origin.

second geometry of prime electrochemical concern has **planar symmetry**, meaning that uniformity of properties exists in planes. The space of interest lies between two parallel planes separated by a distance, L, the planes being very much larger than L in their linear dimensions. Again, there is only one coordinate to consider, now represented by x, where $0 \leq x \leq L$. Each of these two geometries is simple in that there is only one relevant distance coordinate. Thus, when we discuss the field, we mean implicitly the field strength in the direction of increasing r or x.

Coulomb's law tells us that the electric field strength falls off with distance according to the **inverse-square law**: at double the distance from a point source the field is

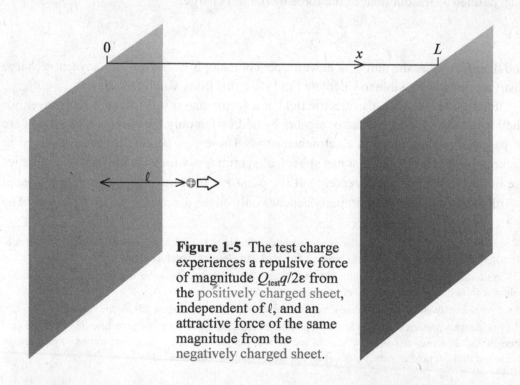

Figure 1-5 The test charge experiences a repulsive force of magnitude $Q_{test}q/2\varepsilon$ from the positively charged sheet, independent of ℓ, and an attractive force of the same magnitude from the negatively charged sheet.

one-fourth. Thus, it is evident that at a distance R from the point charge Q

1:5 $$X(R) = \frac{f}{Q_{\text{test}}} = \frac{Q}{4\pi\varepsilon R^2} \qquad \text{inverse-square law}$$

The field is uniform at all points on the sphere shown in Figure 1-4, falling off as $1/R^2$. The inverse-square law does not apply to the field in planar symmetry. In that geometry, electrochemists are interested in the field between two charged planes, such as electrodes. In Figure 1-5 the left-hand plane is uniformly charged such that the **charge density** (measured in coulombs per square meter, C m^{-2}) is q. The field strength caused by that plane, at a distance ℓ, can be shown[113] to be simply $X(\ell) = q/[2\varepsilon]$. Taking into account the second, oppositely charged, plane, the total field is

1:6 $$X(\ell) = \frac{q}{\varepsilon} \qquad \text{planar symmetry}$$

Provided that the charged sheets are large enough and parallel, the adjacent field doesn't depend on location. The field strength[114] is constant!

Figure 1-6 A test charge moves a short distance δr from point A to point B towards the source of an electric field. It experiences a field of strength X acting in the direction of increasing r.

The concept of a small "test charge" is a valuable fiction; it is also used to define electrical potential. Imagine that we place a test charge at point A, and move it a small distance δr towards a much larger fixed charge as in Figure 1-6. It needs the expenditure of **work** $w_{A \to B}$ for the test charge to reach its destination, point B. Work (measured in **joules**[115], J) can be calculated as *force* × *distance* or, in this case:

1:7 $$w_{A \to B} = f \times [-\delta r] = -Q_{\text{test}} X \delta r$$

The negative sign arises because the journey occurs in the negative r direction. It is said that an **electrical potential** exists at each of points A and B and we define the difference between these potentials as the coulombic work needed to carry a test charge between the two points divided by the magnitude of the test charge. Hence,

---

[113] See Web#113 for the derivation from Coulomb's law. It involves an integration in polar coordinates.

[114] Two square metal plates, each of an area of 6.25 cm^2 are separated by 1.09 cm. They are oppositely charged, each carrying 2.67 nC. The pair is immersed in acetonitrile, a liquid of permittivity 3.32×10^{-10} C^2 N^{-1} m^{-2}. Calculate the field (strength and direction) at a point 500 µm from the negatively charged surface. Check your answer at Web#114.

[115] James Prescott Joule, 1818 – 1889, English scientist and brewer.

1:8
$$\phi_B - \phi_A = \frac{w_{A \to B}}{Q_{test}} = -X \delta r$$

Notice that the definition defines only the *difference* between two potentials, and not the potential, ϕ, itself[116,117]. In differential notation, the equation becomes[118]

1:9
$$\frac{d\phi}{dr} = -X \qquad \text{definition of potential}$$

The unit of electrical potential is the **volt**[119]. The first equality in equation 1:8 shows that one volt equals one joule per coulomb ($V = J \, C^{-1}$).

The situation depicted in Figure 1-6 is simple because the distance moved, in that case, was small enough that the field could be treated as constant. For a longer journey one finds, making use of equation 1:5,

1:10
$$\phi_B - \phi_A = -\int_A^B X(r)dr = \frac{-Q}{4\pi\varepsilon} \int_{r_A}^{r_B} \frac{dr}{r^2} = \frac{Q}{4\pi\varepsilon}\left(\frac{1}{r_B} - \frac{1}{r_A}\right) \qquad \begin{array}{l} \text{spherical} \\ \text{symmetry} \end{array}$$

Moreover, the situation depicted in Figure 1-6 is especially simple in that the journey was along a radial direction. A geometry like that in Figure 1-7 is more general. The force on the moving test charge now varies along the journey, not only because the field strength changes, but also because the angle θ constantly alters as the charge moves. The potential difference between points A and B in this geometry can be calculated from the following chain of equalities

1:11
$$\phi_B - \phi_A = \frac{w_{A \to B}}{Q_{test}} = \frac{-1}{Q_{test}} \int_A^B f \cos\{\theta\} \, d\ell = -\int_A^B X \cos\{\theta\} \, d\ell$$

Both X and θ change as the distance ℓ traveled by the test charge increases. Remarkably, the result of the integration does not depend on the route that the test charge travels on its journey from A to B. The work, and therefore the potential change, is exactly the same for the direct route as for the circuitous path via point C in Figure 1-7, and this fact greatly simplifies the calculation of the potential difference[120]. In fact, equation 1:10 applies.

Equation 1:9 shows the electric field strength to be the negative of the gradient of the electrical potential. In electrochemistry, electrical potential is a more convenient quantity than electric field, in part because it is not a vector. It does have the disadvantage, though,

---

[116] Calculate the electrical potential difference between the 75% and 50% points in the problem cited in Footnote 112. Which is the more positive? Check your answer at Web#116.

[117] Refer to the problem in Footnote 114 and find the potential difference between the point cited and a point at the surface of the nearby electrode. Web#117 has the answer.

[118] From Coulomb's law derive an expression for the electrical potential difference between a point at a distance r from an isolated proton and a point at infinity. Our derivation will be found at Web#118.

[119] Alessandro Guiseppe Antonio Anastasio Volta, 1745–1827, Italian scientist.

[120] See Web#120 for the derivation of the potential change accompanying the $A \to B$ journey in Figure 1-7.

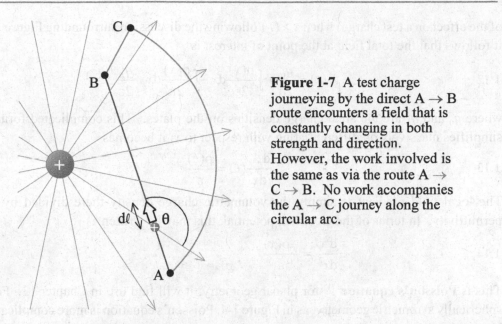

Figure 1-7 A test charge journeying by the direct A → B route encounters a field that is constantly changing in both strength and direction. However, the work involved is the same as via the route A → C → B. No work accompanies the A → C journey along the circular arc.

of being a relative, rather than an absolute quantity. For this reason we more often encounter the symbol $\Delta\phi$ rather than ϕ itself. In this book the phrase **electrical potential difference** will often be replaced by the briefer term **voltage**.

We can define only *differences* in electrical potential. Worse, we can define differences in the electrical potential only between points that lie within phases of the *same* (or very similar) compositions. The essential reason for this is that we do not have innocuous "test charges" at our disposal. We only have electrons, protons and ions. So when we try to measure the coulombic work in moving such charged particles from one phase to another, we inevitably encounter other energy changes arising from the change in the chemical environment in which the particle finds itself. Such **chemical work** is absent only if the departure and arrival sites have similar chemical compositions.

The electroneutrality principle prevents charges accumulating, other than near boundaries. A need exists to investigate the distribution of charge in spaces adjacent to boundaries, because it is at the junctions of phases that electrochemistry largely occurs. The symbol ρ is generally used to represent **volumetric charge density** (unit, coulombs per cubic meter, $C\ m^{-3}$). Do not confuse this quantity with q, the *areal* charge density ($C\ m^{-2}$).

Let us first consider the presence of space charge in the geometry of Figure 1-5. Imagine that, in addition to areal charge densities on the plates, there is a space charge in the region between the plates, its magnitude being $\rho(x)$ at any distance x from the left-hand plate. We seek the field at some point $x = \ell$. The space charge can be regarded as being made up of many thin wafers, each of areal charge density $\rho(x)dx$. Each of these will contribute to the field, positively for the thin wafers to the left of $x = \ell$, negatively (think of the effect on a test charge) when $x > \ell$. Following the discussion surrounding Figure 1-5, it follows that the total field at the point of interest is

1:12
$$X(\ell) = \frac{q_0}{2\varepsilon} + \int_0^\ell \frac{\rho(x)}{2\varepsilon}\,dx - \int_\ell^L \frac{\rho(x)}{2\varepsilon}\,dx - \frac{q_L}{2\varepsilon}$$

where q_0 and q_L are the areal charge densities on the plates. This complicated formula simplifies massively on differentiating with respect to x; it becomes

1:13
$$\frac{dX}{dx}(\ell) = \frac{\rho(\ell)}{\varepsilon}$$

The local field gradient is simply the volumetric charge density there divided by the permittivity. In terms of the electrical potential, this may be written

1:14
$$\frac{d^2\phi}{dx^2} = \frac{-\rho(x)}{\varepsilon} \qquad \begin{array}{l} \text{Poisson's equation} \\ \text{planar symmetry} \end{array}$$

This is **Poisson's equation**[121] for planar geometry; it will find use in Chapter 13. For a spherically symmetric geometry, as in Figure 1-4, Poisson's equation is more complicated, but can be shown[122] to be

1:15
$$\frac{1}{r^2}\frac{d}{dr}\left\{r^2\frac{d\phi}{dr}\right\} = \frac{-\rho(r)}{\varepsilon} \qquad \begin{array}{l} \text{Poisson's equation} \\ \text{spherical symmetry} \end{array}$$

This law finds application in the Debye-Hückel theory discussed in the next chapter.

Capacitance and Conductance: the effects of electric fields on matter

Materials may be divided loosely into two classes: electrical **conductors** that allow the passage of electricity, and **insulators** that do not. The physical state is irrelevant to this classification; both classes have examples that are solids, liquids and gases. Conductors themselves fall into two main subclasses according to whether it is electrons or ions that are the **charge carriers** that move in response to an electric field.

$$\text{materials}\begin{cases} \text{insulators} \\ \text{conductors}\begin{cases} \text{electronic conductors} \\ \text{ionic conductors} \end{cases} \end{cases}$$

Electronic conductors owe their conductivity to the presence of **mobile electrons**. All **metals** are electronic conductors, but some solid inorganic oxides and sulfides (e.g.

---

[121] Siméon-Denis Poisson, 1781–1840, French mathematical physicist.

[122] This is derived in Web#122, preceded by an important lemma.

PbO_2 and[123] Ag_2S) also conduct electricity by virtue of electron flow. These, and most other **semiconductors**[124], owe their conductivity to an excess (**n-type**) or a deficit (**p-type**) of electrons compared with the number required to form the covalent bonds of the semiconductor's crystal lattice. In p-type semiconductors, the missing electrons are known as **holes** and solid-state physicists speak of the conductivity as being due to the motion of these positively-charged holes. Of course, it is actually an electron that moves into an existing hole and thereby creates a new hole at its former site. **Pi-electrons**[125] are the charge carriers in some other materials, of which graphite is the best known, but which also include newer synthetic **conductive polymers**. An example is the cationic form of poly-pyrrole, which conducts by motion of π-electron holes, through a structure exemplified by

Certain crystalline organic salts[126], known as **organic metals**, also conduct by virtue of π-electron motion. Yet another exotic electronic conductor is the tar-like material (see page 94 for an application) formed when the polymer of 2-vinylpyridine reacts with excess iodine to form a so-called "charge transfer compound".

The second class of materials that conduct electricity comprises the **ionic conductors**, which possess conductivity by virtue of the motion of anions and/or cations. Solutions of electrolytes (salts, acids and bases) in water and other liquids are the most familiar examples of ionic conductors, but there are several others. **Ionic liquids** resemble electrolyte solutions in that the motion of both anions and cations contributes to their electrical conductivity[127]: an example is 1-butyl-3-methylimidazolium hexafluorophosphate,

typical
ionic liquid

---

[123] These two solids, and others, are slightly "nonstoichiometric". **Stoichiometric compounds** contain two or more elements in atom ratios that are stricly whole numbers. Water for example, contains *exactly* twice as many H atoms as O atoms. **Nonstoichiometric solids**, in contrast, depart often only slightly from this whole-number rule. Such an abnormality may occur naturally, being associated with defects in the crystal lattice, or be introduced artificially by admixture with a small quantity of a **dopant**.

[124] Read more about semiconductor conductivity at Web#124.

[125] In organic compounds having alternating single and double bonds in chains or rings, the electrons confer unusual properties, including abnormal electrical conductivity. Such electrons are described as "pi-electrons".

[126] Rich in π-electrons, the organic compound tetrathiafulvalene (TTF) readily forms the cation TTF^+. Another organic compound, tetracyanoquinodimethane (TCNQ), conversely forms the anion $TCNQ^-$. Accordingly, a mixture of these two compounds is in equilibrium with the salt $(TTF^+)(TCNQ^-)$ and the mixture, known as an "organic metal", has high electrical conductivity.

[127] Conductivities of various materials, including a similar ionic liquid are listed in the table on page 384.

Ionic liquids are, in fact, **molten salts**, but inorganic salts generally have much high melting points and conduct only at elevated temperatures. **Solid ionic conductors**[128], on the other hand, usually have only one mobile ionic species that may be either an anion (as in zirconia, ZrO_2, which, at high temperatures, allows oxide ions, O^{2-}, to migrate through its lattice[129]) or a cation (as in silver rubidium iodide, $RbAg_4I_5$, in which Ag^+ is mobile even at room temperature). An interesting case is provided by lanthanum fluoride, LaF_3, crystals that have been "doped" by a very small addition of europium fluoride, EuF_2. Because the **dopant** contributes fewer F^- ions to the lattice than its host, the crystal has "fluoride ion holes" which can move exactly as do electron holes in p-type semiconductors. Such crystals find applications in the fluoride ion sensor described on page 121.

A few materials permit the flow of electricity by both electronic and ionic conduction. An example of such mixed conduction is provided by the hot gases known as **plasmas**, which contain positive ions and free electrons[130]. A second example is the solution formed when sodium metal dissolves in liquid ammonia. Such a solution contains sodium Na^+ cations and solvated electrons (see page 41), both of which are mobile and share duties as charge carriers. Yet another example of mixed conduction is provided by hydrogen dissolved in palladium metal; here there is conduction by the migration of protons (hydrogen ions) as well as by electrons. In summary:

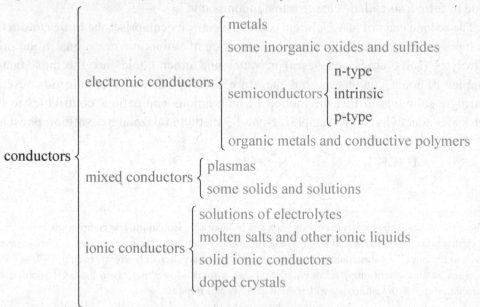

Though we shall not go into details, technological devices exist that produce a constant

---

[128] also known as **solid electrolytes**, but this can be a misleading name.

[129] See pages 68 and 173 for an application. Zirconia exists in two forms, only one of which allows oxide ion motion; to stabilize this form, a small quantity of yttrium is added.

[130] Fluorescent lights and "neon lights" are examples.

difference of electrical potential. Such a device is called a **voltage source** and it has two terminals, one of which (often colored red) is at a more positive electrical potential than the other. There are other devices, named **voltmeters**, that can measure electrical potential differences. Both these devices are electronic; that is, they produce or measure an electrical potential difference by virtue of a deficit of electrons on their red terminals compared with the other. We do *not* have devices able directly to produce or measure deficits or excesses of other charged species, such as protons or ions, so studies on these latter charge carriers are conducted through the medium of electronic devices. Much of the later content of this book is devoted to experiments carried out to investigate the behavior of *ions*, via measurements made with *electronic* devices.

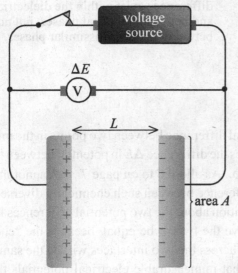

Figure 1-8 Parallel plates store electric charge, and retain the charge when the switch is opened.

Figure 1-8 shows a voltage source connected by wires and a switch to a voltmeter[131] and to a pair of parallel metal sheets, often called **plates**. On closing the switch, a brief surge of electrons occurs and causes charges to appear on the plates. Because of the electroneutrality principle, electrons arrive on the inward-facing surface of the right-hand metal plate. There is a complementary withdrawal of electrons from the inward-facing surface of the left-hand metal plate, leaving a positive charge on that surface.

We have seen in equation 1:6 that such a parallel distribution of charges produces a uniform electric field of strength $X = q/\varepsilon$ in the space between the plates. Here ε is the permittivity of the medium between the plates and, if this is air, it differs only marginally from ε_0. The direction of the field is towards the negative plate, rightwards in Figure 1-8. To carry a test charge a distance L, from a point adjacent to the negatively charged plate to a second point adjacent to the positive plate, will require work w equal to $XQ_{test}L$ or $qQ_{test}L/\varepsilon$ and, accordingly, the potential difference

---

[131] An ideal voltmeter measures the voltage between its terminals while preventing any charge flow through itself. Modern voltmeters closely approach this ideal.

1:16
$$\Delta\phi = \phi_{\text{close to positive plate}} - \phi_{\text{close to negative plate}} = \frac{w}{Q_{\text{test}}} = \frac{qL}{\varepsilon}$$

exists between the destination and starting points. This shows, reasonably, that the medium close to the positive plate is at a more positive electrical potential than is the medium close to the negative plate. And of course, because the field is uniform, the potential changes linearly with distance between the two points as illustrated in Figure 1-9.

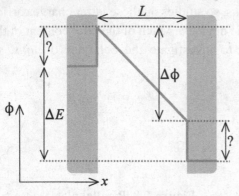

Figure 1-9 Measurable potential differences exist within the dielectric and between the metal phases, but not between points in dissimilar phases.

The $\Delta\phi$ in equation 1:16 is the potential difference between two points in the medium. The voltmeter shown in Figure 1-8 measures the difference ΔE in potential between the two metal plates. We now assert that $\Delta E = \Delta\phi$. As alluded to on page 7, we cannot measure (or even usefully define!) the potential difference between such chemically diverse media as metal and air, and so we have no information about the two potential differences labeled "?" in Figure 1-9. However, we do believe the two to be equal, because the "chemical work" in transferring any charged particle across the two interfaces will be the same. We shall continue to follow the convention that immeasurable electrical potentials that are imagined to exist within phases are symbolized ϕ, whereas electrical potential differences that can be measured by a voltmeter are denoted ΔE.

Return to Figure 1-8 and note that, on reopening the switch, the charges remain on the plates; electric charge is stored. A device, such as the parallel plates[132] just described, that is able to store electric charge is called a **capacitor**. The stored charge is

1:17
$$Q = \frac{-A\varepsilon}{L}\Delta E$$

The ratio of the potential difference across a capacitor to the charge it stores is called the **capacitance** of the capacitor and is given the symbol C

1:18
$$\frac{-Q}{\Delta E} = C = \frac{\varepsilon A}{L}$$
definition of capacitance

planar symmetry

---

[132] A capacitor can have a shape other than the parallel-plate configuration discussed here. Read about another important capacitor – the isolated sphere – at Web#132.

The unit of capacitance[133] is the **farad**, F. One farad equals one coulomb per volt. The negative sign in the last two equations, which you will often find missing from other texts, arises because positive charge flowing into a capacitor produces a negative charge on the remote plate.

We now turn to discuss what happens when an insulator is placed between the parallel plates of a capacitor. Equations 1:17 and 1:18 still apply, with ε becoming the permittivity[134] of the insulator. Permittivities vary greatly, as evidenced in the table on page 382. Notice that the listed permittivities[135] always exceed ε_0 so that the capacitor now has a larger capacitance and stores more charge[136] for a given voltage. The explanation for this is especially easy to understand when the insulator has a dipolar molecule, such as the organic liquid, acetonitrile, CH_3CN. Like the water molecule shown in Figure 1-2, this molecule has a positive end and a negative end and, in an electric field, such molecules tend to align themselves as illustrated in Figure 1-10. The effect is to create localized fields within the insulator that oppose, and partially neutralize, the imposed field, so that more external charge is required to reach the applied voltage ΔE. Insulators that behave in this, and similar[137], ways are often called **dielectrics**.

Figure 1-10 In an electric field, dipoles become aligned, to some extent, so that the dipole field opposes the field applied by the plates.

It is quite a different story if we place an electronic conductor between the plates. The electrons are now able to pass freely from the negative plate into the conductor, and from the conductor into the positive plate, as in Figure 1-11. Electrons being negative, their passage from right to left through the conductor in this figure corresponds to electric charge flowing from left to right. We say an **electric current**, I, flows through the conductor: it

---

[133] Copper sheets, each of 15.0 cm² area are separated by an air gap of 1.00 mm. Calculate the capacitance, checking at Web#133. Also find the charge densities on the plates and the field within the capacitor, when 1000 V is applied.

[134] Confusingly ε is sometimes used to denote **relative permeability** (or **dielectric constant** or **dielectric coefficient**), which is the ratio of the permittivity of the material in question to the permittivity of free space.

[135] These are listed in farads per meter, the conventional unit of permittivity. Show that this is equivalent to the unit given in equation 1:3. See Web#135.

[136] and also more energy. The energy stored by a capacitor is $Q\Delta E/2$. Can you explain the reason for the divisor of 2? If not, consult Web#136.

[137] Even if the material is not dipolar, the presence of the imposed field can induce a temporary dipole. Such **polarizability** is how a material such as tetrachloromethane, CCl_4, exhibits an elevated permittivity.

expresses the rate at which charge passes through the conductor:

$$1:19 \qquad I = \frac{dQ}{dt} \qquad \text{definition of current}$$

Electric current is measured in the **ampere**[138] unit, one ampere corresponding to the passage of one coulomb in a time t of one second[139] ($A = C \cdot s^{-1}$). The flow of electricity through the conductor is continuous, unlike the case of an insulator, in which there is only a brief transient passage of electricity.

Of course, the same flow of electricity that occurs in the conductor is also experienced in the wires and plates that constitute what is known as the **circuit**, the pathway through which the charge flows. That is why an **ammeter**, an electronic device[140] that measures electric current, can be positioned, as in Figure 1-11, remote from the conductor and yet measure the current flowing through it. An important quantity, equal to the current divided by the cross-sectional area through which it flows, is the **current density**, i:

$$1:20 \qquad i = \frac{I}{A} \qquad \text{definition of current density}$$

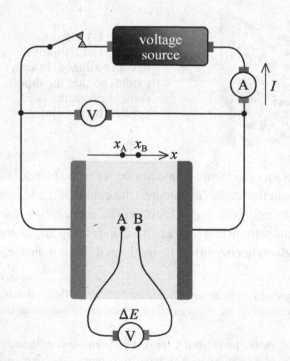

Figure 1-11 Arrangement for measuring the conductivity of an electronic conductor. The method is sometimes called the **4-terminal method** because there are four connections to the conductor. The sample of conductor is of length L and cross-sectional area A.

---

[138] André Marie Ampère, 1775 – 1836, French physicist.

[139] Some pacemaker batteries (see Chapter 5) are required to generate 29 μA of electricity at 2.2 V and to operate reliably for 8 years. What is total charge delivered in the battery's lifetime? How many electrons is that? What energy is liberated? What average power? See Web#139.

[140] An ideal ammeter measures the current flow without producing any voltage across its terminals. Modern ammeters closely approach this ideal.

It is measured in amperes per square meter, A m^{-2}. Unlike the current itself, the current density does differ in different sections of a circuit.

The ratio of the current density flowing in a conductor to the field that creates it is called the **conductivity** κ of the material[141]

1:21
$$\kappa = \frac{i}{X} = \frac{I/A}{-\Delta\phi/L}$$
definition of conductivity

planar symmetry

It has the unit (A m^{-2})/(V m^{-1}) = A V^{-1} m^{-1} = S m^{-1}. S symbolizes the **siemens**[142] unit.

One way in which the conductivity of an electronically conducting material can be measured[143] is illustrated in Figure 1-11. A voltmeter measures the electrical potential difference ΔE between two points A and B on the conductor, which has a uniform cross-sectional area A, and through which a known current is flowing. Then, using 1:9,

1:22
$$\kappa = \frac{i}{X} = \frac{I/A}{-d\phi/dx} = \frac{I(x_B - x_A)}{-A\Delta E}$$

The relationship $i = \kappa X$ is one form of **Ohm's law**[144]. Another is $-\Delta E/I = R$ which defines the **resistance** R, measured in the **ohm** (Ω) unit, equal to S^{-1}. The second equality in the formula

1:23
$$R = \frac{-\Delta E}{I} = \frac{L}{\kappa A}$$
definition of resistance

planar symmetry

applies only to conductors of a simple cuboid[145,146] or cylindrical shape. The resistances of conductors of these and some other geometries are addressed in Chapter 10.

The negative sign in the three most recent equations arose because the current flow occurred in the direction of the coordinate in use. However, Ohm's law is often used in contexts in which there is no clear coordinate direction and accordingly the equation $\Delta E = IR$ is often written, in this book and elsewhere, without a sign. The issue originates because, strictly, I is a vector whereas ΔE and R are not. Exactly the same ambiguity arises in equations 1:17 and 1:18. Just remember that the flow of current through a resistor or capacitor is accompanied by a *decrease* in electrical potential.

Thus far, we have investigated what happens when we apply an electric field to an

---

[141] The reciprocal $1/\kappa$ of the conductivity is known as the **resistivity**. Conduct*ivity* and resist*ivity* are properties of a *material*. In contrast conduct*ance* and resist*ance* are properties of a particular *sample* of material.

[142] Ernst Werner von Siemens, 1816 – 1892, German engineer.

[143] Another uses alternating current, as discussed on page 104.

[144] Georg Simon Ohm, 1789 – 1854, German physicist.

[145] Show that the resistance, measured between opposite edges of a square of a thin conducting film, does not depend on the size of the square. For this reason, the resistances of thin films are often expressed in **ohms per square**. A thin copper film has a resistance of 0.6 Ω per square. Using the table entry on page 385, find its thickness. Compare you answer to that in Web#145.

[146] Find the resistance between opposite faces of a cube of pure water of edge length 1.00 cm. See Web#146.

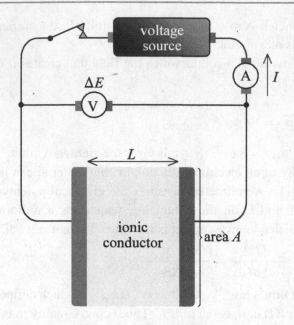

Figure 1-12 In the absence of chemical reaction, current flows transiently when a field is applied to an ionic conductor.

insulator or to an electronic conductor. What occurs when an electric field is applied to an ionic conductor, for example by applying a voltage between two plates that sandwich the conductor? Often a chemical reaction occurs and we enter the realm of electrochemistry. Sometimes, however, if the applied voltage ΔE is small enough, conditions[147] are such that no chemical reaction can occur. In such circumstances, when the switch shown in Figure 1-12 is closed, a current flows that, unlike the case of the electronic conductor, declines in magnitude and eventually becomes immeasurably small. The quantity of charge, $Q(t)$, that has passed increases with time in the manner described by the curves in Figure 1-13. As in the insulator case, the charge passed ultimately, $Q(\infty)$, is proportional to the area A of the plates and (approximately at least) to the applied voltage ΔE. However, it is entirely independent of the separation L. Evidently there are factors at play in the case of an ionic conductor that have no parallel in the other two classes of materials, but the novel behavior is readily explained.

If the ionic conductor contains mobile ions of two types, cations and anions, then the effect of the field is to cause these ions to move, anions leftwards in Figure 1-14 and cations rightwards. As the moving ions approach the impenetrable plates, they are halted and accumulate there. The two sheets of accumulating ions themselves create a field that opposes that caused by the plates, decreasing the field experienced by the moving ions and slowing their motion. Eventually the motion ceases because the two fields entirely cancel and leave the interior of the conductor field-free.

---

[147] Electrochemists refer to such a circumstance as "an electrochemical cell under **totally polarized** conditions". See Chapter 10.

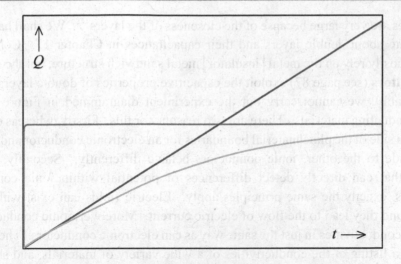

Figure 1-13 How the charge passed varies with time following the imposition of an electric field on three classes of material. For the **insulator**, a charge of magnitude $A\varepsilon\Delta E/L$ passes almost immediately. For the **electronic conductor**, the charge passed increases linearly as $A\kappa t\Delta E/L$. For the **ionic conductor**, the charge accumulates at an ever-decreasing rate.

We now have four sheets of charge: two electronic and two ionic. At the surface of each plate a layer of ions confronts a layer of electronic charge of equal magnitude but opposite sign. This is called a double layer. Just as the two layers of charge in Figure 1-8 constitute a capacitor, so do the layers at each plate in the present case. Such **double layer**

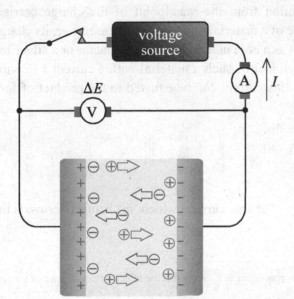

Figure 1-14 In the absence of chemical reaction, ions move and accumulate at the interfaces when a field is applied to an ionic conductor.

capacitances are very large because of the closeness of the layers[148]. We shall have much more to write about double layers and their capacitances in Chapter 13. Commercial capacitors mostly rely on the metal│insulator│metal sandwich structure, but the so-called **supercapacitors** (see page 87) exploit the capacitive properties of double layers.

Regrettably, we cannot carry out the experiment diagrammed in Figure 1-11 on ionically conducting materials. There are two reasons for this. Firstly, whereas electrons exist on each side of the plate│material boundaries for an electronic conductor and can pass from one side to the other, ionic conductors behave differently. Secondly, we lack voltmeters that can directly detect differences of potential within ionic conductors. Nevertheless, exactly the same principles apply. Electric fields can exist within ionic conductors and they lead to the flow of electric current. Moreover, ionic conductors can be assigned conductivities in just the same way as can electronic conductors. The table on page 384 is a listing of the conductivities of a wide variety of materials, and shows the charge carriers responsible in each case.

In truth, the classification of materials into insulators and conductors, though convenient conceptually, is not a rigid one. Most "conductors" exhibit some dielectric behavior and most "insulators" do conduct to some small extent. Water is a good example of a substance that displays both types of behavior. Its small content of hydronium H_3O^+ and hydroxide OH^- ions gives it conductivity, yet the considerable dipole moment of its majority H_2O molecules (Figure 1-2) causes them to orient in an electric field in a manner analogously to that illustrated in Figure 1-10.

Mobilities: the movement of charged particles in an electric field

Let us look at electrical conduction from the standpoint of the charge carriers themselves. Initially, consider the case of a material with a sole charge carrier, its charge being a single positive charge Q_0, such as a hole in a p-type semiconductor or a silver ion in $RbAg_4I_5$. Figure 1-15 pictures a cylinder of such a material with a current I flowing through it. Some thought shows that this current can be equated to the product of four terms

$$1:24 \quad \begin{array}{c} \text{electric} \\ \text{current} \end{array} = \begin{array}{c} \text{rate at} \\ \text{which charge} \\ \text{crosses} \\ \text{any plane} \end{array} = \left(\begin{array}{c} \text{number of} \\ \text{carriers} \\ \text{per unit} \\ \text{volume} \end{array} \right) \left(\begin{array}{c} \text{cross-} \\ \text{sectional} \\ \text{area} \end{array} \right) \left(\begin{array}{c} \text{charge} \\ \text{on each} \\ \text{carrier} \end{array} \right) \left(\begin{array}{c} \text{average} \\ \text{carrier} \\ \text{speed} \end{array} \right)$$

The **first** of these is the number density of charge carriers which, in chemical terms, is the

---

[148] There is such a capacitor at each plate. If its capacitance is C, show that the effective capacitance of the unit as a whole in $C/2$. See Web#148.

concentration c of the carrier multiplied by **Avogadro's constant**[149], N_A, equal to 6.0221 $\times 10^{23}$ mol^{-1}. The **second** and third right-hand terms in 1:24 are simply A and Q_0 respectively. The fourth is the average velocity $\overline{v}$ with which the carrier moves in the x direction. Thus[150], with c representing the concentration of the charge carrier,

$$1{:}25 \qquad\qquad I = N_A c A Q_0 \overline{v} = F A c \overline{v}$$

The product of Avogadro's number and the elementary charge is a quantity that crops up repeatedly in electrochemistry. This product is given the symbol F and the name **Faraday's constant**[151].

$$1{:}26 \qquad F = N_A Q_0 = (6.0221 \times 10^{23}\,\text{mol}^{-1})(1.6022 \times 10^{-19}\,\text{C}) = 96485\ \text{C mol}^{-1}$$

Faraday's constant provides the quantitative connection between chemistry and electricity. In one mole of sodium chloride, the total charge on the sodium ions, Na^+, is 96485 coulombs and, of course, there are -96485 C on the chloride ions, Cl^-.

Equation 1:25 shows that the velocity of the charge carrier is proportional to the current, and therefore to both the electric field and the conductivity

$$1{:}27 \qquad\qquad \overline{v} - \frac{I}{FAc} = \frac{i}{Fc} = \frac{\kappa X}{Fc} \qquad \text{average velocity}$$

The ratio of the average velocity of the moving charge carriers to the field that causes that motion is known as the **mobility**[152] u of the carrier. Thus, we see that the conductivity of the material shown in Figure 1-15 is related to the carrier's mobility by the simple relation

$$1{:}28 \qquad\qquad \kappa = \frac{Fc\overline{v}}{X} = Fuc \qquad \text{single charge carrier}$$

This simple relationship needs modification, of course, if there is more that one charge carrier, or if a carrier has a charge of other than $+Q_0$. A carrier i that has a charge of Q_i is

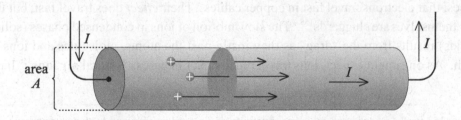

Figure 1-15 Positive charge carriers move in response to a field, leading to the flow of current I.

---

[149] Amadeo Carlo Avogadro, 1776 – 1856, Italian count, lawyer and natural philosopher.

[150] Confirm that the units in equation 1:25 are consistent. Check at Web#150.

[151] Michael Faraday, 1791 – 1867, Englishman and "father of electrochemistry". **Faraday's law** (Chapter 3) establishes the proportionality between electric charge and the amount of chemical reaction during electrolysis.

[152] Mobilities of anions are negative in this book, but elsewhere you may find the absolute value reported. Moreover, mobility is sometimes defined as $\overline{v}_i/(z_i X)$.

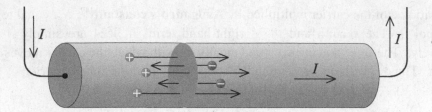

Figure 1-16 Cations, moving to the right with the field, and anions, moving leftwards against the field, both contribute to the current.

said to have a **charge number**

1:29 $$z_i = \frac{Q_i}{Q_0}$$ definition of charge number

For example, electrons e^- and calcium ions Ca^{2+} have charge numbers of -1 and $+2$ respectively. The replacement for equation 1:28 that is true generally is

1:30 $$\kappa = F\sum_i z_i u_i c_i$$ linking mobilities to conductivity

Note that carriers of either sign contribute positively to the conductivity[153] because, if z_i is negative, so is u_i, as clarified in Figure 1-16.

Of course, mobilities are affected by such factors as temperature and the medium in which the charge carriers find themselves, but some representative values are given in the table on page 388. The listed values for ions relate to aqueous solution[154], because this is the most widely studied electrochemical medium. For the most part, mobilities are surprisingly small. One gets the impression, from the "instantaneous" response of electrical appliances, that electrons travel fast in copper cables. Their *effect* does travel fast, but the particles themselves are sluggards[155]. The slow motion of ions in condensed phases (solids and liquids) results from the "drag" as they jostle past the atoms, molecules and ions in their path. As one might expect, ions travel much faster in gases. Natural **air ions**[156] have

---

[153] See Web#153 for the related terms **molar conductivity**, **ionic conductivity**, and **transport number**.

[154] Water purity is often gauged from its conductivity, though **conductimetry** measures only ionic impurities. Assuming the impurity to be sodium chloride (Na^+ and Cl^- ions in solution), assess the purity (in moles per liter and as a percentage) of a water sample of conductivity $22\ \mu S\ m^{-1}$. See Web#154. The conductivity of seawater is measured to assess its **salinity**.

[155] In silver, the most conductive metal, electrons travel at about 6 millimeters per second in a field of one volt per meter, on the assumption of a single mobile electron per Ag atom. The high conductivity of metals arises from the high electron concentration rather than high mobility. In contrast, electrons in silicon travel about twenty times faster.

[156] or **cluster ions**, H^+, NH_4^+, OH^-, and NO_3^- ions with attached H_2O molecules, having an average mass of about $160\ g\ mol^{-1}$.

mobilities close to 1.5×10^{-4} m^2 V^{-1} s^{-1}. In vacuum, charge carriers do *not* travel with a constant speed in a uniform field; they accelerate, negating the concept of mobility.

We have been discussing the movement of charged particles caused by an electric field. Such motion is called **migration**. Motion may occur from other causes, notably by **diffusion** and by **convection**. All three modes are important in electrochemistry and their interplay is the subject of Chapter 8.

Electrical Circuits: models of electrochemical behavior

A device that is fabricated to have a stable resistance is known as a **resistor**[157]; it is represented in circuit diagrams by ‑\/\/\/‑. Similarly, ‑||‑ is used to symbolize a **capacitor**. Resistors and capacitors are examples of **circuit elements**. There are others, but resistors and capacitors are the ones of most interest to electrochemists. When two or more circuit elements are connected so that they experience the same voltage, they are said to be in **parallel**; conversely they are in **series** if they experience the same current[158]. The circuit shown on the left-hand side of Figure 1-17 overleaf has a resistor and a capacitor in parallel: it provides a model of materials that have both dielectric and conductive properties. One or more circuit elements connected into such a circuit as those in the figure are often referred to as the circuit's **load**. How does the circuit behave in response to the load when the switch is closed? The resistor experiences a constant electrical potential difference ΔE, so the current has a constant value, equal in magnitude to $\Delta E/R$. The capacitor charges, in principle immediately, to attain the charge $C\Delta E$. Energy, provided ultimately by the voltage source, has been stored in the capacitor. The resistor, in contrast, has destroyed electrical energy by converting it to heat; that is how electric heaters work. The **power** dissipated by the resistor, measured in joules per second or **watts**[159], is the product of the voltage ΔE and the current I flowing through the conductor[160].

More interesting, and more relevant to electrochemistry, is the right-hand circuit in Figure 1-17 overleaf, in which the resistor and capacitor are in series. The capacitor is uncharged at time $t = 0$, when the switch is closed. What will the current be subsequently?

---

[157] A graphite cylinder, of 2.40 cm length and 0.450 cm diameter, was found to have a resistance of 0.0347 Ω. Calculate the conductivity of this particular graphite, checking your result against Web#157.

[158] Show that if several resistances are connected in series, the overall resistance is the simple sum of all the individual resistors, but that if several capacitors are connected in series, the reciprocal of the overall capacitance is the sum of the reciprocals of the individual capacitors. What are the corresponding rules for a parallel connection? See Web#158.

[159] James Watt, 1736 – 1819, Scottish engineer, after whom the unit (symbol W) is named.

[160] The resistor in left-hand diagram of Figure 1-17 overleaf has a resistance of 425 Ω and the voltage applied is 1.15 V. Calculate the current through the resistor and the power dissipated by it, comparing your results with Web#160. Also, select a voltage and a resistance, such that the current through the resistor is 13 milliamperes and the power is 150 microwatts.

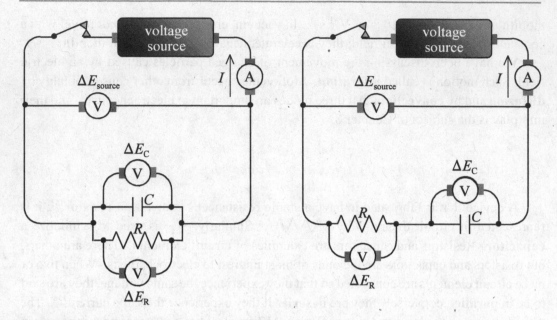

Figure 1-17 Circuits to evaluate the effect of a voltage step on a resistor and a capacitor in parallel (left) and series (right). In the parallel case, ΔE_C and ΔE_R are identical; when the components are in series, the same current I flows through R and C.

There are three voltmeters in the circuit and it is evident[161] that the sum of their readings must be zero:

1:31 $$\Delta E_{source} + \Delta E_R + \Delta E_C = 0$$

The voltage across the resistor, and that across the capacitor, are calculable via Ohm's law and equation 1:18, respectively; and therefore

1:32 $$\Delta E_{source} = -\Delta E_R - \Delta E_C = RI + \frac{Q}{C} = R\frac{dQ}{dt} + \frac{Q}{C}$$

Definition 1:19 was used in the final step. Equation 1:32 is a first-order differential equation that can be solved[162] to give

1:33 $$Q = C\left[1 - \exp\left\{\frac{-t}{RC}\right\}\right]\Delta E_{source}$$

[161] because, after taking three steps around the circuit, you eventually return to your initial point.

[162] It is easy to see how equation 1:32 follows from 1:33; less easy to derive 1:33 from 1:32. See Web#162 for a discussion of **Laplace transformation**, (Pierre Simon de Laplace, 1747–1827, French mathematician) a technique useful for solving such differential equations as 1:32.

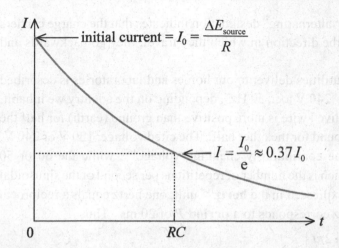

$$\text{initial current} = I_0 = \frac{\Delta E_{source}}{R}$$

$$I = \frac{I_0}{e} \approx 0.37 I_0$$

Figure 1-18 How the current changes following the imposition of a constant voltage on a series arrangement of a resistor and a capacitor.

Differentiation with respect to time gives[163]

1:34 $$I = \frac{1}{R} \exp\left\{\frac{-t}{RC}\right\} \Delta E_{source}$$

which describes how the current falls off exponentially[164] with time, as illustrated in Figure 1-18. Notice that the product RC is a time[165]; it is called the **time constant** or **decay time** of the series circuit. Electrochemical cells have time constants, as we shall discover in Chapter 13.

Connected circuit elements find use in electrochemistry. A frequently used model of an electrochemical cell has one resistor R_s in series with a parallel combination of a capacitor C and a second resistor R_p. In response to a voltage step, it allows the current

1:35 $$I = \frac{\Delta E_{source}}{R_s + R_p} \left[1 + \frac{R_s}{R_p} \exp\left\{ \frac{-(R_s + R_p)}{R_s R_p C} t \right\} \right]$$

to flow[166].

Alternating Electricity: sine waves and square waves

Up to this point we have been discussing electric currents that do not change sign; these are said to be **direct currents** or d.c. Of equal importance, however, is **alternating**

---

[163] Demonstrate that, at sufficiently short times, the series arrangement behaves as a resistor whereas, at long times, the behavior is as if the resistor were not present. See Web#163.

[164] In formulating an exponential function, you may be used to the notation $e^{-t/RC}$ instead of $\exp\{-t/RC\}$. The latter style is used throughout this book.

[165] Demonstrate that multiplying the farad and ohm units give the unit of time, the second. See Web#165.

[166] Derive equation 1:35. Check at Web#166.

current electricity or a.c.. The "alternating" designation indicates that the charge carriers (electrons or ions) alternate in the direction in which they travel: they go backwards and forwards.

The electricity that power utilities deliver to our homes and laboratories is described either as "120 V a.c., 60 Hz" or "240 V a.c., 50 Hz", depending on the country we inhabit. The electrical potential of the "live" wire is more positive than ground (earth) for half the time and more negative than ground for the other half. The cited voltage, 120 V or 240 V, is the **root-mean-square** of the continuously changing voltage[167], while the 60 or 50 designates the **frequency** ω, which is the number of repetitions per second of the sinusoidal voltage pattern. Frequency is expressed in the **hertz**[168] unit; one hertz equals a reciprocal second, so a frequency of 50 Hz corresponds to a **period** P of 20 ms. Thus

1:36 $$E(t) = \sqrt{2}E_{\text{rms}} \sin\left\{\frac{2\pi t}{P}\right\} = (170\,\text{V}) \sin\left\{\frac{2\pi t}{\frac{1}{60}\,\text{s}}\right\} \text{ or } (340\,\text{V}) \sin\left\{\frac{2\pi t}{\frac{1}{50}\,\text{s}}\right\}$$

describes the electricity supply. Figure 1-19 illustrates these two alternating voltages.

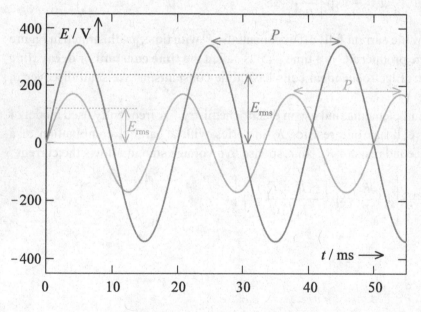

Figure 1-19 Waveforms of the domestic electricity supply: the green and violet curves respectively illustrate 120 V, 60 Hz and 240 V, 50 Hz supplies.

For scientific purposes, E_{rms} and P are less convenient than two alternative parameters: the a.c. **voltage amplitude** $|E|$, equal to the maximum voltage achieved by $E(t)$; and the **angular frequency** ω, equal to $2\pi/P$. When these parameters are adopted, the first equality in equation 1:36 becomes replaced by the equivalent expression

---

[167] The reason for choosing the root-mean-square as the voltage of record is that such an a.c. voltage generates the same power (applied across a resistor, for example) as a d.c. voltage of the same magnitude.

[168] Heinrich Rudolf Hertz, 1857 – 1894, German physicist.

1:37
$$E(t) = |E| \sin\{\omega t\}$$

The a.c. electricity supplied by most American utilities, for example, has a nominal voltage amplitude $|E|$ of 170 V and an angular frequency ω of 376.99 rad s^{-1}. To validate equation 1:37, the zero of time must be chosen to correspond to one of the instants at which the electrical potential is zero and increasing, as in Figure 1-19. When other choices of $t = 0$ are made, the more general equation

1:38
$$E(t) = |E| \sin\{\omega t + \varphi_E\} \qquad \text{a.c. voltage}$$

applies, in which φ_E is the **phase angle** of the a.c. voltage. Thus three parameters are required to describe an arbitrary a.c. voltage: the amplitude $|E|$, the angular frequency ω, and the phase angle φ_E.

The imposition of an a.c. voltage of angular frequency ω often causes an alternating flow of electricity. This is described as an **a.c. current**[169] and it generally has the same frequency as the applied voltage and obeys the equation

1:39
$$I(t) = |I| \sin\{\omega t + \varphi_I\} \qquad \text{a.c. current}$$

which shows that an a.c. current possesses the same three attributes – amplitude, frequency and phase – as an a.c. voltage. The relationship between $I(t)$ and $E(t)$ reflects the nature of the load and incorporates two aspects: how $|E|$ depends on $|I|$, and how the phase angles φ_E and φ_I are related. The ratio $|E|/|I|$ is called the **impedance**, Z, of the load[170] and is measured in ohms. The difference $\varphi_I - \varphi_E$ is the **phase shift** and is measured in radians or degrees.

In Figure 1-20, the $\perp$ symbol indicates a connection to ground (earth). This figure is of a simple circuit, equipped with an a.c. voltmeter and a.c. ammeter, that can measure

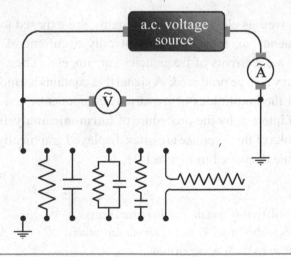

Figure 1-20 An a.c.circuit for measuring the impedance of five alternative loads.

---

[169] Literally, "a.c. current" (alternating current current) contains a redundancy. Nonetheless, this expression is commonly used.

[170] Alternatively, $|I|/|E|$ is the **admittance** Y of the load, measured in siemens. Read more about impedances on pages 305–322.

$|I|$ and $|E|$, and hence the impedance Z of the load. Other devices[171], not shown, can also measure the phase shift. The voltage applied to the load is that in equation 1:38. When the load is a single resistor, the resulting current $|I|\sin\{\omega t + \varphi_I\}$ can be found directly from Ohm's law

1:40 $$I(t) = \frac{E(t)}{R} = \frac{|E|}{R}\sin\{\omega t + \varphi_E\} \quad \text{so that} \quad |I| = \frac{|E|}{R} \text{ and } \varphi_I = \varphi_E$$

The impedance equals the resistance and there is no phase shift. When the load is a capacitor, however, one has

1:41 $$I(t) = \frac{d}{dt}Q(t) = C\frac{d}{dt}E(t) = C|E|\omega\cos\{\omega t + \varphi_E\} = C|E|\omega\sin\left\{\omega t + \varphi_E + \frac{\pi}{2}\right\}$$

so that the impedance and phase shift are as listed in the second row of the table below. This table also has entries[172] for other loads of electrochemical interest[173].

Load	Impedance Z	Phase shift $\varphi_I - \varphi_E$
Resistor R	R	0
Capacitor C	$1/\omega C$	$\pi/2$
R and C in parallel	$R/\sqrt{1 + \omega^2 R^2 C^2}$	$\arctan\{\omega RC\}$
R and C in series	$\sqrt{1 + \omega^2 R^2 C^2}/\omega C$	$\operatorname{arccot}\{\omega RC\}$
Warburg element	$\sqrt{R/\omega C} = W/\sqrt{\omega}$	$\pi/4$

When some circuit elements, as well as electrochemical systems, are exposed to an alternating voltage of angular frequency ω, they generate not only a current of the "fundamental frequency" ω but also a.c. currents of frequencies 2ω, 3ω, etc. These are called **harmonics**[174]. A d.c. current may also be produced. A signal that contains harmonic frequencies may be analyzed to find the amplitudes of its various components, and the phase angles, too, should those be of interest, by the procedure of **harmonic analysis** or **Fourier transformation**[175]. The results of this exercise are often displayed graphically as a **Fourier spectrum**[176], such as the one displayed in Figure 1-21.

---

[171] such as **lock-in amplifiers, phase-sensitive detectors**, and their digital counterparts.

[172] Derive the entries for the parallel and series loads. Or see Web#172 for our derivations.

[173] See Web#173 and Chapter 15 for information on the Warburg element.

[174] Musicians and some others refer to the harmonic of frequency 2ω as the *first* harmonic; others, including ourselves, call it the *second* harmonic. In our terminology, the first harmonic is the fundamental frequency, ω.

[175] Jean Baptiste Joseph Fourier, 1768 – 1830, renowned French mathematician.

[176] or as a **power spectrum**, as on page 325.

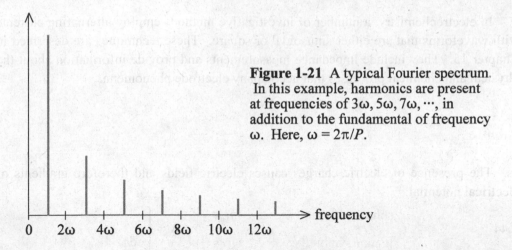

Figure 1-21 A typical Fourier spectrum. In this example, harmonics are present at frequencies of $3\omega, 5\omega, 7\omega, \cdots$, in addition to the fundamental of frequency ω. Here, $\omega = 2\pi/P$.

Alternating currents and voltages may have waveforms other than sinusoidal. One waveform that finds use electrochemically is the square wave, shown in Figure 1-22 and representable algebraically by[177]

1:42
$$E(t) = (-)^{\text{Int}\{2t/P\}} |E| \qquad \text{square wave}$$

Like many other waveforms, the square wave may be represented by the sum of a set of sine waves. In the case of the square wave, the first few terms are

1:43
$$(-)^{\text{Int}\{2t/P\}} |E| = \frac{4|E|}{\pi} \sin\left\{\frac{2\pi t}{P}\right\} + \frac{4|E|}{3\pi} \sin\left\{\frac{6\pi t}{P}\right\} + \frac{4|E|}{5\pi} \sin\left\{\frac{10\pi t}{P}\right\} + \cdots \qquad \text{square wave}$$

In fact, the Fourier spectrum shown in Figure 1-21 is that of the square wave 1:43. Modern electrochemical instrumentation applies waveforms as a discontinuous set of brief constant segments, as illustrated in Figure 1-22, rather than continuously. Likewise, the current response is measured stepwise.

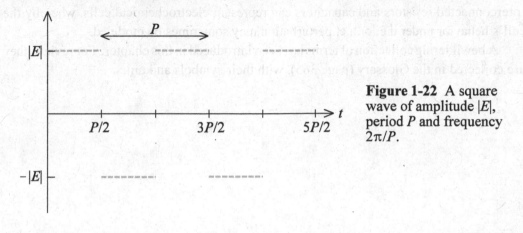

Figure 1-22 A square wave of amplitude $|E|$, period P and frequency $2\pi/P$.

---

[177] Int{ } denotes the integer-value function. Int{y} means the largest integer that does not exceed y; thus Int{8.76} = 8 and Int{−8.76} = −9.

In electrochemistry, a number of investigative methods employ alternating currents with waveforms that are either sinusoidal or square. These techniques are described in Chapter 15. They include impedance measurements and provide information about the physical and chemical processes that accompany electrode phenomena.

Summary

The presence of electric charges causes electric fields and therefore gradients of electrical potential:

1:44
$$\left.\begin{array}{l} \text{spherical symmetry: } Q/(4\pi\varepsilon r^2) \\ \text{planar symmetry : } \qquad q/\varepsilon \end{array}\right\} = X = \begin{cases} -\mathrm{d}\phi/\mathrm{d}r \\ -\mathrm{d}\phi/\mathrm{d}x \end{cases}$$

The imposition of an electric field on an insulator rearranges the charges within it, causing a transitory passage of charge:

1:45
$$-Q = C\Delta E = \frac{\varepsilon A}{L}\Delta E$$

In a conductor, the mobile charges, which may be electrons or ions, move in response to a potential difference leading to a steady current:

1:46
$$-I = \frac{\Delta E}{R} = \frac{\kappa A}{L}\Delta E$$

The conductivity of a conductor, electronic or ionic, reflects the charge numbers, mobilities, and concentrations of all the charge carriers present:

1:47
$$\frac{i}{X} = \kappa = F\sum_i z_i u_i c_i$$

Interconnected resistors and capacitors can represent electrochemical cells, whereby the cell's behavior under d.c. or a.c. perturbation may sometimes be modeled.

A bewildering collection of terms has been introduced in this chapter. For review, they are collected in the Glossary (page 365), with their symbols and units.

2

Chemistry

In Chapter 1 we introduced those aspects of electricity that have close relevance to electrochemistry. In the present chapter our subject is those fundamental aspects of physical chemistry on which much of electrochemistry is based.

Chemical Reactions: changes in oxidation state

In this book we need not go deeply into the subject of **valency** – how atoms bond together to form chemical compounds – but the term **oxidation state** (or **oxidation number**[201]) is relevant both to chemistry and electrochemistry. As a working definition applied to a metallic element, it refers to the number of positive charges on the bare ion, or the number of chloride atoms bonded to each atom of the element in its chloride, or to twice the number of oxygen atoms per atom of the element in its oxide. Thus, aluminum is in oxidation state +3 in Al^{3+}, in $AlCl_3$, and in Al_2O_3. Less obviously it is also in oxidation state +3 in $AlOOH$ and in AlF_4^-. In fact, aluminum is rarely in an oxidation state other than +3, except in its elemental form when its oxidation state is 0. Other metals, however, wantonly adopt a variety of oxidation states; iron, for example, is found in states[202] 0, +2, +3, and +4. In such cases it is common to add a roman numeral to specify the oxidation state. Thus "copper(II) sulfate" implies that the copper in this salt is in oxidation state +2.

The main focus of chemistry, and of electrochemistry, is on **reactions**, in which one form of matter, the **reactants** or **substrates**, is converted into different substances, the **products**. A **stoichiometric equation**[203], or **chemical equation**, of which

2:1 $\qquad Hg_2Cl_2(s) + 2H_2O(\ell) + H_2(g) \rightarrow 2Hg(\ell) + 2H_3O^+(aq) + 2Cl^-(aq)$

---

[201] A table of oxidation numbers will be found on page 386. Read more about oxidation numbers at Web#301.

[202] In magnetite, Fe_3O_4, iron's oxidation state is $8/3$. Or one might prefer to say that this oxide is composed of one-third Fe(II) and two-thirds Fe(III).

[203] **Stoichiometry** is the study of the relative amounts of reactants consumed and products created in a chemical reaction. More on this topic on page 46.

Electrochemical Science and Technology: Fundamentals and Applications, First Edition. Keith B. Oldham, Jan C. Myland, Alan M. Bond.

is an example[204], provides a shorthand way of representing a reaction[205]. A stoichiometric equation must be "balanced"; that is, equal numbers of all atoms must appear on each side of the → symbol. Moreover, the charge on each side must be equal.

Though many chemical reactions do not involve any change in oxidation state, many others do. *When one element undergoes a change in oxidation state during a chemical reaction, another element necessarily changes its oxidation state too.* In reaction 2:1, for example, the elements mercury Hg and hydrogen H both change their oxidation states. When, later in this book, we address electrochemical reactions, you will find that they invariably involve changes in oxidation state, and that usually it is a single element that changes its oxidation state.

Gibbs Energy: the property that drives chemical reactions

Energy can exist in many forms. All nonthermal forms of energy may be converted, often rather easily, into an equivalent quantity of heat. If a reaction generates heat

2:2   Reactants → Products + *heat*

the reactants evidently contained more energy than do the products. Energy associated with chemicals is called **enthalpy** H. The change in enthalpy[206] accompanying reaction 2:2, $\Delta H = H_{products} - H_{reactants}$, is negative. Most, but not all, chemical reactions that proceed spontaneously have negative ΔH values, and liberate heat.

That most chemical reactions liberate heat is a manifestation of a general law of nature: *processes that lead to a lowering of nonthermal energy are favored.* Another rule is: *processes that lead to an increase in disorder are favored.* The chemical property that reflects the change in disorder accompanying a reaction is the **entropy** change ΔS. Depending on the temperature T, either the change in H or the change in S is the more important influence on the reaction. The **Gibbs energy**[207] G takes both factors into account appropriately by defining[206]

2:3 $$\Delta G = \Delta H - T\Delta S$$

Just as a boulder can tumble downhill but not uphill, chemical changes can occur if G decreases, but not if it increases. G is the "chemical energy" that governs the feasibility of chemical reactions. Physical chemists have accurately measured the **standard Gibbs**

---

[204] Find the oxidation states of each element, in each compound, in reaction 2:1. See Web#204.

[205] The parenthesized italic "(s)", "(aq)", "(g)", etc. indicate the state (solid, in aqueous solution, gas, etc.). They are informative, but optional, additions to a stoichiometric equation. Likewise, (ℓ), "(ads)", "(fus)", etc. imply in the liquid state, the adsorbed state, the molten (or fused) state, etc.

[206] The reaction must be conducted at, or corrected to, constant temperature and pressure. We assume constant temperature and constant ambient pressure throughout this section, and indeed most of the book.

[207] Josiah Willard Gibbs, 1839 – 1903, U.S. physical chemist. G is also known as **free energy**.

energy of very many chemicals and a list will be found on page 390; those of the six substances in reaction 2:1 are tabulated as G° values below. To give this property its full name and fully embellished symbol, the tabulated values are the molar Gibbs free energies of formation at 25.00°C of each substance in its standard state, $(\Delta G_f^{\ominus})_{298.15}$. The "of formation" in this name means "during the process of formation from its elements in their standard states", which is why G° is zero for Hg and H_2. Because the Gibbs energies of individual ions cannot be measured, one ion – the hydronium ion H_3O^+ – is chosen as a standard and assigned the same G° value as water[208]. From tabulated values of G°, one can calculate the change in Gibbs energy accompanying any reaction[206,209]. Thus, for reaction 2:1, the value calculated for the change in Gibbs energy is

2:4 $\Delta G^{\circ} = 2G_{Hg}^{\circ} + 2G_{H_3O^+}^{\circ} + 2G_{Cl^-}^{\circ} - G_{Hg_2Cl_2}^{\circ} - 2G_{H_2O}^{\circ} - G_{H_2}^{\circ} = -51.7 \text{ kJ mol}^{-1}$

No reaction that has a positive ΔG° will occur chemically under standard conditions. Processes with negative ΔG°, such as reaction 2:1, are said to be **feasible**[210] and *may* occur.

Substance	State	G° / kJ mol^{-1}
Hg	pure liquid mercury	0
H_3O^+	hydronium ion, in aqueous solution at unit activity	-237.1
Cl^-	chloride ion, in aqueous solution at unit activity	-131.2
Hg_2Cl_2	pure solid mercury(I) chloride	-210.7
H_2O	pure liquid water	-237.1
H_2	hydrogen gas at standard pressure	0

     Thermodynamicists use the term "standard state". By this they mean not only that the substance should be in the physical state usually encountered in the laboratory[211] – for example H_2O as liquid water, not ice – but also that it be in a prescribed condition described as **unit activity**. Precisely what "activity" means is discussed in the following section.

     The statement that no reaction with a positive ΔG° will occur chemically "under standard conditions" means "with all of the reactants *and products* present in their standard states". Under those conditions and with $\Delta G^{\circ} > 0$, the *reverse* reaction will inevitably be

---

[208] Earlier, H^+ was used as the symbol for what is now called the hydronium ion, and its standard Gibbs energy was defined as zero.

[209] What change in standard Gibbs energy accompanies the $Hg_2Cl_2(s) \rightarrow 2Hg(\ell) + Cl_2(g)$ reaction? See Web#209.

[210] or **spontaneous** or **exergonic**.

[211] Standard laboratory conditions are a temperature of 298.15 K and an ambient pressure of 100.00 kPa. These conditions will be assumed throughout this book, unless otherwise stated.

feasible. Except for the unlikely event that $\Delta G° = 0$, either the forward reaction *or* the reverse reaction will always be feasible for any conceivable process. Of course, only to be feasible is not a guarantee that the reaction will, in fact, occur. Moreover, even if a feasible reaction does occur, it will often fail to go to completion; that is, it may cease before all the reactant is consumed.

Associated with $\Delta G°$ is the **equilibrium constant** K for a reaction. A negative value of $\Delta G°$ implies an equilibrium constant of greater than unity. In fact, the important relationship

2:5 $$K = \exp\left\{\frac{-\Delta G°}{RT}\right\} \qquad \text{or} \qquad \Delta G° = -RT\ln\{K\}$$

provides a quantitative link between the two properties[212,213]. Here R is the **gas constant**.

2:6 $$R = 8.3145 \text{ J K}^{-1} \text{ mol}^{-1}$$

Note that $RT = 2.4790$ kJ mol^{-1} at the standard temperature $T° = 298.15$K.

The statement above – that no reaction will occur unless $\Delta G° < 0$ – requires the reactants and products to be in their standard states. For states other than standard, the requirement is that $\Delta G < 0$, the distinction between $\Delta G°$ and ΔG hinging on the *activities* of the products and reactants, a subject addressed in the next section. A chemical reaction ceases when the change in Gibbs energy becomes zero and equilibrium prevails.

2:7 $$\Delta G = 0 \qquad \text{equilibrium}$$

A caveat is that the process be purely chemical. If other kinds of work can be brought into play to foster the reaction, then the requirement is that

2:8 $$\Delta G - W < 0 \qquad \text{feasibility}$$

whereas

2:9 $$\Delta G = W \qquad \text{equilibrium}$$

Here W is the work[214] performed by some external agency when the Reactants → Products reaction takes place. By supplying external work, you *can* make that boulder travel uphill! It is the facility to couple electrical work into a chemical reaction that makes electrochemistry into a powerful synthetic tool, creating a possibility to drive reactions that would otherwise be unfeasible. Examples of **electrosynthesis**, as the generation of chemicals with electrochemical assistance is called, will be found in Chapter 4. Other examples of external work being coupled to chemical processes include metabolically powered physiological "pumps" (Chapter 9), photosynthesis, and some gravitationally powered processes.

---

[212] Calculate the equilibrium constant of reaction 2:1. Check your answer at Web#212.

[213] The equilibrium constant of the reaction $2Ag_2O(s) \rightleftarrows 4Ag(s) + O_2(g)$ is 1.19×10^{-4} at 25°C. Find $G°$ for silver oxide. Compare your answer with that at Web#213 or the tabulated value on page 391.

[214] In contrast to the w encountered in Chapter 1, W is a *molar* work term, in joules per mole, J mol^{-1}.

Activity: restlessness in chemical species

The concept of activity is useful in chemistry and especially in electrochemistry. Before encountering a definition of the term, it is valuable to acquire a qualitative appreciation of what activity implies. In daily life, the terms "unsettled" and "restless" are used to describe discontented persons who are eager to leave their present situation. When a similar characteristic applies to a molecule or ion, we say the substance has a high "activity". A high activity may be manifested in several ways:

$$\left.\begin{array}{c}\text{a substance}\\ \text{with a high}\\ \text{activity}\end{array}\right\}\text{tends to}\left\{\begin{array}{l}\text{react more readily (it has a high Gibbs energy)}\\ \text{react more rapidly (rates of reaction are faster)}\\ \text{transfer to another phase (e.g. by evaporating or dissolving)}\\ \text{diffuse into a more dilute region}\end{array}\right.$$

To pursue the anthropomorphic analogy further, substances with high activity have unsociable or even suicidal tendencies. Low activity confers contentment.

Quantitatively, the **activity** a_i measures the restlessness of a substance i in some condition of interest compared with that in its standard state. Thus, being ratios, activities have no units; they are positive numbers. Activities are affected by temperature and, to a lesser extent, by the ambient pressure, but we shall not explore these dependences, because they seldom have great importance in electrochemistry.

The activity a_i of a gaseous species i depends on its partial pressure. As Figure 2-1 illustrates, the activity of a gas varies linearly with its partial pressure p_i up to very high pressures. The standard state of a gas occurs when it is at the standard pressure, 1 bar, so:

2:10　　　　$a_i = \dfrac{p_i}{p^\circ}$　　　where $p^\circ = 1.0000 \times 10^5$ Pa　　　when i is a **gas**

the **pascal** (Pa) being the *SI* unit in which pressure is measured. Under all conditions, except extreme pressures, the activity of a gas may be accurately replaced by p_i/p°.

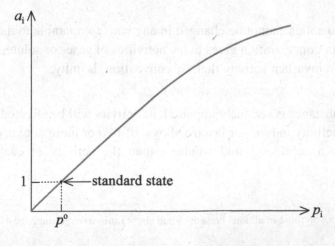

Figure 2-1 Up to very high pressures, the activity of a gas is proportional to its partial pressure.

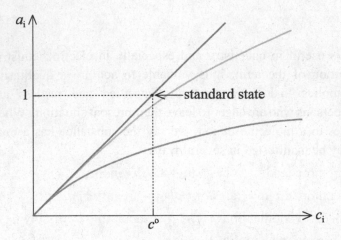

Figure 2-2 The activity of a solute behaving ideally would equal c_i/c^o at all concentrations. Ionic solutes depart from ideality much sooner than do nonionic solutes.

The story for solutes is rather similar. The graphs in Figure 2-2 show how the activities of solutes depend on their concentration. For nonionic solutes, the linearity of the graph extends almost to the standard condition which, for solutes, is chosen as one mole per liter, so that, except for the most concentrated solutions, the approximation

2:11 $a_i \approx \dfrac{c_i}{c^o}$ where $c^o = 1.0000 \times 10^3$ mol m^{-3} when i is an **nonionic solute**

is adequate for most purposes. When the solute is an ion, however, linearity ceases well before the standard condition, as Figure 2-2 demonstrates, so that a relationship like 2:11 serves only as a very crude description of behavior. Instead, an empirical factor called an **activity coefficient**, with the symbol γ, is introduced; it is not a constant, itself depending on concentration[215]

2:12 $a_i = \dfrac{\gamma_i c_i}{c^o}$ where $c^o = 1.0000 \times 10^3$ mol m^{-3} when i is an **ionic solute**

As you may have guessed, the reason that dissolved ions behave differently from dissolved molecules is because of their electric charges; we shall say more about ionic activity coefficients on pages 41–45.

The activities of liquids and solids cannot be changed in any way comparable to the variability that partial pressure or concentration gives to the activities of gases or solutes. These condensed phases have an invariant activity that, by convention, is unity:

2:13 $a_i = 1$ when i is a pure **liquid** or **solid**

Of course, if the purity of the substance is seriously impaired, its activity will be affected; copper has a lower-than-unity activity in brass or bronze alloys. If two or more not very dissimilar substances compose a solid or liquid solution, then the activity of each

---

[215] not only on its own concentration, but on those of all ions present. Read about **ionic strength** on page 41.

component will be close to its mole fraction[216] x_i.

2:14 $\qquad a_i \approx x_i \qquad$ when i is a **component** of a solid or liquid solution

As would be expected, the activity of water is generally somewhat lower than unity[217] when it serves as a solvent. A relation analogous to 2:14 occurs in films that are one molecule thick. If the fraction (by area) occupied by a particular component of that film is θ_i, then

2:15 $\qquad a_i \approx \theta_i \qquad$ when i is a component of a **unimolecular film**

What about the activity of electrons? This takes us outside the realm of traditional thermodynamics, but there are good reasons for ascribing an activity of

2:16 $$a_{e^-} = \exp\left\{\frac{-F(\phi - \phi^\circ)}{RT}\right\} \qquad \text{for an \textbf{electron}}$$

to an electron at a point where the electrical potential is ϕ. We learned in Chapter 1 that only differences in potential can be measured and only then between two phases of the same (or very similar) compositions. Let I and II signify two such similar phases within which electrons exist, then the electron activities in the two phases are in the ratio

2:17 $$\frac{a_{e^-}^{II}}{a_{e^-}^{I}} = \exp\left\{\frac{-F\Delta E}{RT}\right\} \qquad \Delta E = \phi^{II} - \phi^{I}$$

where ΔE is the measurable voltage between phases II and I.

As we saw on page 33, activity influences the behavior of species in several realms of chemistry and physics. Its effect on Gibbs energy is through the relationship

2:18 $$G_i = G_i^\circ + RT \ln\{a_i\}$$

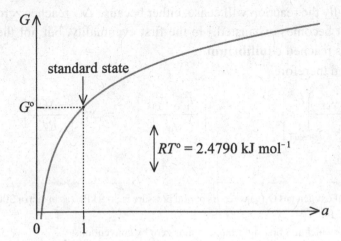

Figure 2-3 The Gibbs energy of a substance varies logarithmically with its activity, equaling its standard value G° at unit activity.

In figure: G, standard state, G°, $RT^\circ = 2.4790 \text{ kJ mol}^{-1}$, 0, 1, a

---

[216] The **mole fraction**, x_i, of a component in a solution is the ratio of the number of moles of that component to the total number of moles of all components in the solution.

[217] though sometimes larger; for example $a_{H_2O} = 1.004$ in a 2.0 mol L^{-1} aqueous solution of KCl.

The logarithmic increase in the Gibbs energy of substance i with its activity[218] is illustrated in Figure 2-3. The first right-hand term in equation 2:18 usually[219] dominates the free energy of a substance at normal temperatures.

Recall that ΔG° is the Gibbs energy change accompanying a reaction *when all reactants and products are in their standard states*. In other circumstances, when activities are not all equal to unity, we use the unsuperscripted symbol ΔG to denote the Gibbs energy change for a reaction. Turning to reaction 2:1 as an example of the way in which the activities of the reactants and products influence the Gibbs energy change, one finds

$$\Delta G = \Delta G^\circ + 2RT \ln\{a_{\text{Hg}}\} + 2RT \ln\{a_{\text{H}_3\text{O}^+}\} + 2RT \ln\{a_{\text{Cl}^-}\} - RT \ln\{a_{\text{Hg}_2\text{Cl}_2}\}$$

2:19

$$- 2RT \ln\{a_{\text{H}_2\text{O}}\} - RT \ln\{a_{\text{H}_2}\} = \Delta G^\circ + RT \ln\left\{\frac{a_{\text{Hg}}^2 a_{\text{H}_3\text{O}^+}^2 a_{\text{Cl}^-}^2}{a_{\text{Hg}_2\text{Cl}_2} a_{\text{H}_2\text{O}}^2 a_{\text{H}_2}}\right\}$$

Notice how the stoichiometric coefficients from the reaction's equation have appeared as powers in the argument of the logarithm, with the products in the numerator and the reactants in the denominator. The activities of the condensed phases (solids and liquids) can be omitted, because they are unity, and the other three activities may be replaced by recourse to equations 2:10 and 2:12, so that[220]

2:20 $$\Delta G = \Delta G^\circ + RT \ln\left\{\frac{a_{\text{H}_3\text{O}^+}^2 a_{\text{Cl}^-}^2}{a_{\text{H}_2}}\right\} = \Delta G^\circ + RT \ln\left\{\frac{p^\circ \gamma_{\text{H}_3\text{O}^+}^2 \gamma_{\text{Cl}^-}^2 c_{\text{H}_3\text{O}^+}^2 c_{\text{Cl}^-}^2}{p_{\text{H}_2} (c^\circ)^4}\right\}$$

As the reaction proceeds, the concentrations of the hydronium H_3O^+ and chloride Cl^- ions will steadily increase and the partial pressure of hydrogen H_2 may[221] decrease. These activity changes cause the $RT \ln\{\ \}$ term to increase steadily in magnitude, making ΔG steadily less negative. Eventually the reaction will cease, either because ΔG reaches zero or because the Hg_2Cl_2 reactant becomes exhausted. In the first eventuality, but not the second, we say the reaction has reached **equilibrium**.

At equilibrium, $\Delta G = 0$ and therefore

2:21 $$\Delta G^\circ = -RT \ln\left\{\left(\frac{a_{\text{H}_3\text{O}^+}^2 a_{\text{Cl}^-}^2}{a_{\text{H}_2}}\right)_{\text{equil}}\right\} \quad \text{or} \quad \left(\frac{a_{\text{H}_3\text{O}^+}^2 a_{\text{Cl}^-}^2}{a_{\text{H}_2}}\right)_{\text{equil}} = \exp\left\{\frac{-\Delta G^\circ}{RT}\right\}$$

[218] At 25°C, what is the Gibbs energy of oxygen gas $\text{O}_2(g)$ when its partial pressure is 20.9 kPa (as in air) or 300 bar (as in a gas cylinder)? See Web#218.

[219] Except, of course, for elements, for which the first right-hand term is zero by convention.

[220] The p° and c° terms are frequently omitted from equations such as this, on the understanding that pressures and concentrations are the unit-less values of these properties measured in bars and moles per liter respectively. Such concentration values are sometimes represented by the formula of the solute in brackets, as in [Cl⁻].

[221] Most likely, the reaction would be carried out by bubbling hydrogen through a suspension of mercury(I) chloride in water, in which case the hydrogen activity would soon remain constant.

the subscript "equil" serving as a reminder that this is the value of the activity grouping only at equilibrium. Note an important and valuable feature of equations 2:21: one side of the relationship relates to standard conditions, the other to equilibrium conditions. On comparing this result with equation 2:5, we find

2:22
$$\left(\frac{a_{H_2}}{a_{H_3O^+}^2 \, a_{Cl^-}^2}\right)_{equil} = K \qquad \text{law of chemical equilibrium}$$

which is an example of the usual formulation of the **law of chemical equilibrium**.

In addition to affecting the Gibbs energy change in reactions, activities also govern the ΔG accompanying the transfer of a substance from one phase to another, as in Figure 2-4. Imagine that such a small amount of a nonionic substance transfers from phase L to phase R that no significant change occurs in the composition of either phase. Then the Gibbs energy change is

2:23
$$\Delta G = RT \ln\left\{\frac{a_i^R}{a_i^L}\right\} \qquad \text{for i neutral}$$

However, if the transferred species is ionic, we may think of the transfer taking place in two steps, via the intermediate reservoir labeled I in Figure 2-4. This has the same electrical potential as Reservoir L but the same composition as Reservoir R. Thus the first half of the journey, $L \rightarrow I$, encounters no electrical discontinuity and the change in Gibbs energy is as in equation 2:23. The remainder of the journey, $I \rightarrow R$ involves no chemical discontinuity and is assumed to leave the potentials of phases I and R unchanged. The second half of the journey requires electrical work to be performed of a magnitude that equation 1:8 shows to be $Q_i(\phi^R - \phi^I)$ for an individual ion or $z_iF(\phi^R - \phi^I)$ on a molar basis. Accordingly, the total Gibbs energy change is

2:24 $$\Delta G = \Delta G^{L \rightarrow I} + \Delta G^{I \rightarrow R} = RT \ln\left\{\frac{a_i^I}{a_i^L}\right\} + z_iF(\phi^R - \phi^I) = RT \ln\left\{\frac{a_i^R}{a_i^L}\right\} + z_iF(\phi^R - \phi^L)$$

The equations refer to the transfer of one mole of i but, of course, much smaller transfers would be demanded to satisfy the stipulation of unchanged composition and (especially) potential.

If phases R and L are at equilibrium with respect to species i, there is no change in Gibbs energy accompanying the transfer and equation 2:24, which of course incorporates

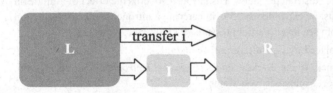

Figure 2-4 The transfer of an uncharged substance i, between Reservoirs L and R, can be made directly but, for a charged species, it is convenient to imagine travel through an intermediate Reservoir I.

2:23, may be rearranged to

2:25
$$a_i^R \exp\left\{\frac{z_i F}{RT}\phi^R\right\} = a_i^L \exp\left\{\frac{z_i F}{RT}\phi^L\right\}$$

which identifies a condition of **transfer equilibrium** for charged (or uncharged) species. Clearly the concept may be extended to multiple phases or to a continuum. If a gradient of potential (an electric field) exists in a fluid at equilibrium, ions i will distribute themselves so that $a_i \exp\{z_i F\phi/RT\}$, which is known as the **electrochemical activity**, is the same at all points. Equation 2:25 finds application in the theories of Gouy-Chapman (pages 262–264) and of Debye-Hückel (pages 41–45). This equilibrium is one example of the **Boltzmann's distribution law**[222], which in a more general form asserts that, at equilibrium, the concentrations at two sites are related by[223]

2:26
$$\frac{c_i^R}{c_i^L} = \exp\left\{\frac{-N_A}{RT}w_i^{L\to R}\right\}$$ Boltzmann's law

where $w_i^{L\to R}$ is the work needed to move a particle of i from L to R.

Ionic Solutions: the behavior of dissolved ions

Studies of solutions of inorganic chemicals in water have provided most of our knowledge of the behavior of ions. Even though electrochemists now use many others, water remains the most important solvent[224] and aqueous solutions of ions are the traditional, and best understood, examples of ionic conductors. Accordingly, it is *aqueous* solutions that are mostly discussed in this section.

A word about **concentration**. The *SI* unit of concentration is mol m^{-3}, but chemists[225] prefer to use moles per liter (mol L^{-1}), which is spoken of as **molar** and abbreviated M. As we have seen, the thermodynamic standard concentration c° is one molar. For compatibility, this book often cites millimolar (mM) concentrations, this being equal to the *SI* unit.

A valid, but nonthermodynamic, interpretation of equilibrium is that the forward and reverse reactions are proceeding at equal rates. This idea is conveyed by the reversed arrows that chemists use to indicate the existence of equilibrium. For example, the equilibrium that exists in water between molecular water and its ions is described by the

---

[222] Ludwig Eduard Boltzmann, 1844–1906, Austrian physicist. His law also governs the effect of altitude on atmospheric pressure. How much work must be expended at 25°C in carrying a nitrogen molecule to the top of a mountain (partial pressure 57 kPa), from sea level (partial pressure 79 kPa)? See Web#222.

[223] The law is usually written with R/N_A replaced by k_B, **Boltzmann's constant**.

[224] In fact, until the seminal work of Wazonek, Blaha, Berkey, and Runner (*J. Electrochem. Soc.* 102, 1955, 235), the electrochemical solvent used almost exclusively was water.

[225] Another unit favored by physical chemists is **molality**; the molar amount of solute per kilogram of solvent.

stoichiometric equation[226]

2:27
$$2\,H_2O(\ell) \rightleftarrows H_3O^+(aq) + OH^-(aq)$$

Note that, at equilibrium, the concepts of reactant and product are meaningless and we could equally write the equation the other way round[227]. Equilibrium 2:27 is addressed by the law of chemical equilibrium in the form

2:28
$$K = \left(\frac{a_{H_3O^+}\,a_{OH^-}}{a_{H_2O}^2} \right)_{equil} \qquad \text{equilibrium constant}$$

The value of K can be calculated from G° data for the $2\,H_2O(\ell) \to H_3O^+(aq) + OH^-(aq)$ reaction[228] and is

2:29
$$1.005 \times 10^{-14} = K = a_{H_3O^+}\,a_{OH^-} = \frac{\gamma_{H_3O^+}\,\gamma_{OH^-}\,c_{H_3O^+}\,c_{OH^-}}{(c^\circ)^2} \qquad \begin{array}{l}\text{ionic}\\ \text{product}\end{array}$$

The equilibrium constant for this reaction is so important that it has been given the special symbol K_w and the name **ionic product for water**[229]. In formulating equation 2:29, we have dropped the "equil" subscript[230], and taken the activity of water to be unity, which is valid for dilute aqueous solutions. If the water is free of other ions, then electroneutrality requires that $c_{H_3O^+} = c_{OH^-}$. Moreover, activity coefficients can be treated as unity at the low concentrations involved, so that taking the square-root of equation 2:29 gives

2:30
$$c_{H_3O^+} = c_{OH^-} = c^\circ \sqrt{K_w} = 1.003 \times 10^{-7}\,\text{mM} \qquad \text{in pure water}$$

In aqueous solution, the activities (and, approximately, the concentrations) of the H_3O^+ and OH^- ions are in inverse relationship to each other, as illustrated in Figure 2-5 overleaf.

Water is by far the most common solvent and the activity of the hydronium ion is important in determining the properties of aqueous solutions. The usual way of reporting this property is through the *p*H, defined[231] by

2:31
$$pH = -\log_{10}\left\{ a_{H_3O^+} \right\} \qquad \text{definition of } pH$$

---

[226] This equilibrium is often written $H_2O(\ell) \rightleftarrows H^+(aq) + OH^-(aq)$ but, because a naked proton cannot exist in aqueous solution, we prefer the representation in 2:27. There is evidence from the work of Eigen[251], however, that the formula $H_9O_4^+(aq)$ might be more appropriate than $H_3O^+(aq)$.

[227] whereupon the equilibrium constant would become the reciprocal of that given in 2:28 and 2:29. The standard change in Gibbs energy would become the negative of that reported in the following footnote.

[228] From the data on page 391 show that $\Delta G^\circ = 79.9$ kJ mol^{-1}. Then, go on to calculate the K value given in equation 2:29, as in Web#228.

[229] Other solvents have ionic products; for sulfuric acid it is 4.7×10^{-3}, for dimethyl sulfoxide 5.0×10^{-34}. The ions involved, in each case, result from **autoprotolysis**, the transfer of a proton from one solvent molecule to another. When its ionic product is very large, a solvent is classed as an **ionic liquid** (page 9).

[230] These reactions are so fast that equilibrium is almost always established.

[231] The problematic definition of *p*H as $-\log_{10}\{c_{H_3O^+}\}$ or $-\log_{10}\{c_{H^+}\}$ may be encountered.

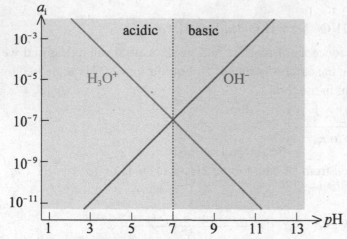

Figure 2-5 In water, the activities of the hydronium and hydroxide ions are interrelated by $a_{H_3O^+} \cdot a_{OH^-} = K_w$. The pH is defined as $-\log_{10}\{a_{H_3O^+}\}$.

pH values of less than 7 have a preponderance of hydronium ions over hydroxide ions and are said to be **acidic**. When pH values exceed 7, $c_{OH^-} > c_{H_3O^+}$, and the solution is **basic** or alkaline. Only when the pH is very close to 7 is the solution **neutral**. Following the groundbreaking concepts of Arrhenius[232], solutes that increase the hydronium ion content, and thereby decrease the pH, of water are said to be **acids**. Carbon dioxide is a case in point; it sets up the equilibrium[233]

2:32 $$CO_2(g) + 2H_2O(\ell) \rightleftarrows H_3O^+(aq) + HCO_3^-(aq)$$

The equilibrium constant for this process[234] is known as the **acidity constant** of carbon dioxide[235]. Conversely, a species such as sodium oxide that increases the pH of water is a **base**. It reacts with water completely[236]

2:33 $$Na_2O(s) + H_2O(\ell) \rightarrow 2Na^+(aq) + 2OH^-(aq)$$

to produce two hydroxide ions for each sodium oxide molecule.

---

[232] Svante August Arrhenius, 1859 – 1926, Swedish chemist and 1903 Nobel laureate. To Arrhenius we owe present-day concepts of ions, and the **Arrhenius equation**, describing how rate constants depend on temperature via the factor $\exp\{-\Delta G^{\ddagger}/RT\}$ where $\Delta G^{\ddagger}$ is the Gibbs **energy of activation**.

[233] Equation 2:32 is a simplification of what actually occurs. The species $CO_2(aq)$, $H_2CO_3(aq)$ and $CO_3^{2-}(aq)$ are involved in the full story.

[234] Its value is 4.7×10^{-7}. Calculate the approximate pH of "soda water" (H_2O in equilibrium with CO_2 at one bar partial pressure). See Web#234.

[235] Weak bases similarly have **basicity constants**. For example, ammonia NH_3 has a basicity constant defined by $a_{NH_4^+} a_{OH^-} / a_{H_2O} a_{NH_3}$. How is the acidity constant of the ammonium ion, NH_4^+ related to this basicity constant? See Web#235.

[236] How is the pH of one liter of water affected when one gram of sodium oxide, Na_2O, dissolves? See Web#236.

Electrolytes are compounds that dissolve (usually in water) to produce ions[237]. The adjectives **weak** and **strong** are applied according to whether the electrolyte does, or does not, establish an equilibrium: carbon dioxide is a weak acid; sodium oxide is a strong base. There are numerous exceptions, but generally inorganic electrolytes are strong, whereas organic electrolytes are weak. When an electrolyte is only partially soluble, as is lead sulfate in water, an equilibrium is established between the solid and the ions it forms, as in

2:34 $$PbSO_4(s) \rightleftarrows Pb^{2+}(aq) + SO_4^{2-}(aq)$$

The equilibrium constant that describes this equilibrium

2:35 $$K = \frac{a_{Pb^{2+}} a_{SO_4^{2-}}}{a_{PbSO_4}} = \frac{\gamma_{Pb^{2+}} \gamma_{SO_4^{2-}} c_{Pb^{2+}} c_{SO_4^{2-}}}{(c^o)^2}$$ solubility product

is given the special name **solubility product**[238].

The properties of the solvent are profoundly affected by the presence of solutes, and particularly ionic solutes. The molecules of dipolar solvents, such as water (page 2) cluster around dissolved ions, becoming oriented so that the end of opposite charge is closest to the ion. The phenomenon is known as **solvation**, or specifically **hydration** if the solvent is water. For example, four water molecules cluster around a dissolved copper(II) ion; they are bound so strongly that $[Cu(H_2O)_4]^{2+}(aq)$ is perhaps a more appropriate formula than $Cu^{2+}(aq)$ for the dissolved ion.

Ionic Activity Coefficients: the Debye-Hückel model

The approximation $a \approx c/c^o$ is usually acceptable for nonionic solutes, but for ions only at unusually low concentrations. The distinction in behavior between ions and nonions is easy to understand: nonionic solutes encounter each other only when they happen to collide, but ions feel coulombic forces even from their remote neighbors. Any one ion experiences the forces from all the ions present, those from multiply charged ions being particularly strong. Accordingly it is the **ionic strength**[239], defined by

2:36 $$\mu = \tfrac{1}{2} \sum_i z_i^2 c_i$$ definition of ionic strength

that affects the activities of individual ions. For example, in a solution prepared by

---

[237] In the context of electrochemical cells (Chapter 3), "electrolyte" has another meaning.

[238] and sometimes the special symbol K_{sp}. Though tabulations of solubility product values are usually given without units, they are often based on concentrations, not activities.

[239] The concentrations of the major ions in ocean water are: Cl⁻ 554.1 mM, Na⁺ 469.7 mM, Mg²⁺ 47.0 mM, SO_4^{2-} 15.3 mM, K⁺ 9.9 mM, Ca²⁺ 9.5 mM, and the **ion pair** $NaSO_4^-$ 6.1 mM. Calculate the ionic strength of seawater. See Web#239. This Web also discusses ion association generally.

dissolving 0.174 grams (one millimole) of K_2SO_4 in water to make one liter of solution, there will be the ionic[240] concentrations: $c_{K^+} = 2.00$ mM and $c_{SO_4^{2-}} = 1.00$ mM, whence the ionic strength is easily calculated as $\mu = [(+1)^2 c_{K^+} + (-2)^2 c_{SO_4^{2-}}]/2 = 3.00$ mM. The ionic strength is one component of an important parameter that occurs repeatedly in the theory of ionic solutions. This is known as the **Debye length**[241] and is defined by[242]

$$2:37 \qquad \beta = \sqrt{\frac{RT\varepsilon}{2F^2\mu}} \qquad \beta^{aq} = \frac{9.622 \text{ nm}}{\sqrt{\mu/\text{mM}}} \quad \text{at } 25°C \qquad \text{Debye length}$$

For the aqueous potassium sulfate solution just discussed, the Debye length is 5.56 nm, a small distance, but considerably larger than most ionic sizes.

A famous model, devised in 1923 by two European collaborators, is designed to explain how the electrical interactions between ions lead to their less-than-unity activity coefficients. In their treatment, Debye and Hückel[243] considered a solution containing any number of ions of various species, but focused on one ion only: the **central ion**, as illustrated in Figure 2-6. All others are not treated as individuals, but as contributors to a cloud-like **ionic atmosphere** surrounding the central ion. If the central ion is a cation, coulombic forces cause the spherically symmetrical cloud to be more heavily populated by anions than cations. The presence of this negatively charged cloud makes the central cation more stable and therefore less active than it otherwise would be.

The mathematics of the Debye-Hückel theory is quite elaborate and is given in detail elsewhere[244]. Here we shall merely sketch the salient steps in the treatment and cite the results. The solution is at equilibrium and therefore the Boltzmann distribution formula (page 38) holds for each species of ion

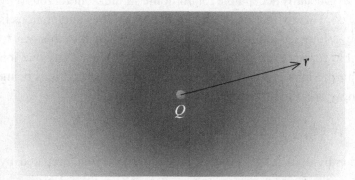

Figure 2-6 The ionic atmosphere is pictured as a **cloud of electricity**, of sign opposite to that of the central ion.

---

[240] Like most salts, potassium sulfate is a strong electrolyte, fully ionized both in the solid state and in solution.

[241] Peter Joseph Wilhelm Debye, 1884 – 1966, Dutch physicist and 1936 Nobel laureate.

[242] Verify that the units of β are those of length and confirm the value of β^{aq} given in equation 2:37. See Web#242.

[243] Erich Armand Joseph Hückel, 1896 – 1980, German physicist, also known for his molecular orbital work.

[244] A full treatment will be found in Web#244.

2:38
$$a_i(r)\exp\left\{\frac{-z_i F}{RT}\phi(r)\right\} = \text{an } r\text{-independent constant for each i}$$

where $\phi(r)$ is the local electrical potential. By making plausible approximations, and recognizing that the charge density at any distance r from the central ion is caused by a cation/anion disparity, so that

2:39
$$\rho(r) = F\sum_i z_i c_i(r) \qquad \text{local charge density}$$

Debye and Hückel established the result

2:40
$$\frac{-\rho(r)}{\varepsilon} = \frac{\phi(r)-\phi(\infty)}{\beta^2} = \frac{\Delta\phi(r)}{\beta^2}$$

This linear relationship between charge density and potential, coupled with the requirement that the overall charge in the ionic atmosphere must balance the charge Q on the central ion,

2:41
$$Q + 4\pi\int_0^\infty r^2 \rho(r)\,dr = 0 \qquad \text{charge balance}$$

is all that is needed to solve Poisson's law, equation 1:15. The solution is

2:42
$$\Delta\phi(r) = \frac{Q}{4\pi\varepsilon r}\exp\left\{\frac{-r}{\beta}\right\}$$

Without the ionic cloud, the potential would have been $\Delta\phi_{naked}(r) = Q/(4\pi\varepsilon r)$ and so the potential $\phi_{cloud}(r)$ due to the cloud is the difference $\Delta\phi(r) - \Delta\phi_{naked}(r)$. Therefore, at the site of the central ion, the potential caused by the ionic cloud is found to be

2:43
$$\phi_{cloud}(0) = \lim_{r\to 0}\left[\frac{Q}{4\pi\varepsilon r}\left(\exp\left\{\frac{-r}{\beta}\right\}-1\right)\right] = \frac{-Q}{4\pi\varepsilon\beta}$$

Next, there is a need to assess the stabilization energy that the central ion possesses by virtue of its ionic atmosphere. The calculation of this quantity resembles finding the energy stored by a capacitor, which is one-half of the product of the charge stored and the potential across the plates, being $Q^2/(8\pi\varepsilon\beta)$ in the case of the ionic atmosphere. This energy stabilizes the central ion and represents external work W that must be supplied to overcome the presence of the ionic atmosphere. As on page 32, external work subtracts from ΔG and hence modifies the activity. The resultant effect emerges as the activity coefficient

2:44
$$\gamma_i = \exp\left\{-z_i^2\sqrt{\mu/\mu_{DH}}\right\} \qquad \text{where } \mu_{DH} = \left(2\pi N_A\right)^2\left(2RT\varepsilon/F^2\right)^3$$

μ_{DH} is a constant that equals 727 mM for aqueous[245] solutions at 25°C. For example, equation 2:44 predicts that, in our exemplary potassium sulfate solution, $\gamma_{K^+} = 0.938$ and $\gamma_{SO_4^{2-}} = 0.773$.

---

[245] Recalculate β and μ_{DH} for acetonitrile, CH_3CN, at 0.00°C. Compare with Web#245.

Electroneutrality demands that anions always accompany cations, so we have no way of measuring the properties of individual ions. Instead, for an electrolyte solution composed of one anionic and one cationic species, we define a **mean ionic activity** by

2:45
$$a_\pm = \left(\frac{a_-^{z_+}}{a_+^{z_-}}\right)^{1/(z_+ - z_-)}$$
mean activity

Here z_+ and z_- are the charge numbers of the cation and anion, the latter being negative. For our potassium sulfate solution, the relation is $a_\pm = (a_{K^+})^{2/3}(a_{SO_4^{2-}})^{1/3}$. Likewise, the **mean ionic activity coefficient** is a composite defined in a similar way and therefore

2:46 $\gamma_\pm = \exp\left\{z_+ z_- \sqrt{\mu/\mu_{DH}}\right\}$ $\mu_{DH}^{aq} = 727$ mM at $25°$C   limiting Debye-Hückel law

generally, according to the simple Debye-Hückel model. The prediction for our millimolar potassium sulfate solution example is $\gamma_\pm = 0.879$. Notice that μ_{DH} is the ionic strength at which an aqueous solution of a $z_+ z_- = -1$ strong electrolyte (such as NaCl) is predicted by equation 2:46 to have a mean ionic activity coefficient of $1/e$ or 0.368. These $\gamma_\pm$ values may be measured by electrochemical and other[246] methods and have been tabulated[247]. For our millimolar K_2SO_4 case, the measured value of 0.885 is not too far removed from the 0.879 prediction, but at higher ionic strengths the divergence worsens. Examples of experimental data for potassium fluoride KF and calcium chloride $CaCl_2$ are graphed as green and violet points respectively in Figure 2-7 and compared with the prediction of equation 2:46, illustrated in red. Evidently the theory becomes increasingly unsuccessful as μ increases. Nevertheless it does hold in the limit of sufficiently low ionic strengths and, for this reason, equation 2:46 (or 2:44) is often known as the **Debye-Hückel limiting law**.

It is easy to level criticisms against the Debye-Hückel model and there have been many attempts to improve it, or to adjust the limiting law empirically. One obvious flaw in the model is that ions are accorded no size. It is unreasonable to have the ion cloud extending to $r = 0$. When an inner limit of $r = R_c$ is placed on the ion cloud, as in Figure 2-8, the treatment[244] yields the **Debye-Hückel extended law**

2:47 $\gamma_\pm = \exp\left\{\frac{z_+ z_- \sqrt{\mu/\mu_{DH}}}{1 + (R_c/\beta)}\right\}$ $\mu_{DH}^{aq} = 727$ mM, $\beta^{aq} = \dfrac{9.62\,\text{nm}}{\sqrt{\mu/\text{mM}}}$ extended D-H law

The difficulty with the extended law is in assigning a value to R_c. Should this be the radius of the central ion? The mean of the radii of the cation and anion? With or without an adjustment for solvation? In practice, R_c is often regarded as an adjustable parameter,

---

[246] The concentrations of its ions in a saturated solution of lead sulfate, $PbSO_4$, are 0.14 mM. Assuming the Debye-Hückel limiting law, calculate the mean ionic activity coefficient. Also calculate the solubility product of lead sulfate and hence find the change in standard Gibbs energy that accompanies the reaction $PbSO_4(s) \rightarrow Pb^{2+}(aq) + SO_4^{2-}(aq)$. Does the table on page 391 confirm this? See Web#246.

[247] notably by Petr Vanýsek, in *Handbook of Chemistry and Physics*, CRC Press, annual edns.

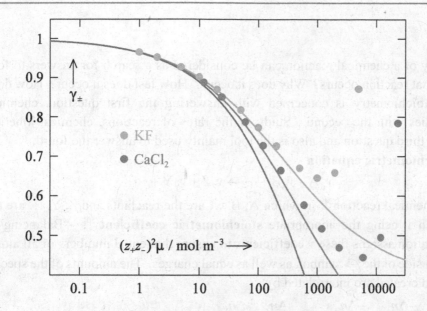

Figure 2-7 The Debye-Hückel model in its limiting version predicts that the mean ionic activity coefficients of salts depend on $(z_+z_-)^2\mu$ according to the red line. The points show experimental data for potassium fluoride and calcium chloride solutions. The predictions of the extended law for $z_+z_- = -1$ cases are shown in green and those for $z_+z_- = -2$ are in violet.

chosen to fit experimental data. Often a value close to 0.39 nm, about the diameter of a water molecule[248], is used for aqueous solutions of simple electrolytes. This was the choice used in calculating the green and violet lines in Figure 2-7. As the figure illustrates, the extended law provides a distinct improvement over the limiting law. At the highest ionic strengths, the mean ionic activity coefficients of individual electrolytes behave idiosyncratically, often increasing above unity.

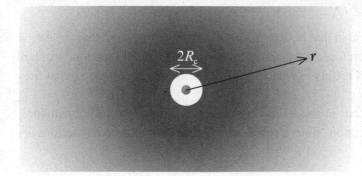

Figure 2-8 In the extended version of the Debye-Hückel model the ion cloud extends inwards only as far as a sphere of radius R_c.

---

[248] The calculation of this diameter will be found at Web#248.

Chemical Kinetics: rates and mechanisms of reactions

The study of a chemical reaction can be considered as a search for answers to four questions: What reaction occurs? Why does it occur? How fast does it occur? How does it occur? Stoichiometry is concerned with answering the first question, chemical thermodynamics with the second. Study of the rates of reactions, chemical kinetics, addresses the third question and also is the tool mainly used to answer the fourth.

The **stoichiometric equation**

2:48
$$v_A A + v_B B + \cdots \rightarrow v_Z Z + v_Y Y + \cdots$$

describes a chemical reaction[249] in which A, B, ... are the reactants and Z, Y, ... are the products, each v being the appropriate **stoichiometric coefficient**[250]. "Balancing" a chemical equation assigns these v coefficients to ensure that equal numbers of all atoms occur on each side of the $\rightarrow$ symbol, as well as equal charges. The amounts of the species consumed and created are interrelated by

2:49
$$\frac{-\Delta n_A}{v_A} = \frac{-\Delta n_B}{v_B} = \cdots = \frac{\Delta n_Z}{v_Z} = \frac{\Delta n_Y}{v_Y} = \cdots \qquad \text{amounts destroyed}$$
$$\text{or created}$$

where Δn_A, Δn_B, Δn_Z, ... are changes in the amounts (moles) of A, B, Z, ... during some interval of time. In an infinitesimal length of time, we have

2:50
$$\frac{-1}{v_A}\frac{d n_A}{dt} = \frac{-1}{v_B}\frac{d n_B}{dt} = \cdots = \frac{1}{v_Z}\frac{d n_Z}{dt} = \frac{1}{v_Y}\frac{d n_Y}{dt} = \cdots$$

This equation, based on purely stoichiometric principles, provides the basis for the definition of reaction rate.

Before leaving the topic of stoichiometry note that, although we usually think of stoichiometric coefficients as being positive integers, equations 2:49 and 2:50 remain valid if species are transferred, in whole or in part, from one side of the chemical equation to the other, provided that a change of sign accompanies the transfer. For example, recognize that the reactions described by the equations

2:51
$$2A + B \rightarrow Z + Y \qquad \text{and} \qquad 2A + \tfrac{1}{2}B - Z \rightarrow Y - \tfrac{1}{2}B$$

are stoichiometrically equivalent in that both correspond to the requirement that $-\tfrac{1}{2}\Delta n_A = -\Delta n_B = \Delta n_Z = \Delta n_Y$ and are therefore "balanced" correctly. Later in this section, we shall make use of such transfers as a means of enriching stoichiometric equations so that they also convey kinetic information.

Chemical reaction may occur in spaces having one, two or three spatial dimensions. One-dimensional reactions occur on a line, a **three-phase junction**, but such reactions are

---

[249] The stoichiometric equation of an *electrochemical* reaction, many of which will be encountered in future chapters, differs from a chemical equation only in including electrons on one side of the equation or the other.
[250] Some authorities define the stoichiometric coefficients of reactants as negative numbers; we do not.

among the least studied, and least understood, chemical processes. Two-dimensional reactions, of which reaction 2:1 is an example, occur on surfaces or phase boundaries and are referred to as **heterogeneous reactions**. The reactions leading to equilibrium 2:27 provide examples of a chemical reaction taking place in three spatial dimensions, a class to which the name **homogeneous reaction** is applied. This latter class is the simplest and the one on which, historically, the study of chemical kinetics has been based.

If reaction 2:48 is homogeneous, it will be occurring uniformly in some space of volume V. Then, any of the terms in equation 2:50, divided by V, defines the reaction rate v, which implies

$$2:52 \qquad \frac{-1}{v_A}\frac{dc_A}{dt} = \frac{-1}{v_B}\frac{dc_B}{dt} = \cdots = \frac{1}{v_Z}\frac{dc_Z}{dt} = \frac{1}{v_Y}\frac{dc_Y}{dt} = \cdots = v \qquad \text{definition of reaction rate}$$

Rates of homogeneous reactions thus have *SI* units of mol m^{-3} s^{-1}.

As noted on page 33, reaction rates respond to the activities of the reactants. As an example, one reaction and the corresponding **rate law** are[251]

$$2:53 \qquad 2H_2O(\ell) \rightarrow H_3O^+(aq) + OH^-(aq) \qquad \vec{v} = \vec{k}a_{H_2O}^2 = \vec{k}$$

Here, the proportionality constant $\vec{k}$ is an activity-based **rate constant**. In this case, the reactant activity in unchangeable, so the reaction rate is a constant. The rate of the backward reaction

$$2:54 \qquad 2H_2O(\ell) \leftarrow H_3O^+(aq) + OH^-(aq) \qquad \overleftarrow{v} = \overleftarrow{k}a_{H_3O^+}a_{OH^-}$$

is proportional to the product of the activities of the reacting ions, through a second rate constant $\overleftarrow{k}$. Because activities are pure numbers, all activity-based rate constants for homogeneous reactions take the *SI* unit mol m^{-3} s^{-1}. When the reaction is at equilibrium, the forward and backward rates are equal

$$2:55 \qquad 2H_2O(\ell) \rightleftarrows H_3O^+(aq) + OH^-(aq) \qquad \vec{v} = \overleftarrow{v}$$

and it therefore follows that the ratio of the rate constants is equal to an activity quotient. This ratio, by the law of chemical equilibrium (see equation 2:28), is the **equilibrium constant** for the reaction

$$2:56 \qquad \frac{\vec{k}}{\overleftarrow{k}} = \left(\frac{a_{H_3O^+}a_{OH^-}}{a_{H_2O}^2} \right)_{equil} = K \qquad \text{equilibrium}$$

in this case, the ionic product of water. This is a remarkable relationship in that it links two major branches of physical chemistry: chemical thermodynamics and chemical kinetics.

The proton-exchange reactions 2:53 and 2:54 are particularly simple because the entire reaction appears to occur in a single step; generally several steps are required. The

---

[251] Manfred Eigen (German physical chemist, 1967 Nobel laureate for his work on fast reactions) reported the rate constant values of 2.5×10^{-5} s^{-1} and 1.1×10^{11} L mol^{-1} s^{-1} for reactions 2:53 and 2:54 respectively. Are these values consistent with the known ionic product of water? See Web #251.

mechanism of a chemical reaction is an inventory of the **elementary steps** by which the reaction is believed to proceed. Rarely is it necessary to invoke more than two types of elementary step[252]. A **bimolecular step**, of which reaction 2:54 is an example, requires an interaction between two particles[253] and has the following model equation and rate law:

2:57 $\qquad A + B \xrightarrow{k_1} \text{product(s)} \qquad v_1 = k_1 a_A a_B \qquad$ Step (1)

and therefore two activities are pertinent. A **unimolecular step**

2:58 $\qquad\qquad C \xrightarrow{k_2} \text{product(s)} \qquad v_1 = k_2 a_C \qquad$ Step (2)

involves a single particle (atom, molecule or ion). The number of activity terms entering the rate law is called the reaction's **order**; reaction 2:57 is of order two, reaction 2:58 of order one. The possibility of the step, say Step (3), proceeding in both directions simultaneously must usually be countenanced, as in

2:59 $\qquad A + B \underset{k_{-3}}{\overset{k_3}{\rightleftarrows}} Z \qquad v_3 = \vec{v}_3 - \bar{v}_3 = k_3 a_A a_B - k_{-3} a_Z \qquad$ Step (3)

the rate law of which represents the net effect of a bimolecular forward component and a unimolecular backward component. Mechanisms are, by their nature, putative, but an acceptable mechanism must correctly predict the stoichiometric, kinetic, and equilibrium properties of the overall reaction:

2:60 $\qquad \left.\begin{array}{r}\text{a satisfactory mechanism}\\[4pt]\text{must meet three criteria}\end{array}\right\rbrace \left\{\begin{array}{l}\text{(i) satisfy the stoichiometry}\\[4pt]\text{(ii) predict the experimental rate law}\\[4pt]\text{(iii) be compatible with the equilibrium law}\end{array}\right.$

Multistep mechanisms generally involve one crucial step, known as the **rate-determining step**. This is the "slow" step[254] of the mechanism, other steps being rapid in comparison. Rapid bidirectional steps are effectively at equilibrium. Thus if 2:59 is rapid, its effect is to make the two terms $k_3 a_A a_B$ and $k_{-3} a_Z$ large in comparison to their difference, so that $k_3 a_A a_B \approx k_{-3} a_Z$ and therefore $a_Z / a_A a_B \approx k_3 / k_{-3} = K_3$, the equilibrium constant of Step (3).

The following is one example[255] of a reaction with a stoichiometry that is far too complicated for the reaction to occur in a single step

$$2Br^-(aq) + H_2O_2(aq) + 2H_3O^+ \rightleftarrows Br_2(aq) + 4H_2O(\ell)$$

2:61 $$v = \vec{k} a_{Br^-} a_{H_2O_2} a_{H_3O^+} - \bar{k}\, \frac{a_{Br_2}}{a_{Br^-} a_{H_3O^+}}$$

---

[252] The reaction $2NO(g) + Cl_2(g) \rightarrow 2NOCl(g)$ could be explained by a termolecular step. Suggest an alternative mechanism that does not involve the simultaneous collision of three molecules? See Web#252.

[253] The particles may be the same $A + A \xrightarrow{k_1} \text{product(s)} \qquad v_1 = k_1 a_A^2$

[254] It is common practice to refer to the rate-determining step as the "slow step", though the various steps in a mechanism usually proceed at the same speed, just as do ships in convoy.

[255] A second may be found at Web#255.

The arrows of unequal length signify that both the forward and backward reactions occur, but at unequal rates. The experimental rate law[256] of reaction 2:61, also bespeaks mechanistic complexity. A four-step mechanism proposed for this reaction is

2:62
$$\begin{cases} \text{Br}^-(aq) + \text{H}_3\text{O}^+(aq) \underset{k_{-1}}{\overset{k_1}{\rightleftarrows}} \text{HBr}(aq) + \text{H}_2\text{O}(\ell) & \text{Step (1), rapid} \\[2mm] \text{HBr}(aq) + \text{H}_2\text{O}_2(aq) \underset{k_{-2}}{\overset{k_2}{\rightleftarrows}} \text{HOBr}(aq) + \text{H}_2\text{O}(\ell) & \text{Step (}\hat{2}\text{), rate-determining} \\[2mm] \text{Br}^-(aq) + \text{HOBr}(aq) \underset{k_{-3}}{\overset{k_3}{\rightleftarrows}} \text{Br}_2(aq) + \text{OH}^-(aq) & \text{Step (3), rapid} \\[2mm] \text{H}_3\text{O}^+(aq) + \text{OH}^-(aq) \underset{k_{-4}}{\overset{k_4}{\rightleftarrows}} 2\text{H}_2\text{O}(\ell) & \text{Step (4), rapid} \end{cases}$$

Note our use of the circumflex, as in $(\hat{2})$, to indicate the rate-determining step. Species such as HBr and HOBr that appear in the mechanism, but not in the overall reaction, are known as **intermediates**. Inasmuch as adding the four mechanistic steps in 2:62 yields equation 2:61, the first of the three criteria listed above in 2:60 is satisfied by this mechanism. That criteria (ii) and (iii) are also met is less obvious, though it is true[257]. Accordingly 2:62 *may* be the correct mechanism.

There is a concise method of predicting the kinetic consequences of a mechanism. The method involves creating a so-called "stoichiokinetic equation" by implementing the following rules. Mechanism 2:62 is used to illustrate the rules.

(a) Write the equations of all the steps that precede the rate-determining step by transferring all species from the right- to the left-hand side, introducing negative signs on transfer, and writing the reaction as an equilibrium. In the mechanistic example 2:62, only Step (1) precedes the rate determining step and it becomes

2:63 $\qquad \text{Br}^- + \text{H}_3\text{O}^+ - \text{HBr} - \text{H}_2\text{O} \rightleftarrows \qquad$ Step (1)

(b) Write the rate-determining step unchanged. In the example, this is

2:64 $\qquad\qquad\qquad \text{HBr} + \text{H}_2\text{O}_2 \rightleftarrows \text{HOBr} + \text{H}_2\text{O} \qquad$ Step $(\hat{2})$

(c) Modify all the steps that follow the rate-determining step by transferring all species from the left to the right, introducing negative signs on transfer. Steps (3) and (4) in the example become

2:65 $\qquad\qquad\qquad\qquad\qquad \rightleftarrows \text{Br}_2 + \text{OH}^- - \text{Br}^- - \text{HOBr} \qquad$ Step (3)

2:66 $\qquad\qquad\qquad\qquad\qquad \rightleftarrows 2\text{H}_2\text{O} - \text{H}_3\text{O}^+ - \text{OH}^- \qquad$ Step (4)

(d) Leaving the rate-determining step unchanged, multiply or divide other equations, if

---

[256] The role of the activity of water in the rate law cannot be established experimentally, because water's activity is never significantly different from unity in aqueous solutions.

[257] See Web#257 for an algebraic proof that criteria (ii) and (iii) are met, without invoking the stoichiokinetic route. Thereby the stoichiokinetic approach is substantiated.

necessary, by a small integer (seldom other than 2), so that all intermediates disappear on implementing rule (e). This is unnecessary in the example.

(e) Add the modified equations to generate the stoichiokinetic equation. Cancel terms on addition, but do *not* combine terms from one side with those from the other.
The example of mechanism 2:62 is summarized in the following table:

Mechanistic equations	Step	Modified equations
$Br^- + H_3O^+ \rightleftharpoons HBr + H_2O$	(1)	$Br^- + H_3O^+ - HBr - H_2O \rightleftharpoons$
$HBr + H_2O_2 \rightleftharpoons HOBr + H_2O$	(2̂)	$HBr + H_2O_2 \rightleftharpoons HOBr + H_2O$
$Br^- + HOBr \rightleftharpoons Br_2 + OH^-$	(3)	$\rightleftharpoons \begin{cases} Br_2 + OH^- \\ -Br^- - HOBr \end{cases}$
$H_3O^+ + OH^- \rightleftharpoons 2H_2O$	(4)	$\rightleftharpoons \begin{cases} 2H_2O \\ -H_3O^+ - OH^- \end{cases}$
$\left.\begin{array}{l} 2Br^- + H_2O_2 \\ +2H_3O^+ \end{array}\right\} \rightleftharpoons Br_2 + 4H_2O$	sum	$Br^- + H_3O^+ + H_2O_2 - H_2O \rightleftharpoons \begin{cases} 3H_2O + Br_2 \\ -Br^- - H_3O^+ \end{cases}$

The result of this exercise,

2:67
$$Br^-(aq) + H_3O^+(aq) + H_2O_2(aq) - H_2O(\ell)$$
$$\rightleftharpoons 3H_2O(\ell) + Br_2(aq) - Br^-(aq) - H_3O^+(aq)$$

is the **stoichiokinetic equation** of the mechanism. It earns this name because it not only correctly displays the stoichiometry of the reaction, but it also reveals the overall kinetics implicit in the mechanism. The link between the stoichiokinetic equation and the mechanism is the correspondence between the stoichiometric coefficients and the powers to which the activities are raised in the rate law. The implied rate law is

2:68
$$v = \vec{v} - \overleftarrow{v} = \vec{k} a_{Br^-} a_{H_3O^+} a_{H_2O_2} a_{H_2O}^{-1} - \overleftarrow{k} a_{H_2O}^3 a_{Br_2} a_{Br^-}^{-1} a_{H_3O^+}^{-1}$$

for our exemplary mechanism. Apart from the immeasurable dependences on water activity, this mechanistic rate law agrees perfectly with the experimental finding in 2:61, thereby satisfying criterion (ii)[258]. Moreover, imposing the equilibrium condition $v = 0$ leads immediately to the conclusion that the equilibrium condition

2:69
$$\frac{\vec{k}}{\overleftarrow{k}} = \left(\frac{a_{Br_2} a_{H_2O}^4}{a_{Br^-}^2 a_{H_2O_2} a_{H_3O^+}^2} \right)_{equil} = K$$

---

[258] What rate law is predicted by a mechanism identical with 2:62 except that Step (3) is rate determining? Experimentally, how may this mechanism be discounted? Compare your answer with that at Web#258.

satisfies criterion (iii). Elsewhere[257] you will find a proof that the rate law 2:68 is an algebraic consequence of mechanism 2:62; the same source shows how the overall rate constants $\vec{k}$ and $\overleftarrow{k}$ are related to those of the mechanistic steps.

Unlike the foregoing discussion in terms of activities, chemical rate laws are customarily written in terms of concentrations and some consider this to be a more accurate description of kinetic behavior. Equation 2:59 would be replaced by

2:70
$$A + B \underset{k_{-3}}{\overset{k_3}{\rightleftharpoons}} Z \qquad v_3 = \vec{v}_3 - \overleftarrow{v}_3 = k'_3 c_A c_B - k'_{-3} c_Z$$

for example, where each primed k is a concentration-based rate constant, which incorporates an activity coefficient and the standard concentration. Thus the k'_3 in 2:70 replaces and equals $k_3 \gamma_A \gamma_B / (c^\circ)^2$. Chemical kineticists have a strong preference[259] for concentration-based rate laws because their use leads more easily to integrated rate laws. One penalty paid for this is that the units of concentration-based rate constants depend on the order of the reaction[260]; thus, whereas k'_{-1} has units of s^{-1}, the *SI* unit of k'_3 becomes $m^3 \, mol^{-1} \, s^{-1}$. Though concentration-based k's are not primed in the literature of chemical kinetics, we shall do so in this book, because the distinction between k and k' is important in *electro*chemical kinetics.

During the last few paragraphs, we have been addressing homogeneous reactions. Heterogeneous reactions, which include electrochemical reactions, differ in several respects from the simpler homogeneous reactions. Apart from a few very special cases, reactions occurring at surfaces require the **transport** of reactants to, and/or products from, the surface. Thus, especially in electrochemistry, concern is for processes in which reaction and transport are intimately linked. Whereas the rate of a second-order homogeneous reaction is measured in the mol $m^{-3} \, s^{-1}$ unit, this is replaced by mol $m^{-2} \, s^{-1}$ for a heterogenous reaction. Rate constants, therefore, also have different units. For a first-order reaction involving the reactant A, the heterogeneous rate law is

2:71
$$\vec{v} = \vec{k} a_A^s \qquad \text{or} \qquad \vec{v} = \vec{k}' c_A^s$$

where the unit of $\vec{k}'$ is m s^{-1}. The superscript ^{s} serves as a reminder that it is the activity or concentration of the reactant *at the surface* that is relevant. This concentration at the surface[261] is often very different from the concentration at a more distant point.

Another phenomenon encountered in heterogeneous reactions that has no parallel in homogeneous reactions is **adsorption**, illustrated in Figure 2-9 overleaf. Species reaching the surface at which a heterogeneous reaction occurs may form a bond with the surface so that a temporary "compound" – an **adsorbate** – is created.

2:72
$$A(g \text{ or } soln) \rightleftharpoons A(ads)$$

---

[259] So strong, in fact, that kinetics texts seldom mention activities at all.

[260] Worse, concentration-based equilibrium constants, that we denote K', sometimes have units, sometimes not.

[261] Not to be confused with the "surface (or superficial) concentration", which is a quantity with the mol m^{-2} unit used to quantify the extent of adsorption.

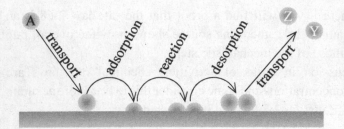

Figure 2-9 Some of the processes that may be involved in a heterogeneous reaction.

If the adsorbate is too strongly bonded, it will soon cover the surface preventing any further event. Otherwise, the adsorbate may desorb or it may decompose or undergo some other reaction, such as

2:73 $A(ads) \rightleftarrows Z(ads) + Y(ads)$ or $A(ads) + A(g \text{ or } soln) \rightleftarrows Z(ads)$

Again, it is necessary for a continuing reaction that the product of such a reaction not itself adsorb too strongly, or the surface will become blocked[262]. The final step again involves transport.

2:74 $Z(ads) \rightleftarrows Z(g \text{ or } soln)$

Figure 2-9 shows some of the processes that may be involved. Reactions of this sort, often occurring at a metal/gas interface, are of great importance, not only in electrochemistry but also in the petrochemical industry where they are described as **heterogeneous catalysis** because the metal[263] plays the role of a catalyst in what would otherwise be a feasible, but elusive, gas-phase homogeneous reaction. Enzymes, too, are heterogeneous catalysts.

Summary

The molar "chemical energy" of a compound or ion, compared with its elements, is represented by its Gibbs energy. This equals its standard Gibbs energy plus a term that reflects its activity:

2:75 $G_i = G_i^\circ + RT \ln\{a_i\} = \begin{cases} G_i^\circ & \text{pure solid, liquid or solvent of a dilute solution} \\ G_i^\circ + RT \ln\{p_i / p^\circ\} & \text{gas except at very high pressure} \\ G_i^\circ + RT \ln\{c_i / c^\circ\} & \text{nonionic solute in a dilute solution} \\ G_i^\circ + RT \ln\{\gamma_i c_i / c^\circ\} & \text{ionic solute, even in dilute solution} \end{cases}$

The activity coefficient of an ion may differ markedly from unity for reasons that the Debye-Hückel theory helps us understand. The feasibility of a chemical reaction such as

[262] sometimes usefully, as when corrosion (Chapter 11) is thereby inhibited.

[263] or other solid. Carbon surfaces catalyze the $C_2H_4(g) + H_2(g) \rightarrow C_2H_6(g)$ reaction, for example.

2:76
$$v_A A + v_B B + \cdots \rightarrow v_Z Z + v_Y Y + \cdots$$

is determined by the sign of the Gibbs energy change, with ΔG being zero at equilibrium. The *standard* Gibbs energy change is related through activities to the equilibrium constant:

2:77
$$\exp\left\{\frac{-\Delta G^\circ}{RT}\right\} = \frac{a_Z^{v_Z} a_Y^{v_Y} \cdots}{a_A^{v_A} a_B^{v_B} \cdots} = K = \frac{\vec{k}}{\overleftarrow{k}} \qquad \text{equilibrium}$$

Equilibrium is also characterized by the equality of the rates of the forward and backward reactions. Those rates depend on the concentrations (or more precisely on the activities) of the participants, A, B,..., Z, Y, ..., in the reaction, raised to powers that are not predictable from the stoichiometry but which must conform to the requirements of relationship 2:77. It is generally possible to rearrange the stoichiometric equation into a stoichiokinetic equation that matches the kinetics of the forward and reverse reactions and can suggest likely mechanisms. Chemical reactions may be homogeneous or heterogeneous, the latter class including electrochemical reactions.

3

Electrochemical Cells

Two classes of conductor – electronic and ionic – were identified in Chapter 1. The junction between an ionic conductor and an electronic conductor is named an **electrode** and it is here that the chemistry of electrochemistry occurs. Chemical processes that occur at electrodes are termed **electrochemical reactions** or **electrode reactions**, and they are described by equations that differ from ordinary chemical equations only by having electrons, in addition to chemical species, as participants; equations 3:2 and 3:3 below are examples. The simplest kind of **electrochemical cell** incorporates two electrodes.

Equilibrium Cells: two electrochemical equilibria generate an interelectrode voltage

When an electronic conductor is brought into contact with an ionic conductor, a reaction may occur. If so, this is a heterogeneous chemical reaction as discussed in Chapter 2. For example, when the electronic conductor lead, $Pb(s)$, is placed in contact with a solution of sulfuric acid, an ionic conductor containing principally the $H_3O^+(aq)$ and $HSO_4^-(aq)$ ions, the reaction[301]

$$3:1 \qquad Pb(s) + H_3O^+(aq) + HSO_4^-(aq) \rightarrow PbSO_4(s) + H_2O(\ell) + H_2(g)$$

which has a ΔG° of -47.1 kJ mol^{-1}, occurs slowly. This is *not* usually treated as an electrochemical reaction, though it could be regarded as the composite of the following two electrochemical processes:

$$3:2 \qquad\qquad 2H_3O^+(aq) + 2e^- \rightarrow 2H_2O(\ell) + H_2(g) \qquad \text{reduction}$$

$$3:3 \qquad Pb(s) + HSO_4^-(aq) + H_2O(\ell) \rightarrow 2e^- + PbSO_4(s) + H_3O^+(aq) \qquad \text{oxidation}$$

These occur simultaneously on the metal surface. A reaction such as 3:3, in which electrons are generated is termed an **oxidation**; whereas reaction 3:2, in which electrons are consumed, is an example of a **reduction**.

---

[301] Some chemists analyze reactions, chemical and electrochemical, in terms of oxidation numbers. Read about this approach at Web#301.

Electrochemical Science and Technology: Fundamentals and Applications, First Edition. Keith B. Oldham, Jan C. Myland, Alan M. Bond.
© 2012 John Wiley & Sons, Ltd. Published 2012 by John Wiley & Sons, Ltd.

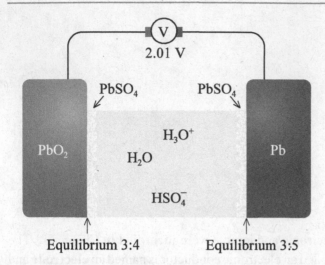

Figure 3-1 An example of a simple electrochemical cell is the lead-acid cell, here shown at rest.

   No electrical measurements can be made on a single electronic|ionic junction, such as the one we have just discussed, because there is no electric current flowing[302] and no pair of sites between which an electrical potential difference might be measured. The minimal configuration needed for an electrochemical measurement is an **electrochemical cell**, as exemplified in Figure 3-1, in which a voltmeter measures the electrical potential difference between two (different) electronic conductors, linked by a single ionic conductor. The illustrated example is an electrochemical cell familiar to most of us: the **lead-acid cell**[303]. Six such cells in series make up the common 12 V car battery. One of the electronic conductors in the lead-acid cell is lead, Pb, the other is lead dioxide, PbO_2. The ionic conductor is a concentrated aqueous solution of sulfuric acid in which the dominant species, apart from water, are the hydronium H_3O^+ and bisulfate HSO_4^- ions[304]. On discharge, a layer of lead sulfate forms on each electrode. Beneficially, these lead sulfate layers are porous, allowing the solution to penetrate to the underlying electrode.

   Notice our inconsistent usage of the word "electrode". Confusingly, electrochemists use this word with diverse meanings. Depending on the context, **electrode** may refer to (a) the two-dimensional interface between an electronic and an ionic conductor, (b) that interface plus the adjacent layers of each conductor, (c) half of the cell, and (d) the electronic conductor itself. Likewise, the term "electrolyte" is used with two meanings in this book and elsewhere. As on page 41, **electrolyte** means a substance that dissolves to

---

[302] other than internally. In corrosion reactions, of which 3:1 is an example, oxidation and reduction may occur at different sites on the metal surface, with the electrons traveling through the bulk metal.

[303] There is further discussion of the lead/acid cell on page 94. Write the equation $2PbO_2(s) + 2H_3O^+(aq) + 2HSO_4^-(aq) \rightarrow 2PbSO_4(aq) + 4H_2O(\ell) + O_2(g)$ as two electrochemical reactions proceeding concurrently. Compare your equations with those in Web#303.

[304] "Hydrogen ion" or "proton" and "H^+" are the names and symbol sometimes used to replace "hydronium ion" and "H_3O^+". An alternative, and better, name for HSO_4^- is "hydrogen sulfate" ion.

form ions. However, as a contraction of "electrolyte solution", it is also used to mean the solution thus formed. Thus, "electrolyte" is often used as an alternative to "ionic conductor" when the latter is a solution.

Figure 3-1 shows the chemical components of the lead-acid cell though not, of course, the architecture of an automotive battery. As shown, the included voltmeter would read a value close to 2.0 V, the PbO_2 being positive with respect to the Pb. The sign of the voltage reflects the greater activity of electrons on the lead electrode. The magnitude of the voltage can, and will now, be explained using the principles addressed in Chapters 1 and 2.

Heterogeneous electrochemical equilibria are established at each electrode interface. At the lead dioxide electrode the equilibrium is

$$3:4 \qquad PbO_2(s) + HSO_4^-(aq) + 3H_3O^+(aq) + 2e^-(PbO_2) \rightleftarrows PbSO_4(s) + 5H_2O(\ell)$$

while that at the lead electrode is[305]

$$3:5 \qquad PbSO_4(s) + H_3O^+(aq) + 2e^-(Pb) \rightleftarrows Pb(s) + HSO_4^-(aq) + H_2O(\ell)$$

Any actual reaction would require electrons to cross the electrode interfaces and no such motion is possible without a circuit being established[306]. In each case, the forward and reverse processes are taking place at equal rates. So no net reactions occur; the equilibria are undisturbed. Notice that we have written $e^-(Pb)$ or $e^-(PbO_2)$ to indicate that the electrons in question are "dissolved" in the parenthesized phase, though we shall discontinue to do this later in the book.

We can conjoin equations 3:4 and 3:5 by subtracting the latter from the former and then splitting the difference into a chemical part

$$3:6 \qquad PbO_2(s) + Pb(s) + 2HSO_4^-(aq) + 2H_3O^+(aq) \rightleftarrows 2PbSO_4(s) + 4H_2O(\ell)$$

that we call the **cell reaction**, and an electrical part

$$3:7 \qquad 2e^-(PbO_2) \rightleftarrows 2e^-(Pb)$$

that shows an exchange of electrons between the two electrodes. Though we have written them as equilibria, processes 3:6 and 3:7 cannot occur with the arrangement in Figure 3-1 because the voltmeter does not allow the passage of electrons.

The change in Gibbs energy accompanying the forward direction of process 3:6 can be calculated[307] according to the principles described in Chapter 2, as

$$3:8 \ \Delta G = \Delta G^\circ + RT \ln \left\{ \frac{a_{PbSO_4}^2 \, a_{H_2O}^4}{a_{Pb} a_{PbO_2} a_{HSO_4^-}^2 a_{H_3O^+}^2} \right\} = -375.1 \text{ kJ mol}^{-1} + RT \ln \left\{ \frac{(c^\circ)^4 a_{H_2O}^4}{\gamma_\pm^4 c_{H_3O^+}^2 c_{HSO_4^-}^2} \right\}$$

---

[305] The convention adopted for electrode reactions at equilibrium is to write the electrons on the left-hand side. Analyze reactions 3:4 and 3:5 in terms of changes in oxidation numbers[301] and compare with Web#305.

[306] Recall that the voltmeter measures an electrical potential difference without allowing the passage of electrons. To achieve this goal in practice, a voltage follower is used; see Web #1029.

[307] Use data from the table on page 391 to confirm the value of ΔG° reported in 3:8. See Web#307.

Under the conditions typical of the operation of a lead-acid battery, each of the two ions, has an average concentration of about 3000 mM and the final term in 3:8 has a value close to -16 kJ mol^{-1}. Notice that, as is usual in such calculations, the $\Delta G°$ term numerically dominates the $RT\ln\{\ \}$ term in leading to an overall ΔG of about -387 kJ mol^{-1}. What is the significance of "mol^{-1}" here? Per mole of what? It can be thought of as "per mole of the reaction as written". That is, in this case, per mole of lead destroyed or per *two* moles of lead sulfate created.

Despite its thermodynamic favorability, the forward direction of reaction 3:6 cannot occur unless 3:7 occurs in the reverse direction and this is thwarted by the absence of an electron pathway between the two electrodes[305]. The eagerness of cell reaction 3:6 to occur in the forward direction is manifested by the higher electron activity in the Pb compared to the PbO$_2$. Reaction 3:6 "wants" to go in the forward direction; reaction 3:7 "wants" to go backwards. If these processes were permitted, work could be performed by the passage of electrons from the Pb to the PbO$_2$. How much work? The passage of two electrons, each of which carries a charge of $-Q_0$, down a voltage drop of ΔE could perform work of $2Q_0\Delta E$. Our choice of signs in Chapter 1 was that w represented the *external* work that needed to be performed. Here no coercion is needed, w is negative and equal to $-2Q_0\Delta E$. On a molar basis, the external work would be

$$3:9 \qquad\qquad W = N_A w = -2N_A Q_0 \Delta E = -2F\Delta E$$

In accord with the equilibrium principles discussed on page 32, $\Delta G = W$ and therefore

$$3:10 \qquad\qquad \Delta E = \frac{-\Delta G}{2F} = \frac{-(-387\text{ kJ mol}^{-1})}{2\times(96485\text{ C mol}^{-1})} = 2.01\text{ V}$$

in agreement with experiment. Here ΔE is the potential difference between the two electrode potentials[308].

The procedure for finding the voltage of a simple electrochemical cell at equilibrium may be generalized as follows[309]:

(a) Write down the equilibrium reaction at electrode I, and that at electrode II, with electrons on the left-hand side of each equation

---

[308] You might be concerned that, whereas in Chapter 1 we stated that electrical potential differences could only be measured between phases of similar composition, here we appear to be measuring the potential difference between two dissimilar electronic conductors: PbO$_2$ and Pb in our example. In fact, what the voltmeter measures is the electrical potential difference between its two terminals, both of which are copper! Our interest, however, is in the activities of electrons and when two electronic conductors, such as lead and copper, are in contact and at equilibrium, then the electron activities in (not the potentials of) the two conductors will be the same.

[309] Using data from the Appendix, page 391, follow this protocol to calculate the equilibrium voltage of the cell in which the electrode reactions are Ag$_2$O(s)+H$_2$O(ℓ)+2e$^-$(Ag) $\rightleftarrows$ 2Ag(s)+2OH$^-$(aq) and ZnO(s)+H$_2$O(ℓ)+2e$^-$(Zn) $\rightleftarrows$ Zn(s)+2OH$^-$(aq), reporting which of the electrodes is positive. Explain why the only data you need for this calculation are the standard Gibbs energies of silver oxide and zinc oxide. Compare you answers with those at Web#309.

(b) If necessary, double or triple either equation to make the numbers of electrons match. The matched number of electrons is n.

(c) After any such modification, subtract equation II from equation I, to obtain a purely chemical equation. Calculate the Gibbs energy change accompanying that reaction, allowing for any nonunity activities.

(d) Then the potential difference between the two electrodes is

$$3:11 \qquad \Delta E_{\text{I versus II}} = \phi_{\text{I}} - \phi_{\text{II}} = -\frac{\Delta G}{nF} \qquad \text{cell voltage}$$

where ΔG is the Gibbs energy change accompanying the overall cell reaction.

Of course, the cell voltage depends on the activities of the chemicals participating in the cell reaction. When these are all unity, the interelectrode potential difference is called the **standard cell voltage**, denoted ΔE°.

$$3:12 \qquad \Delta E^{\circ} = \frac{-\Delta G^{\circ}}{nF} \qquad \text{standard cell voltage}$$

For the lead-acid cell reaction, it follows that

$$3:13 \qquad \Delta E^{\circ} = \frac{-(-371.4 \text{ kJ mol}^{-1})}{2 \times (96485 \text{ C mol}^{-1})} = 1.925 \text{ V}$$

This is the voltage that the lead-acid cell would develop if the hydronium and bisulfate ions were in their standard states. Notice that the standard cell voltage, the change in standard Gibbs energy, and the equilibrium constant[310] of the cell reaction are equivalent concepts, each of which is a way of expressing the equilibrium ratio of the activities of products and reactants. The relationships that apply for the general reaction 2:48 are:

$$3:14 \qquad \exp\left\{\frac{nF}{RT}\Delta E^{\circ}\right\} = \exp\left\{\frac{-\Delta G^{\circ}}{RT}\right\} = K = \left(\frac{a_{Z}^{v_Z} a_{Y}^{v_Y} \cdots}{a_{A}^{v_A} a_{B}^{v_B} \cdots}\right)_{\text{equilib}} \qquad \text{equilibrium}$$

Even though there is no chemistry actually taking place, equilibrium cells are an important source of chemical data. The measurement of cell voltages, which can be carried out with great precision, has provided most of the exact thermodynamic data that chemists need. Not only standard Gibbs energies, which are accessible directly from measured ΔE° values, but also enthalpies and entropies can be found from the way in which ΔE° changes with temperature[311]. Moreover, activity coefficients are mostly measured by studying the effect of concentration on cell voltages, a procedure that also yields E° values[312].

---

[310] From this standard cell voltage reported in 3:13, calculate the equilibrium constant of reaction 3:6. What are the equilibrium ion concentrations if samples of the three solids are brought into mutual contact under water? Is this realistic? See Web#310.

[311] The standard molar enthalpy change accompanying the cell reaction can be found as $nFT^2\text{d}(\Delta E^{\circ}/T)/\text{d}T$ and the corresponding standard molar entropy change is $nF\text{d}(\Delta E^{\circ})/\text{d}T$.

[312] An example of such a study is described in Web#312.

Cells not at Equilibrium: interchanges of chemical and electrical energy

In the previous section, we saw that a lead-acid cell could potentially perform work if an electron path were to be provided between the two electronic conductors. Figure 3-2 shows such a pathway established through a "load" and with an ammeter to measure the resulting electric current. The load presents an opportunity to perform useful work, such as starting your car; or the load could be a resistor, in which case energy is released as heat. The ammeter measures the current flowing clockwise[313] through a circuit that is partly made up of electronic conductors and partly of the ionic conductor. Through the latter, electricity flows by the migration of hydronium H_3O^+ and bisulfate HSO_4^- ions. The current crosses each of the electronic|ionic junctions by virtue of electrochemical reactions.

The chemistry at the electrode surfaces is no longer in equilibrium. Interconversions in both directions are still taking place, but they are no longer occurring at equal rates. A convenient way of representing a departure from equilibrium is by using arrows of unequal length in the equation. At the left-hand electrode, the process becomes

3:15 $\qquad PbO_2(s) + HSO_4^-(aq) + 3H_3O^+(aq) + 2e^-(PbO_2) \rightleftharpoons PbSO_4(s) + 5H_2O(\ell)$

with the forward process now outrunning the backward conversion. Therefore a net reduction occurs. An electrode at which a reduction is occurring is termed a **cathode**. Conversely, at the right-hand electrode, the backward process outruns the forward:

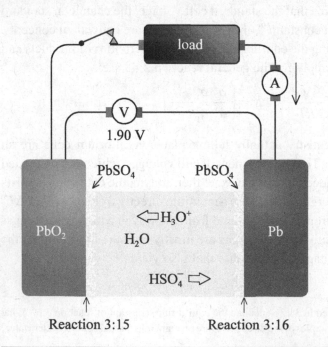

Reaction 3:15   Reaction 3:16

Figure 3-2 In its galvanic mode the lead-acid cell provides energy to a "load". The cell voltage is less than its equilibrium value.

---

[313] Electrons and anions, being negatively charged, circle counterclockwise; cations move clockwise. The ammeter is assumed not to impede the flow of current. To achieve this in practice, a current follower is used. See Web#1029.

3:16
$$PbSO_4(s) + H_3O^+(aq) + 2e^-(Pb) \rightleftarrows Pb(s) + HSO_4^-(aq) + H_2O(\ell)$$

and a net oxidation occurs. An electrode at which an oxidation is occurring is described as an **anode**. Electrons are consumed at a cathode, created at an anode. As Figure 3-2 illustrates, the cell voltage falls somewhat below its equilibrium value during this process – the greater the current, the larger the drop – as a result of electrode **polarization**. This is a phenomenon with several causes, as described in detail in Chapter 10.

As the two electrochemical reactions proceed, the Gibbs energy of the cell's contents steadily declines as chemical energy is converted into work or heat. The conversion of chemical energy, through electrical energy, to work can approach 100% efficiency; there is no Carnot limitation[314,541].

An electrochemical cell, such as that we have been describing, in which chemical energy is being destroyed is named a **galvanic cell**[315]. Conversely, when a cell captures electrical energy and converts it to chemical energy, it is said to be acting as an **electrolytic cell**. Figure 3-3 shows a lead-acid cell in electrolytic mode. Electrical current is being forced by a d.c. voltage source through the cell, which adopts a voltage somewhat higher than its equilibrium value. Many properties of the cell are now opposite to what they were when the cell was behaving galvanically. The current flows counterclockwise. The left-hand electrode, that was a cathode when the cell functioned galvanically, is now an anode at which the reaction is

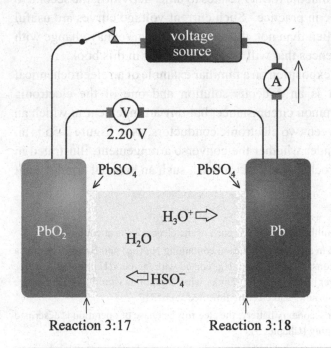

Reaction 3:17   Reaction 3:18

Figure 3-3 In its electrolytic mode the Gibbs energy of the lead-acid cell is increased at the expense of external electrical energy. The cell voltage now exceeds its equilibrium value.

---

[314] Read in thermodynamic texts about this limitation, which is encountered in converting heat to work.

[315] Luigi Galvani, 1737 – 1798, Italian physician and physicist.

3:17 $PbO_2(s) + HSO_4^-(aq) + 3H_3O^+(aq) + 2e^-(PbO_2) \rightleftarrows PbSO_4(s) + 5H_2O(\ell)$

whereas the right-hand electrode, previously the anode, now functions cathodically:

3:18 $PbSO_4(s) + H_3O^+(aq) + 2e^-(Pb) \rightleftarrows Pb(s) + HSO_4^-(aq) + H_2O(\ell)$

Of course, if the d.c. voltage source is carefully adjusted to the value 2.01 V, the current will cease and equilibrium will be restored. Any lower applied voltage leads to galvanic operation. There is a unique voltage at which the cell is at equilibrium[316]. Thus we see that a single cell may function[317] in three modes:

3:19   electrochemical cell $\begin{cases} \text{galvanic mode} \\ \text{equilibrium mode} \\ \text{electrolytic mode} \end{cases}$

all of which are useful[318]. Many practical applications of galvanic and electrolytic cells are discussed in Chapters 5 and 4 respectively. For a variety of reasons, not all electrochemical cells are able to function efficiently in both galvanic and electrolytic modes. When they can, a useful way of describing the electrical properties of the cell is in terms of a **polarization curve**[319]. This is a graph of the cell current I versus the cell voltage ΔE, as in Figure 3-4. The unique point on such a graph, where the current is zero, goes by a variety of names[320], including equilibrium cell voltage and **null voltage**, ΔE_n. The distinction between ΔE° and ΔE_n is that the former relates to unity activities, the second to whatever the activities happen to be in practice. Such current-voltage curves are useful devices, but recognize that, more often than not, the shape of the curve may change with time, for reasons and with consequences that will be explored later in this book.

We chose to base the foregoing exposition on a familiar example of an electrochemical cell in which the ionic conductor is an aqueous solution and one of the electronic conductors is a metal. This is a common circumstance, but *any* arrangement in which an ionic conductor is sandwiched between two electronic conductors, as in Figure 3-5a is an electrochemical cell. You may wonder whether the converse arrangement, illustrated in Figure 3-5b, is not equally an electrochemical cell. Indeed, such an unusual arrangement

---

[316] It can be argued that the cell is *not* at equilibrium; however, each of the electrodes is at equilibrium.

[317] A cell contains two electrodes immersed in an aqueous solution containing $Na^+(aq)$ and $Br^-(aq)$, each at a concentration of 100 mM. One electrode consists of mercury, Hg, coated with mercury(I) bromide, Hg_2Br_2. The second is a silver plate coated with silver bromide, AgBr. Under what conditions would this cell operate in each of the three modes: galvanic? equilibrium? and electrolytic? See Web#317.

[318] In the context of a lead-acid cell, or other secondary battery, the electrolytic phase of operation is described as **charging**, the galvanic phase as **discharging** (Chapter 5).

[319] Such curves are sometimes called voltammograms, but we reserve that name for current-voltage curves resulting from voltammetry (Chapter 12).

[320] also **rest voltage**, **open-circuit voltage** and **reversible voltage**.

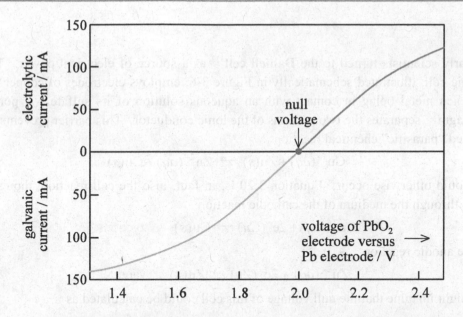

Figure 3-4 In this polarization curve for a lead-acid cell, the colored segments correspond to operation in the galvanic and electrolytic modes.

is possible[321] but, without further junctions being introduced, electrical measurements cannot be made with devices that are available. Yet a further junction of interest to electrochemists is that between two immiscible liquids. Such interfaces, at which electrochemical reactions may be performed without an electrode, are discussed in Chapter 14, where certain other interfaces are also addressed.

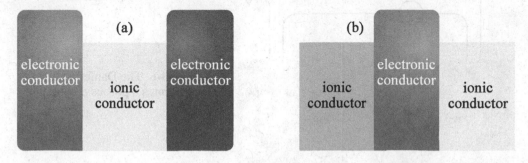

Figure 3-5 (a) a conventional electrochemical cell and (b) the complementary, but impractical, arrangement of conductors.

---

[321] The electronic conductor in such an arrangement is known as a **bipolar electrode**. See Figure 3-8 for an example.

Cells with Junctions: two ionic solutions prevented from mixing

Early scientists turned to the **Daniell cell**[322] as a source of electrical power. This galvanic cell, illustrated schematically in Figure 3-6, employs electrodes of copper and zinc, each metal being in contact with an aqueous solution of its sulfate. A porous diaphragm[323] separates the two portions of the ionic conductor. This barrier prevents the so called "parasitic" chemical reaction

3:20 $$Cu^{2+}(aq) + Zn(s) \rightleftarrows Zn^{2+}(aq) + Cu(s)$$

that would otherwise occur. Equation 3:20 is, in fact, also the cell reaction, though it occurs through the medium of the cathodic reaction

3:21 $$Cu^{2+}(aq) + 2e^-(Cu) \rightleftarrows Cu(s) \qquad \text{cathode}$$

and the anodic reaction

3:22 $$Zn^{2+}(aq) + 2e^-(Zn) \rightleftarrows Zn(s) \qquad \text{anode}$$

One might imagine that the null voltage of this cell could be calculated as

3:23 $$\Delta E_n = \frac{-\Delta G}{nF} = \frac{-1}{2F}\left[G^{\circ}_{Zn^{2+}} - G^{\circ}_{Cu^{2+}} + RT\ln\left\{ \frac{a_{Zn^{2+}}}{a_{Cu^{2+}}} \right\} \right]$$

in the standard way. This gives a value of about 1.102 volts if the activities of the two

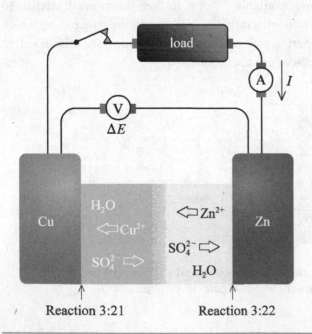

Figure 3-6 The Daniell cell incorporates a porous diaphragm.

---

[322] John Frederic Daniell, 1790 – 1845, English chemist. See also page 89.

[323] Fritted glass, fine capillaries, filter paper, porous porcelain, gels and membranes of various kinds are used as diaphragms, the object being to allow the passage of ions while inhibiting the mixing of the two solutions.

cations are equal[324]. This is close to the voltage of the Daniell cell, but equation 3:23 does not tell the full story. There is a **liquid junction potential difference** associated with the porous diaphragm. To understand the origin of this term, recognize that the flow of current through the cell requires the transport of electricity from right to left through the porous barrier and that two processes – the passage of Zn^{2+} ions from right to left and of SO_4^{2-} ions from left to right – contribute to this. A detailed examination of the voltage of the Daniell cell must take account of this fact.

Liquid junction potential differences[325] exist even in the absence of current flow. Consider the nonelectro-chemical arrangement shown here in which two aqueous solutions of lithium bromide are separated by a permeable barrier. The concentration difference will cause ions to move from the left-hand compartment.

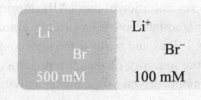

But Br^- anions migrate about twice as fast as Li^+ cations[326] and so have a tendency to arrive earlier in the right-hand compartment. The electroneutrality constraint ensures that the speedy Br^- ions do not significantly outrun their sluggish lithium counterions, but the small effect is enough to give the right-hand compartment a significant negative potential with respect to the left-hand solution. The magnitude of this liquid junction potential difference[327] depends somewhat on the geometry of the barrier's pores and on the degree of stirring of the two solutions, but in the simplest circumstance it is given by the **Henderson equation**[328]

$$3:24 \qquad \Delta\phi = \phi^L - \phi^R = \frac{RT}{F(u_+ - u_-)}\left(\frac{u_+}{z_+} - \frac{u_-}{z_-}\right)\ln\left\{\frac{c^R}{c^L}\right\} \qquad \text{Henderson equation}$$

for the case of a junction between two different concentrations, c^L and c^R, of a single electrolyte. For the fivefold concentration disparity shown in the diagram, the Henderson equation gives a 14 mV potential difference for lithium bromide[329].

The Henderson equation predicts the absence of a liquid junction potential difference

---

[324] Check the voltage calculation using data from the table on page 391. Also use 3:23 to estimate the null voltage after the Daniell cell is half discharged, so that, assuming equal volumes, the activity of the zinc ion will have increased by about 50%, while that of the copper ion has halved. See Web#324.

[325] often simply called a **junction potential**.

[326] See the table of mobilities on page 388. You may be surprised that the small Li^+ ions move more slowly than the larger Br^-. This is because the former are strongly hydrated (page 41) and have to drag as many as four water molecules along with them. Also see page 150.

[327] also called a **diffusion potential** difference because the impetus for the motion of the ions is a concentration gradient.

[328] Lawrence Joseph Henderson, 1878 – 1942, U.S. chemist whose name is also associated with the Henderson-Hasselbalch equation of titrimetry. See Web#328 for a derivation of equation 3:24.

[329] Confirm this value and recalculate $\Delta\phi$ for potassium chloride, KCl, solutions. What significance has the sign of your answers? See Web#329.

between solutions of an electrolyte for which $u_+/u_- = z_+/z_-$; potassium chloride comes very close to meeting that criterion[329]. For this reason, KCl is the most popular electrolyte[330] for use in a **salt bridge**. This device, which may adopt a variety of geometries, is used in cells whenever it is desired to prevent contact between the two ionically conducting liquids. It consists of a concentrated salt solution interposed between the half-cells and separated from them by porous diaphragms.

Figure 3-7 shows a salt bridge linking two half-cells that differ only in their concentration of copper(II) chloride. Despite there now being two barriers, liquid junction potentials have been virtually eliminated by the near equality of the mobilities of the K^+ and Cl^- ions and because the high concentration of these dominant ions means that they carry the majority of the current across the barriers. When this cell operates in the galvanic mode, driven by the higher activities of the ions in the left-hand half-cell, electricity flows around the circuit in a clockwise direction. The carriers of this electric current are diverse:

(a) across the right-hand electrode, by the oxidation $Cu(s) \rightleftarrows 2e^- + Cu^{2+}(aq, c^R)$;

(b) leftwards through the right-hand half-cell, by migration of $Cu^{2+}(aq)$ ions and the countermigration of $Cl^-(aq)$ ions;

(c) across the right-hand barrier, largely by migration of Cl^- ions out of the salt bridge[331];

(d) within the salt bridge by the migration of K^+ and the countermigration of Cl^- ions;

(e) across the left-hand barrier, largely by migration of K^+ ions out of the salt bridge[331];

(f) leftwards through the left-hand half-cell, by migration of the $Cu^{2+}(aq)$ ions and countermigration of $Cl^-(aq)$ ions;

(g) across the left-hand electrode, by the reduction $Cu^{2+}(aq, c^L) + 2e^- \rightleftarrows Cu(s)$; and

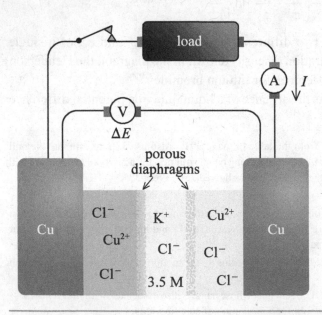

Figure 3-7 This **concentration cell** uses a KCl salt bridge to prevent the transfer of Cu^{2+} ions between the half-cells. The concentration c^L of $CuCl_2$ in the left-hand half-cell exceeds that, c^R in the right-hand half-cell.

---

[330] Ammonium nitrate, NH_4NO_3, is an alternative, in part because it is extremely soluble in water.

[331] There is also an innocuous concentration-gradient-driven diffusion of KCl from the salt bridge.

(h) throughout the remainder of the circuit, by counterclockwise electron flow.

Such a cell is named a **concentration cell without transference** because the two half-cells differ only in concentration, and because there is no transfer of electroactive species between the two half-cells. Subtraction of the reaction described in (g) from that in (a) gives

3:25
$$Cu^{2+}(aq,c^L) \rightleftharpoons Cu^{2+}(aq,c^R)$$

But that is not all! Additional chloride ions have also appeared in the left-hand chamber and disappeared from its right-hand counterpart, so that the total stoichiometry is given by[332]

3:26
$$Cu^{2+}(aq,c^L) + 2Cl^-(aq,2c^L) \rightleftharpoons Cu^{2+}(aq,c^R) + 2Cl^-(aq,2c^R)$$

Thus the chloride ions, too, contribute to the Gibbs energy change. The cell's null voltage, developed after the switch shown in Figure 3-7 is opened, lacks a "standard" term ($\Delta G^o = 0$) and reflects only the activity disparity:

3:27
$$\Delta F_n = \frac{-\Delta G}{2F} = \frac{-1}{2F}\left[\Delta G^o + RT\ln\left\{\frac{a^R_{Cu^{2+}}\left(a^R_{Cl^-}\right)^2}{a^L_{Cu^{2+}}\left(a^L_{Cl^-}\right)^2}\right\}\right] = \frac{-RT}{2F}\ln\left\{\frac{\left(c^R\gamma^R_{\pm}\right)^3}{\left(c^L\gamma^L_{\pm}\right)^3}\right\}$$

This is the equilibrium voltage of the left-hand electrode with respect to the right and will be positive if c^L exceeds c^R.

Figure 3-8 shows another way of preventing transference, dispensing with diaphragms

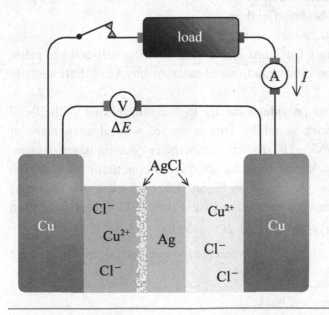

Figure 3-8 A bipolar electrode is an alternative to a salt bridge in avoiding transference in a concentration cell.

---

[332] In practice there is a further complication because copper(II) exists in aqueous solution to a significant extent as $CuCl_4^{2-}$. Moreover, an appreciable concentration of copper(I) chloride is present on account of the equilibria $CuCl_4^{2-}(aq) + Cu(s) \rightleftharpoons 2CuCl_2^-(aq) + 2Cl^-(aq) \rightleftharpoons 2Cu^+(aq) + 4Cl^-(aq)$.

by interposing another electronic conductor between the two half-cells. Here a silver-chloride-coated dual action silver electrode allows the transfer of chloride ions between the two half-cells by virtue of the two reactions:

3:28 $\qquad AgCl(s) + e^-(Ag) \rightleftharpoons Ag(s) + Cl^-(aq)$ $\quad$ right-hand interface

3:29 $\qquad AgCl(s) + e^-(Ag) \rightleftarrows Ag(s) + Cl^-(aq)$ $\quad$ left-hand interface

An electrode behaving in this way is termed a **bipolar electrode**[333].

If the experiment shown in Figure 3-7 is repeated, but with the salt bridge replaced by a single porous barrier, the null cell voltage is quite different. This is now a **concentration cell with transference**, because ions from one half-cell must now migrate, through the porous barrier, to the other half-cell. The null voltage is smaller than that given in equation 3:27 and involves the ionic mobilities[334].

Semipermeable membranes are barriers that allow the passage of some species but not others. Of particular electrochemical interest are membranes that allow the passage of only anions or only cations[335]. They probably function by having small pores with charged walls, the pores being accessible only to ions of opposite sign. Their use in concentration cells with transference has advantages over less discriminating barriers. For example, replacing the porous barrier of the Daniell cell by an anion-selective membrane would be beneficial in allowing the cell to be recharged, which is not 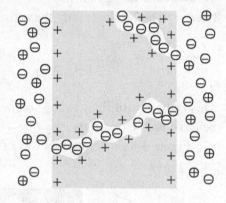 possible with a porous barrier because Cu^{2+} ions would enter the zinc half-cell and cause the parasitic reaction 3:20. On the other hand, membranes of any kind detrimentally increase the resistance of a cell.

Of course, there is no need to provide a barrier to the intermixing of half-cell ingredients when the ionic conductor is solid. This is one of several advantages of solid-phase cells. An example is provided by the high-temperature zirconia-based oxygen-concentration cell[336], in which the electrolyte is the solid ionic conductor ZrO_2 though which oxide ions, O^{2-}, can migrate. As shown in Figure 3-9, each electrode is porous platinum through which oxygen from a gas stream can diffuse and establish the equilibrium

3:30 $\qquad\qquad O_2(g) + 4e^-(Pt) \rightleftharpoons 2O^{2-}(ZrO_2)$

The cell voltage responds to the discrepancy between the oxygen partial pressures at the

---

[333] Why does equation 3:27 apply to this cell? See Web#333.

[334] See Web#334 for more details.

[335] See pages 81–83 for an application in electrodialysis and other industrial application.

[336] See Footnote 129. For a sensor application, see page 173.

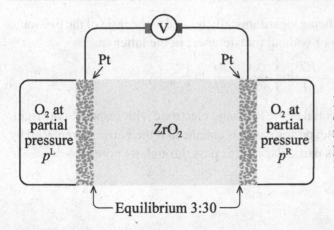

Figure 3-9 The zirconia-based oxygen-concentration cell responds to the ratio of the partial pressures of oxygen at its two electrodes.

left-hand and right-hand electrodes through the equation[337]

$$3:31 \qquad \Delta E_n = \frac{RT}{4F} \ln \left\{ \frac{p_{O_2}^L}{p_{O_2}^R} \right\}$$

Rather similar to this is a cell which uses hydrogen uranyl phosphate tetrahydrate, $HUO_2PO_4 \cdot 4H_2O$, a waxy solid known colloquially as "HUP", that allows proton migration. The voltage of a cell in which HUP is sandwiched between two palladium electrodes responds to differences in hydrogen partial pressure on its two faces.

Summary

The simplest electrochemical cells have an ionic conductor sandwiched between two electronic conductors; the junctions serve as electrodes at which electrochemical reactions occur. The sum of those two reactions is the cell reaction, the ΔG of which is reflected in the null cell voltage (the cell voltage at open-circuit):

$$3:32 \qquad \Delta E_n = \frac{-\Delta G}{nF} = \frac{-\Delta G^\circ}{nF} + \frac{RT}{nF} \ln \left\{ \begin{matrix} \text{activity} \\ \text{term} \end{matrix} \right\} = \Delta E^\circ + \frac{RT}{nF} \ln \left\{ \begin{matrix} \text{activity} \\ \text{term} \end{matrix} \right\}$$

The magnitude of the cell voltage will exceed its null value when the cell operates in an electrolytic mode, current being driven by an external source, but will be less than the null value when the cell delivers current into a load and operates galvanically. To prevent unwanted reactions with a liquid ionic conductor, it is often necessary to interpose a diaphragm, thereby dividing the cell into two. This introduces a liquid junction potential, which can be mitigated by means of a salt bridge or eliminated by a bipolar bridge. A concentration cell is one in which the two half-cells differ only in the concentration of some

---

[337] Calculate the voltage of the concentration cell at 425°C if the oxygen partial pressure on the left-hand side is 1000 Pa and the right-hand side is ambient air (21% oxygen). See Web#337.

electroactive ion i, the cell voltage being logarithmically related to the ratio of the two ionic activities. Such cells may be with or without transference; in the latter case:

3:33
$$\Delta E_{\mathrm{n}} = \frac{RT}{nF} \ln \left\{ \frac{a_i^{\mathrm{L}}}{a_i^{\mathrm{R}}} \right\} \approx \frac{RT}{nF} \ln \left\{ \frac{c_i^{\mathrm{L}}}{c_i^{\mathrm{R}}} \right\}$$

where ΔE_{n} is the difference in potential of the left-hand electrode with respect to the right. Many sensors work on this principle; performance is enhanced if the barrier is replaced by a membrane that selectively allows particular ions to pass through its pores.

4

Electrosynthesis

When a cell operates in its electrolytic mode, the process is termed **electrolysis** and chemicals are produced that have a greater Gibbs energy than the reactants. Many of these electrolysis products have a high commercial value, or are of chemical interest. In this brief chapter, examples are given of processes that are used to manufacture – often industrially in huge tonnages, sometimes on a much smaller scale – products that are valuable economically or scientifically. In several cases, such an **electrosynthesis** is the *only* known route for making a particular element or compound.

Metal Production: many metals are made or purified electrolytically

The metals Li, Na, K, Mg, Ca, Sr, Ba, Ra, Al, and Ta, as well as the gases F_2 and Cl_2, are manufactured almost exclusively by electrolysis of salts of these elements, either in the fused (molten) state or in aqueous solution. For several other metals, including Cr, Mn, Co, Ni, Cu, Ag, Au, Zn, Cd, Ga, In, and Tl, electrolytic methods compete commercially with classical chemical methods of extraction. A metal is said to be "won" from its ore and the electrolytic production of a metal from its ores is therefore known as **electrowinning**. As the most important example, we firstly consider the electrowinning of aluminum.

Though the rival Alcoa process[401] is also in use and other methods are being researched, most of the 12 billion tons of aluminum produced annually still comes from the original Hall-Hèroult process[402]. The raw material from which aluminum is made is the ore **bauxite**, a mixture of $Al(OH)_3$ and $AlOOH$. This is purified and converted to alumina, Al_2O_3 which, in the Hall-Hèroult process, is then dissolved in molten **cryolite**[403] Na_3AlF_6.

---

[401] This has much in common with the Hall-Hèroult process but uses chloride, instead of fluoride, salts.

[402] Charles Martin Hall (1863–1914, U.S. chemist) and Paul Louis Toussaint Hèroult (also 1863–1914, French metallurgist) independently invented the process in the 1880s.

[403] Though cryolite is a naturally occurring mineral, what is actually used is a mixture of NaF and AlF_3. A small percentage of calcium fluoride CaF_2 is added to improve the conductivity.

Electrochemical Science and Technology: Fundamentals and Applications, First Edition. Keith B. Oldham, Jan C. Myland, Alan M. Bond.
© 2012 John Wiley & Sons, Ltd. Published 2012 by John Wiley & Sons, Ltd.

The molten mixture is an ionic liquid containing a variety of ions, including AlF_4^-, F^-, and O^{2-}. Electrolysis is conducted at about 960°C, at which temperature both the electrolyte and the aluminum product are liquid. The cathode reaction is probably

4:1 $$AlF_4^-(fus) + 3e^-(Al) \rightarrow Al(\ell) + 4F^-(fus)$$

and it occurs on the surface of the pool of molten aluminum that rests on graphite at the base of the electrolytic bath, as in Figure 4-1. Continuously formed in situ from pitch and petroleum coke or anthracite, the anode consists of rods of baked carbon, which are consumed by the reaction

4:2 $$2O^{2-}(fus) + C(s) \rightarrow 4e^-(C) + CO_2(g)$$

The overall cell reaction is therefore[404]

4:3 $$2Al_2O_3(s) + 3C(s) \rightarrow 4Al(\ell) + 3CO_2(g)$$

with the applied voltage being about 4 V and the current density close to 5000 A m^{-2}. The heat generated by the massive current flow helps maintain the temperature of the bath.

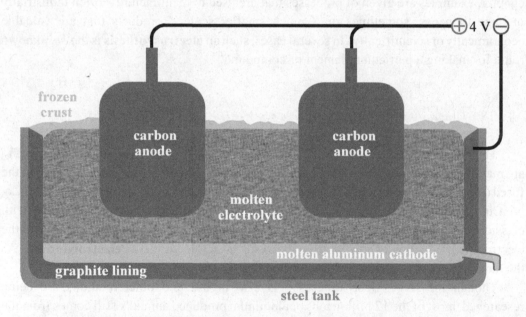

Figure 4-1 Hall-Hèrault cell for aluminum manufacture.

From the increase in standard Gibbs energy accompanying reaction 4:3, one learns that about 20 MJ are required thermodynamically to generate one kilogram of aluminum[405]. However, the electrical energy consumed by the cell is closer to 40 MJ per kilogram,

[404] Construct this overall equation as in Web#404.

[405] Use data from the tables on pages 386 and 391 to confirm this. See Web#405.

largely on account of the heat needed to sustain the high operating temperature. Aluminum manufacture is the largest single user of electricity on the planet. The high electrical demand means that it is more economical to site aluminum smelters close to sources of electricity than adjacent to the bauxite mines. The Hall-Hèroult process is a serious source of pollution, not only on account of the 1.2 kg of carbon dioxide produced[406] for each kilogram of aluminum, but also because of the unwelcome escape of hydrogen fluoride, HF, and carbon monoxide, CO, from the smelters. For both environmental and economic reasons, a better method of making aluminum is sorely needed.

Some copper is electrowon from low-grade ores, but this metal is mostly extracted by traditional smelting. However, regardless of how the copper is won, it is invariably subjected to **electrorefining**[407]. The process is, in principle, supremely simple. The impure copper forms the anode of an electrolytic cell in which the ionic conductor (the "electrolyte") is an aqueous solution containing copper(II) sulfate and sulfuric acid. Pure copper is formed at the cathode. The electrode reactions[408] are

4:4 $$Cu(s, impure) \rightarrow 2e^- + Cu^{2+}(aq)$$

and

4:5 $$Cu^{2+}(aq) + 2e^- \rightarrow Cu(s, pure)$$

The impurities either remain in solution or fall to the floor of the electrolysis vessel as an "anode sludge". The latter is such a valuable source of silver and gold, that the copper refining process pays for itself.

Similar electrorefining treatments, though with diverse electrolytes, are employed to purify the metals cobalt, nickel, tin, and lead. Some forms of **electroplating** use cells similar to refining cells in that metal is transferred from an anode and deposited on the surface of the cathode, which is generally of a different metal. Electroplating is as much an art as a science and unlikely additives are customarily added to the plating bath to achieve desirable properties, such as smoothness, abrasion resistance, and luster, in the electroplated layer or on its surface.

Notice that, because the reactant and product of the cell reaction are almost identical, a refining cell has a null voltage of virtually zero. Nevertheless, some voltage must be applied, primarily to provide the energy for migration. The name **overvoltage** is given to electrical potential differences in excess of those required to overcome the positive Gibbs energy demand of an cell. There are three phenomena that lead to overvoltages: they are called **polarizations** – ohmic, kinetic, and transport – and are examined in detail in Chapter 10.

---

[406] Confirm this statement. Also calculate how much electricity (in coulombs and watts) is used in the generation of one kilogram of aluminum. See Web#406.

[407] The term "**electroraffination**" is also used (from the French *raffiné*, refined or cultivated).

[408] What quantity of electricity is needed to purify one kilogram of copper? Compare your answer with that at Web#408.

The Chloralkali Industry: a bounty of products from salt and water

One of the world's major industries is based on the electrolysis of brine, a concentrated aqueous solution of sodium chloride. There are three rival processes[409], but newer electrolyzers are all of the **membrane cell** type, in which the electrode reactions are[410]:

4:6   **anode :** $2Cl^-(aq) \rightarrow 2e^- + Cl_2(g)$

4:7   **cathode :** $2H_2O(\ell) + 2e^- \rightarrow H_2(g) + 2OH^-(aq)$

The chloralkali membrane cell, diagrammed in Figure 4-2, operates continuously with liquid streams of the compositions shown, and with the anode and cathode compartments separated by a selectively cation-permeable membrane[411]. By inhibiting the migration into the anode compartment of hydroxide ions, the membrane prevents their reaction[412] with the

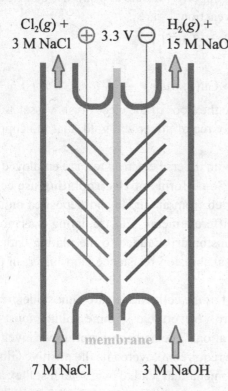

$Cl_2(g)$ +
3 M NaCl $\oplus$ 3.3 V $\ominus$ $H_2(g)$ +
                                      15 M NaOH

7 M NaCl   3 M NaOH

Figure 4-2 A chloralkali membrane cell. The multiplate cathode and anode are louvered as shown to deflect gas bubbles away from the membrane. The bubbles of hydrogen and chlorine become entrained in their respective solution streams, from which they are subsequently separated. Some of the sodium hydroxide produced in the electrolysis is diluted and recycled.

[409] Read about the other two at Web#409.

[410] Use the table on page 391 to estimate the Gibbs energy change for the net cell reaction. Make reasonable assumptions about the activities of the species involved in the chloralkali cell. Check with Web#410. Then calculate the cell's null potential and interpret its sign.

[411] More about selectively permeable membranes will be found on pages 79–83. Membranes manufactured by E. I. du Pont de Nemours and Company use the trade-name **Nafion**.

[412] Some cells operate without a membrane separator, allowing reaction 4:9 to occur, so that hypochlorite is produced in situ.

chlorine. The cathode is iron or steel. Traditionally, graphite was used for the anode, but so-called **dimensionally stable anode** materials offer advantages; these contain ruthenium and titanium oxides which are electrocatalytic for chlorine evolution and thereby diminish the amount of oxygen that is produced by the unwelcome anodic side reaction $4OH^-(aq) \rightarrow 4e^- + 2H_2O(\ell) + O_2(g)$. To decrease cell resistance, the electrodes are positioned close to either side of the membrane. The cell operates at a voltage[413] close to 3.3 V and a current density[414] of about 4.0 kA m^{-2}.

The primary products of the chloralkali process are chlorine gas, hydrogen gas and sodium hydroxide solution[415], for each of which markets exist. The uses of chlorine include water disinfection and remediation, as discussed in Chapter 9. In addition to these, however, there are many other subsequential products of the industry[416]. High purity hydrogen chloride is produced by the mutual combustion of the gaseous products

4:8 $$H_2(g) + Cl_2(g) \rightarrow 2HCl(g)$$

Much of the produced chlorine is reacted with the sodium hydroxide solution:

4:9 $$Cl_2(g) + 2OH^-(aq) \rightarrow Cl^-(aq) + OCl^-(aq) + H_2O(\ell)$$

to produce a solution of mixed sodium chloride NaCl and sodium hypochlorite NaOCl salts, sold as laundry bleach. Sodium hypochlorite, in turn, is anodically oxidized to sodium chlorate NaClO$_3$ for which there is a large market in the pulp and paper industry. This salt is also generated chemically in aqueous solution by the disproportionation of sodium hypochlorite:

4:10 $$3ClO^-(aq) \rightarrow ClO_3^-(aq) + 2Cl^-(aq)$$

a reaction which is very slow in basic solution, but which is catalyzed by hydronium ions. Furthermore, sodium perchlorate NaClO$_4$ is produced electrochemically by anodic oxidation of acidic solutions of sodium chlorate:

4:11 $$ClO_3^-(aq) + 3H_2O(\ell) \rightarrow 2e^- + ClO_4^-(aq) + 2H_3O^+(aq)$$

hydrogen being formed at the cathode.

Organic Electrosynthesis: nylon from natural gas

Though production volumes are small in comparison with those of the large inorganic electrosynthetic enterprises, many electrosyntheses of organic compounds are currently

---

[413] This voltage is to be compared with the null cell voltage of 2.20 V calculated in Web#410.

[414] Use these data, and the assumption of 100% yield, to calculate the rate of chlorine generation in the mol m^{-2} s^{-1} unit. Also compare the energy consumed in generating one mole of chlorine with that calculated in Web#410. Check your answers at Web#414.

[415] known industrially as **caustic soda** or simply "caustic".

[416] Demonstrate that reactions 4:8, 4:9, and 4:10 are thermodynamically feasible. See Web#416.

being exploited commercially worldwide[417]. The **Baizer[418]-Monsanto process** for conversion of acrylonitrile to adiponitrile is the most noteworthy example. The cathodic reaction is both a reduction (adding hydrogen atoms into the organic structure) and a dimerization (combining two acrylonitrile molecules into a single adiponitrile molecule)

4:12 $\qquad 2CH_2CHCN + 2H_3O^+ + 2e^- \rightarrow NCCH_2CH_2CH_2CH_2CN + 2H_2O$

The reaction, the mechanism of which is not firmly established, occurs at a cadmium electrode in an aqueous emulsion containing, in addition to the reactant and product, sodium phosphate and a quaternary ammonium salt.

This electrosynthesis is a key step in the synthesis of nylon from natural gas. The propane $CH_3CH_2CH_3$ fraction of the gas is dehydrogenated to propene CH_2CHCH_3 before being converted to acrylonitrile CH_2CHCN. Following the electrochemical step, adiponitrile is converted in part to adipic acid $HOOC(CH_2)_4COOH$ and in part to 1,6-diaminohexane $H_2N(CH_2)_6NH_2$. These are the two components from which the nylon-66 polymer[419] $[OOC(CH_2)_4COONH(CH_2)_6NH]_\infty$ is formed.

Among smaller scale processes, the electrochemical reduction of bromides is used by organic chemists as a means of synthesis. With R denoting an organic group, the first step in such reductions probably yields a radical

4:13 $\qquad\qquad\qquad RBr(soln) + e^- \rightarrow R^\bullet(soln) + Br^-(soln)$

Depending on the nature of R and of the solvent, the radical may dimerize to R_2, further reduce and abstract a proton from the solvent to form RH, or even attack the electrode (giving HgR_2 or $RHgBr$ with a mercury electrode). With dibromides, double bonds or rings may form, as in the examples[420]

4:14 $\qquad CH_3CHBrCHBrCH_3(soln) + 2e^- \rightarrow CH_3CHCHCH_3(soln) + 2Br^-(soln)$

or

4:15 $\qquad BrCH_2CH_2CH_2CH_2Br(soln) + 2e^- \rightarrow \begin{matrix} CH_2-CH_2 \\ | \qquad\quad | \\ CH_2-CH_2 \end{matrix}(soln) + 2Br^-(soln)$

Another electrosynthetic route of interest in organic chemistry is the **Kolbe synthesis**[421], in which the electrooxidation of carboxylates at a platinum anode yields dimeric hydrocarbons and carbon dioxide:

4:16 $RCO_2^-(soln) - e^- \rightarrow RCO_2^\bullet(ads) \rightarrow R^\bullet(ads) + CO_2(ads) \rightarrow \frac{1}{2}R_2(soln) + CO_2(g)$

Discovered in 1843, this synthesis was the first important organic electrosynthetic reaction.

---

[417] see Chapter 6 in D. Pletcher and F.C. Walsh, *Industrial Electrochemistry*, 2nd edn, Kluwer, 1990.

[418] Manuel M. Baizer, 1914–1988, organic electrochemist.

[419] The number 66 indicates that there are six carbon atoms in the diacid comonomer and six in the diamine.

[420] Predict the product of the electroreduction of the compound $C(CH_2Br)_4$. See Web#420.

[421] Adolph Wilhelm Hermann Kolbe, 1818–1884, German organic chemist, who followed up earlier observations by Faraday on the electrolysis of acetate solutions.

It is used to construct higher saturated hydrocarbons. "Mixed" product formation by joining two different R groups can be accomplished by having one of the reactant carboxylates in excess.

Electrolysis of Water: key to the hydrogen economy?

Proponents of the **hydrogen economy**[422] see the electrolysis of water, in which the electrode reactions are[423]

4:17   **anode :** $6H_2O(\ell) \rightarrow 4e^- + O_2(g) + 4H_3O^+(aq)$

and

4:18   **cathode :** $4H_3O^+(aq) + 4e^- \rightarrow 2H_2(g) + 4H_2O(\ell)$

as the "green" method of generating hydrogen from electricity. Presently, the so-called "reforming" of natural gas[424] is a more cost effective process for manufacturing hydrogen. Accordingly, hydrogen is made electrolytically only in niche applications[425]. It is to be expected, however, that the depletion of fossil fuel reserves will one day tip the economic balance in favor of the electrosynthesis of hydrogen.

As the table on page 384 confirms, water is a poor conductor and therefore a strong electrolyte must be added to the water to provide sufficient conductivity. The identity of the ions provided by this electrolyte is largely immaterial, provided that they do not themselves undergo any competitive electrode reaction; H_2SO_4 is often used as the electrolyte.

The Gibbs energy change accompanying the cell reaction

4:19 $2H_2O(\ell) \rightarrow 2H_2(g) + O_2(g)$

is positive and equal to[426] 474.2 kJ mol^{-1}, from which a null voltage of magnitude 1.229 V may be calculated. It might therefore be expected that if the voltage source shown in Figure 4-3 overleaf is adjusted to any positive value in excess of 1.229 V, current will flow and the reactions 4:17 and 4:18 will occur at the left and right-hand electrodes respectively. In fact, these reactions do not occur significantly until the applied voltage reaches about

---

[422] This is the concept in which hydrogen, H_2, plays a role similar to that of electricity as an "energy currency", being generated at one site, then transported to a remote energy-needful site (such as a home or an automobile engine) where it is combusted with air to liberate energy with the formation of water.

[423] These equations, which of course make no pretense of depicting the mechanisms of the electrode reactions, are applicable to the electrolysis of acidic water. The overall reaction 4:19 applies irrespective of the pH.

[424] This is the catalyzed reaction of methane $CH_4(g)$ with steam $H_2O(g)$ to give hydrogen and oxides of carbon.

[425] As in the prolonged electrolysis of water with a view to increasing its **heavy water** content. D_2O electrolyzes more reluctantly than H_2O and therefore accumulates in the residual electrolyte solution.

[426] Confirm this from the table on page 391 and go on to check the cited null voltage. Rationalize the fact that your calculation resulted in a *negative* voltage. See Web#426.

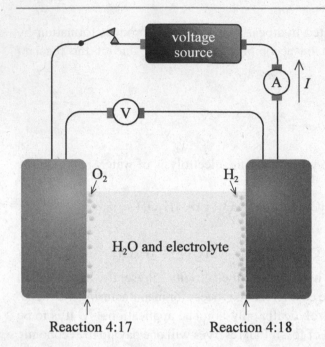

Figure 4-3 The electrolysis of water produces oxygen at the anode and hydrogen at the cathode. Ions from an electrolyte are necessary to provide conductivity but these play no role in the electrode reactions.

2 V. A phenomenon impeding an electrode reaction is known as a **polarization** and three of these are the subject of Chapter 10. In the case of water electrolysis, the polarization can be attributed primarily to the slow kinetics of reaction 4:18 and, especially, reaction 4:17.

Reaction 4:18 provides a powerful example of the finding that the nature of the electrode can profoundly affect the electrochemistry of a reaction even though the electrode material is not stoichiometrically involved in the electrode reaction. Study of the kinetics of this reaction (Chapter 7 and Web#743) shows that the hydrogen-evolution reaction proceeds by three distinct mechanisms, depending on the metal at which the reaction occurs. For most metals, including Hg, Ag, Pb, Cu, and Fe, the mechanism is believed to involve the formation of an adsorbed hydrogen atom in the rate-determining step

4:20
$$\begin{cases} H_3O^+(aq) + e^- \rightarrow H(ads) + H_2O(\ell) & \text{Step } (\hat{1}), \text{ slow} \\ H(ads) + H_3O^+(aq) + e^- \rightarrow H_2(g) + H_2O(\ell) & \text{Step } (2), \text{ fast} \end{cases}$$

On the metals Mo, W, Ti, and Ta, the kinetics can be explained on the basis of the *same* two reactions, but with the second step now being rate determining:

4:21
$$\begin{cases} H_3O^+(aq) + e^- \rightarrow H(ads) + H_2O(\ell) & \text{Step } (1), \text{ fast} \\ H(ads) + H_3O^+(aq) + e^- \rightarrow H_2(g) + H_2O(\ell) & \text{Step } (\hat{2}), \text{ slow} \end{cases}$$

The mechanism for Pd, Rh, and Ir is believed to be

4:22
$$\begin{cases} H_3O^+(aq) + e^- \rightarrow H(ads) + H_2O(\ell) & \text{Step } (1), \text{ fast} \\ 2H(ads) \rightarrow H_2(g) & \text{Step } (\hat{2}), \text{ slow} \end{cases}$$

It is the principles described in Chapter 7 that permit us to speculate on the mechanisms that apply in each case. Several examples are provided there and mechanisms 4:20, 4:21, and 4:22 are referenced elsewhere[743].

Mechanistically, the anodic reaction 4:17 is no more straightforward than the cathodic reaction 4:18, and it is inherently slower. The mechanism at some metal anodes is believed to involve the following steps:

4:23 $$2H_2O(\ell) \rightarrow e^- + HO^{\bullet}(ads) + H_3O^+(aq)$$

4:24 $$HO^{\bullet}(ads) + H_2O(\ell) \rightarrow e^- + O(ads) + H_3O^+(aq)$$

4:25 $$2O(ads) \rightarrow O_2(aq)$$

On other electrodes, hydrogen peroxide may be involved, because this is known to be an intermediate in the reduction of oxygen.

The speed, and even the products, of the reactions described in this section, are determined by the nature of the electrode material, even though the latter is not part of the reaction's stoichiometry. Because of this, such reactions are often described as examples of **electrocatalysis**.

Much research is being conducted aimed at diminishing the voltage needed to effect water electrolysis. Voltages as low as 1.6 V have been achieved by the use of catalytic electrodes. A related research endeavor is to use illuminated semiconductor electrodes. Under suitable conditions, the energy of captured photons can augment the electrical energy (page 291).

The electrolysis of water can yield other products. An electrosynthesis perfected by the Dow company is used to manufacture hydrogen peroxide H_2O_2. The electrolysis is conducted in oxygenated basic solution and electrode reactions are

4:26   **anode:** $2OH^-(aq) \rightarrow 2e^- + H_2O(\ell) + \tfrac{1}{2}O_2(g)$

and

4:27   **cathode:** $H_2O(\ell) + O_2(g) + 2e^- \rightarrow HO_2^-(aq) + OH^-(aq)$

with the overall reaction being

4:28   cell: $OH^-(aq) + \tfrac{1}{2}O_2(g) \rightarrow HO_2^-(aq)$

The hydroperoxide ion HO_2^- is a weak base and, on neutralization, hydrogen peroxide is formed by the proton-transfer reaction

4:29 $H_3O^+(aq) + HO_2^-(aq) \rightarrow H_2O(\ell) + H_2O_2(aq)$

Selective Membranes: a quiet revolution in small-scale inorganic electrosynthesis

The migration of ions through electrosynthetic cells generates heat. Sometimes, as in the Hall-Hèrault cell, this heat provides a valuable contribution to raising the temperature of the ionic conductor. More often, however, it represents wasted energy, and accordingly it is diminished as much as possible. Two ways in which this can be done are to maximize the conductivity and to minimize the distance between the anode and cathode. Of course, if the electrodes are placed too close together, there is a danger of their inadvertently touching and, to prevent this, porous barriers are used, in both electrolytic and galvanic cells. Though the strict distinctions in meaning are not always maintained, it is useful to distinguish between three types of barriers, according to the function they serve:

separator: to physically contain or separate the electrodes

diaphragm: to impede the mixing of two solutions within a cell

membrane: to preferentially allow the passage of certain solutes

To emphasize their selective permeability, membranes are often said to be "semipermeable". The word "membrane" is, of course, also used to describe many different kinds of biological barrier; these too are often discriminatory in their behavior towards ions. Certain biological membranes have "gates" that only specific ions can easily transit; these may be "active" or "passive" according to whether or not the organism needs to expend metabolic energy to "pump" the ions through the gate (page 189). Sadly, we have far to go to even begin to match nature's membrane selectivity.

One characteristic that can be ascribed to barriers of most types is their **permeability**. This is defined as the rate at which a solute i crosses unit area of the barrier in unit time under a unit concentration gradient. Symbolically, the definition of permeability[427] is

4:30
$$P_i = \frac{L}{A\Delta c_i}\frac{dn_i}{dt}$$

where L and A are the thickness and area of the barrier, and dn_i/dt is the rate at which the solute crosses the membrane under a concentration differential of Δc_i. Its unit[428] is $m^2\ s^{-1}$. That P_i for a diaphragm is largely independent of the identity of i, could be considered the distinction between a diaphragm and a membrane.

The prototypical synthetic membrane is Nafion[429]. This is the generic name of a series of polymeric products in which side chains bearing sulfonic (and sometimes also carboxylic $-CO_2H$) acid groups have been introduced into a poly(tetrafluoroethene) backbone. Because the $-SO_3H$ groups ionize in the presence of water, the membrane is hydrophyllic and imbibes water. The tethered $-SO_3^-(aq)$ groups strongly repel anions, so that nafion acts as a cation-only permeable membrane. There are alternative membranes with $-NR_3^+(aq)$ groups, with R denoting an organic moiety such as CH_3, that are preferentially anion permeable, though these have yet to achieve the robustness of their Nafion cousins. The electrosynthetic use of membranes was noted in an earlier section in the context of the

---

[427] The term "permeability" has several different meanings in other areas of science and technology. For example, see Goldman permeability on page 188.

[428] Confirm this allocation of an *SI* unit. See Web#428.

[429] This is the name the DuPont Company gives to its sulfonic acid **ionomer** (a polymer with charged groups).

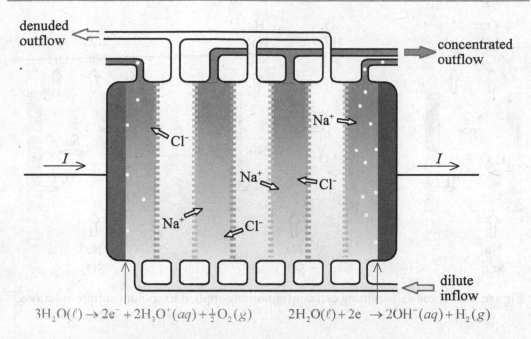

$$3H_2O(\ell) \rightarrow 2e^- + 2H_3O^+(aq) + \tfrac{1}{2}O_2(g) \qquad 2H_2O(\ell) + 2e \rightarrow 2OH^-(aq) + H_2(g)$$

Figure 4-4 In electrodialysis, multiple membranes, alternatingly anion permeable and cation permeable, direct the ions into the salt compartments, denuding the chambers they leave.

chloralkali industry, but the increasing availability and refinement of selectively permeable membranes has considerably enhanced the repertoire of inorganic electrosynthesis. Frequently these cells are of small scale and are sited where a need exists for their product, reducing the cost and hazards of transportation.

Electrodialysis uses an alternating array of anion-permeable membranes and cation-permeable membranes, as in Figure 4-4, to remove most of the salt from brackish[430] water. The process is not economically competitive with reverse osmosis for the desalination of ion-rich waters such as seawater.

A salt is formed spontaneously from the interaction of an acid and a base. The converse process, carried out in an electrosynthetic cell, is called **salt splitting** or **electrohydrolysis**. There are several strategies, illustrated overleaf in the three diagrams of Figure 4-5. Though almost any soluble salt may be split, our illustration uses the example of sodium sulfate. This salt is a valueless byproduct of the viscose[431] process, and of the pulp-and-paper industry, that presents a disposal problem; it can be electrochemically split into sulfuric acid and caustic soda, for both of which a market exists. The first

---

[430] "Brackish" means mildly salty, and thereby unfit for drinking or irrigation. Analyze the chemistry involved in the technique illustrated in Figure 4-4. Ideally, how much electricity per cubic meter would be needed to remove all the sodium chloride from a 10 mM input stream. See Web#430.

[431] Viscose is dissolved cellulose, made from wood or cotton and used to make rayon and cellophane.

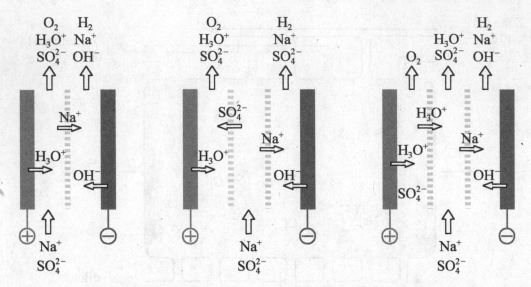

Figure 4-5 Three salt-splitting cell configurations applied to sodium sulfate solutions.

diagram shows a dual-compartment cell with a single membrane. Pure sodium hydroxide is produced by this cell, but the sulfuric acid retains some salt that must be removed by subsequent crystallization. In the second diagram two membranes are used, but the imperfection of the anion-selective membrane, in allowing the passage of some hydronium ions, prevents high concentrations of sulfuric acid being allowed to build up. There are three compartments also in the third diagram, in which the membranes are both cation-selective.

A rather recent innovation in membrane technology is the perfecting of **bipolar membranes**. These are dual membranes with one half being anion permeable, the other cation permeable. Thus, at least in theory, all ions are blocked from passing through the bipolar membrane. However, the immersed membrane readily imbibes water, so that when an electric field of the appropriate polarity is applied, the water ionizes and hydronium ions

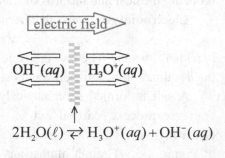

stream out from one face, with hydroxide ions exiting the other. A salt-splitting cell design incorporating both bipolar and monopolar membranes is illustrated in Figure 4-6. Salt solution, S, enters the compartments so marked. Water enters the compartments marked W. The anion of the salt remains within its compartment, becoming partnered by hydronium ions from the bipolar membrane on its left, and exiting as the acid A. The salt's cation travels through the monopolar cation-selective membrane into the adjoining compartment, where it becomes partnered by hydroxide ions from the bipolar membrane

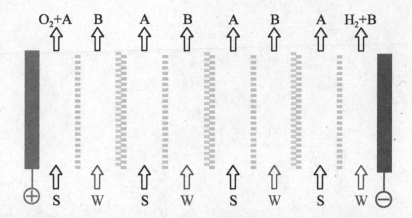

Figure 4-6 Salt-splitting cell incorporating bipolar and monopolar membranes. The letters represent Salt, Water, Acid, and Base.

on its right, and exits as the base B. A cell of similar design is used in Japan for reclaiming lactic acid from sodium lactate solutions.

Ion-selective membranes are not without their problems, but electrochemistry aided by such membranes appears to have a bright future[432]. The use of membranes tailor-made for a specific application will speed the adoption of these economical and eco-friendly techniques.

Summary

Major industrial enterprises, notably aluminum electrowinning and the chloralkali industry, employ electrosynthetic methods to manufacture commercially valuable materials. Electrochemical methods circumvent the thermodynamic $\Delta G < 0$ requirement imposed on purely chemical processes, making electrosynthesis the preeminent route to high-energy chemicals. On a much smaller scale, organic chemists make use of electrolytic steps as components in longer synthetic enterprises. Not presently economical, electrolytic hydrogen, possibly with sunlight contributing some of the required 237.1 kJ mol^{-1} of Gibbs energy, is foreseen by some as filling energy transportation needs of the future. Recent trends are toward the electrosynthesis of valuable chemicals at their point of use, often in cells employing selective membranes.

[432] T.A. Davis, J.D. Genders, and D. Pletcher, *A First Course in Ion Permeable Membranes*, Alesford Press, 1997.

5

Electrochemical Power

In contrast to the electrolytic cells discussed in the previous chapter, which convert electrical energy into chemical energy, the cells addressed here make the reverse conversion: they generate electrical energy from the net Gibbs energy liberated by the reactions at the two electrodes of a galvanic cell.

Types of Electrochemical Power Source: primary or secondary batteries, fuel cells

Batteries[501] and **fuel cells** are galvanic cells designed to be efficient sources of electric power. Whereas a battery is self-contained, a fuel cell employs reactants that are supplied as fluids from outside the cell. Batteries are of two main types: **primary cells** and **secondary cells**; the former operate solely in the galvanic mode, whereas secondary cells also operate electrolytically, and can thereby be "recharged".

$$\text{electrochemical power source}\begin{cases}\text{battery}\begin{cases}\text{primary cell}\\\text{secondary cell}\end{cases}\\\text{fuel cell}\end{cases}$$

Secondary cells are also called **storage cells**, **accumulators** or **rechargeable batteries**.

When a primary battery's initial supply of reactants is exhausted, its useful life is over and it is discarded. A fuel cell, on the other hand, has the advantage that its **fuel** (the chemical consumed at the anode) and the **oxidizer** or **oxidant** (consumed at the cathode) are supplied from outside the cell and the products of the electrode reactions are removed continuously, so the cell can be used almost indefinitely. Though a secondary cell may be discharged and recharged repeatedly, such cells do have a finite **cycle life**, for a variety of reasons. Many batteries also have a limited **shelf life**; that is, their useful life slowly deteriorates even when they are unused.

---

[501] Strictly "battery" refers to several interconnected galvanic cells in a single unit (they were so named by Benjamin Franklin, 1706–1790, one of the founding fathers of the United States, by analogy to a battery of cannons), but nowadays the term is applied also to a single cell.

Electrochemical Science and Technology: Fundamentals and Applications, First Edition. Keith B. Oldham, Jan C. Myland, Alan M. Bond.
© 2012 John Wiley & Sons, Ltd. Published 2012 by John Wiley & Sons, Ltd.

You will be acquainted with the many uses to which batteries and fuel cells are put: starting internal combustion engines, powering vehicles and spacecraft, providing emergency lighting, load leveling by electric utilities, powering pacemakers and hearing aids, energizing all manner of power tools and electronic devices, and so on. Each of these applications has a different electrical requirement and, in response to this variety of needs, a wide array of battery and fuel cell types has been developed, with new ones appearing frequently. The technology of electrochemical power sources is a sophisticated blend of science and engineering; the scientist identifies what might be possible, while the engineer brings that possibility to fruition.

Battery Characteristics: quantifying the properties of batteries

The words "anode" and "cathode" are ambiguous when applied to the electrodes of a secondary battery because what is the anode when the battery is charging becomes the cathode on discharge and is neither when the battery is idle. Therefore the descriptors **positive** and **negative**, even used as nouns, are employed to distinguish the two electrodes of secondary batteries, and this usage often spills over into primary battery terminology too. For example, the lead dioxide of a lead-acid battery is said to be its "positive electrode" or its "positive plate" or simply its "positive". The term **active material** is used to signify the chemicals that are consumed as a battery discharges.

The **nominal voltage** cited for a primary cell (or a fully charged secondary cell) is close to its null (open-circuit) value. As described in Chapter 3, this can often be calculated from standard Gibbs energies and the activities of the species involved[502]. Because of polarization (Chapter 8), the voltage will be less when the cell is delivering charge and, moreover, the voltage will usually decline somewhat with time as the relevant activities within the cell change. The end of the useful life of a primary cell (recharge becomes needed for a secondary cell) occurs at some chosen **cut-off voltage** as shown in Figure 5-1.

The quantity of electricity that a charged battery can deliver before becoming exhausted is known as its **capacity** (not to be confused with *capacitance*, page 12). This is not a well characterized quantity because the charge delivered depends on the magnitude of the load, the duty cycle[503], the cut-off voltage, and the temperature; for specified battery types and uses, protocols govern these factors in the battery industry. The capacity could be expressed in coulombs, but the usual unit in battery parlance is the **ampere-hour** (A h).

5:1 $$1 \text{ A h} = 3600 \text{ C}$$

The total electrical energy (in joules or watt hours) delivered by a battery is the product of its voltage and its capacity.

---

[502] Calculate the voltage of the cell that has an overall reaction as in 5:15 on page 93. Check at Web#502.

[503] That is, whether the load is constant or changing, is applied continuously or intermittently, and so on.

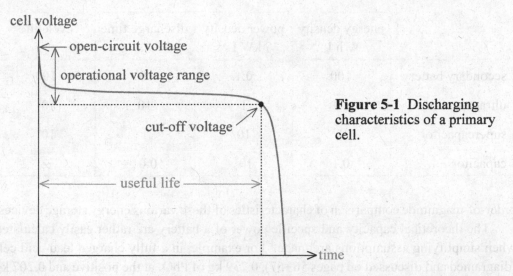

Figure 5-1 Discharging characteristics of a primary cell.

For a particular battery application, the important parameter may be the energy delivered by a given mass of battery or that can be fitted into a given space. These criteria are expressed as the **specific energy** of the battery (watt hours per kilogram) or its **energy density** (watt hours per liter)[504]. For other applications, however, such as starting an engine, it is not so much the energy that is important as how much power the battery can deliver in a brief time period. This is expressed by the **specific power** (watts per kilogram) or **power density** (watts per liter).

Recall from Chapter 1 that capacitors store electrical energy, as batteries do[505]. Which does the job better? From the standpoint of specific energy, batteries are greatly superior, whereas capacitors perform extraordinarily better in terms of specific power. Capacitance is a property of any electrode (page 18 and Chapter 13) and therefore batteries have capacitance too, though the electric charge stored capacitively is generally insignificantly small. Some specialty batteries, however, use highly porous electrodes of abnormally large surface area with surface-confined active materials. Such a device, sometimes called an **ultracapacitor**[506], represents a hybrid between a battery and a supercapacitor[507] (page 18), storing energy both faradaically and capacitively. The table overleaf provides a crude

---

[504] though you may encounter the term "energy density" incorrectly used as a synonym for specific energy. The terms "gravimetric capacity" and "volumetric capacity" are also in use.

[505] A distinction is that, whereas batteries maintain a more-or-less constant voltage as they discharge, the potential difference between the plates of a capacitor steadily declines as it discharges. The name **electrolytic capacitor** is given to a commercial capacitor with plates of Al, Ti or Ta, and an oxide dielectric, formed electrolytically. Despite their unfortunate name, there is nothing faradaic about these devices.

[506] Ultracapacitors are used to store energy from regenerative braking to provide a burst of power for subsequent acceleration. They have also found applications in hybrid vehicles and have been used as the sole power source in some Chinese buses.

[507] The distinction between a supercapacitor and an ultracapacitor is not always made. The name **electrochemical capacitor** is also encountered.

	energy density / W h L^{-1}	power density / kW L^{-1}	discharge time / s	cycle life
secondary battery	100	0.1	10^4	100
ultracapacitor	10	1	10	10^5
supercapacitor	1	10	1	10^6
capacitor	0.1	10^5	0.01	∞

order-of-magnitude comparison of characteristics of these various energy storage devices.

The theoretical capacity and specific power of a battery are rather easily calculated when simplifying assumptions are made. For example, in a fully charged lead-acid cell (diagramed and discussed on pages 56–57), 0.239 kg of PbO_2 at the positive and 0.207 kg of Pb at the negative (one mole of each) would theoretically require only 0.385 liters (weighing 0.499 kg) of 5.2 molar aqueous sulfuric acid to permit the cell reaction

$$5:2 \qquad PbO_2(s) + Pb(s) + 2HSO_4^-(aq) + 2H_3O^+(aq) \xrightarrow{\text{discharge}} 2PbSO_4(s) + 4H_2O(\ell)$$

to proceed to completion. This galvanic process would produce[508]

$$5:3 \qquad (2.00 \text{ mol}) \times (96485 \text{ C mol}^{-1}) = 1.93 \times 10^5 \text{ C} = 53.6 \text{ A h}$$

of electricity, so that our ideal lead-acid cell would have a capacity of 53.6 ampere hours and a mass of $(0.239 + 0.207 + 0.499)$ kg = 0.945 kg. With a cell voltage of 2.01 V, as calculated in equation 3:10, we compute a specific energy of

$$5:4 \qquad \frac{(53.6 \text{ A h}) \times (2.01 \text{ V})}{0.945 \text{ kg}} = 114 \text{ W h kg}^{-1}$$

These calculations make no allowance for incomplete utilization of the electroactive materials[509], for voltage losses arising from polarization, or for the masses of current collectors, terminals, separator, and the battery case; so it comes as no surprise that, in practice, lead-acid batteries have specific energies of only one-quarter to one-third of this theoretical value.

Primary Batteries: the Leclanché cell and its successors

A primary battery generally comprises a negative electrode consisting of a metal low in the electrochemical series (page 113), in contact with an electrolyte, often in the form

---

[508] The "2.00 mol" arises because 2 electrons were cancelled in creating cell reaction 3:6 (or 5:2) from electrode reactions 3:4 and 3:5.

[509] Of course, it is essential for there to be excess sulfuric acid, beyond that required by the stoichiometry, as otherwise the ionic conductor would have no conductivity at full discharge.

of a paste or held within the pores of a separator. The positive electrode is either a metal higher in the electrochemical series or a metal oxide[510]. A primary battery of the latter type is illustrated in the generic diagram to the right. Metal oxides are generally poor electrical conductors and, to overcome this handicap, a **current collector** is often provided in the form of carbon either mixed with the oxide or as a rod providing a conduction path to the battery's positive terminal. Current collectors play no role in the cell's chemistry.

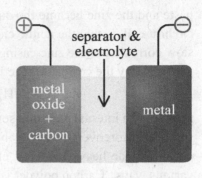

Although archeological finds suggest that batteries may have existed in antiquity[511], it was **Volta's**[119] **pile**, consisting of a stack of alternating zinc and copper disks, each pair being separated from its neighbors by pieces of brine-soaked cloth, that is the first documented device capable of producing a steady flow of current. That was in 1800 but, 36 years later, the **Daniell cell** (pages 64–65) replaced it as a more convenient laboratory device. Primary cells entered commerce in 1868 after **Leclanché**[512] patented the zinc-carbon **wet cell**[513], many thousands of which were used to power early telegraph lines. Each cell contained a carbon rod serving as the positive current collector, surrounded by crushed manganese dioxide in a porous pot beaker. This assembly was immersed in a jar containing an aqueous solution of ammonium chloride, NH_4Cl, into which dipped a zinc rod, the negative electrode. The reaction at the negative electrode is the electrodissolution of the zinc:

5:5 $\qquad$ **negative :** $\qquad$ $Zn(s) \rightarrow 2e^- + Zn^{2+}(aq)$

while at the positive electrode the manganese dioxide is reduced to an oxyhydroxide in a lower oxidation state

5:6 $\qquad$ **positive :** $\qquad$ $MnO_2(s) + NH_4^+(aq) + e^- \rightarrow MnOOH(s) + NH_3(aq)$

The ammonia produced in this cathodic reaction subsequently complexes the zinc ions, to form an ammoniated zinc cation, which precipitates as its chloride, so that the overall cell reaction is

5:7 $\quad Zn(s) + 2MnO_2(s) + 2NH_4^+(aq) + 2Cl^-(aq) \rightarrow 2MnOOH(s) + Zn(NH_3)_2Cl_2(s)$

The wet cell's voltage is close to 1.55 V when the cell is new but, during discharge, it gradually falls to about 1.0 V, beyond which it is exhausted.

The inconvenience of a liquid electrolyte ended, twenty years later, with the invention of the **dry cell** in which the ammonium chloride solution was immobilized in a starch-based

---

[510] The metal is one that has several oxidation states (page 29), the oxide in question being in a higher state.

[511] Read about the Baghdad "battery" at Web#511.

[512] Georges Leclanché, 1839–1882, French inventor.

[513] so called following the 1886 invention of the dry cell by Carl Gassner Jr., 1839–1892, German scientist.

paste and the zinc became the outer casing. A problem with this cell, the so-called **zinc-carbon cell**, is that the acidic character of the ammonium chloride electrolyte leads to a slow corrosion of the zinc casing, causing a decrease in shelf life. As well, the hydrogen generated by the corrosion reaction

$$5\text{:}8 \qquad\qquad Zn(s) + 2NH_4^+(aq) \rightarrow Zn^{2+}(aq) + 2NH_3(aq) + H_2(g)$$

produces an internal pressure, sometimes leading to rupture of the case and leakage of the solution. Moreover, the cell's ability to delivery its full power is jeopardized if it is called on to provide heavy bursts of electricity. These shortcomings have been overcome in various ways. Carbon powder is added to supplement the conductivity of the manganese dioxide. The corrosiveness of the electrolyte is decreased by replacing much of the ammonium chloride by the less acidic zinc chloride $ZnCl_2$. Organic additives are incorporated[514] to lessen corrosion. Modern dry cells[515] have an initial voltage of 1.6 V and maintain a higher voltage throughout their active life than did their forebears. In the battery world, such a cell, illustrated in Figure 5-2, is known as the **zinc chloride cell**.

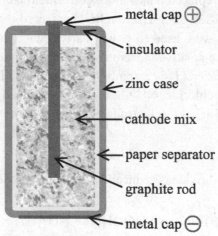

metal cap $\oplus$

insulator

zinc case

cathode mix

paper separator

graphite rod

metal cap $\ominus$

Figure 5-2 Cross-sectional diagram of a modern dry cell. The cathode mix is composed of powdered carbon and manganese dioxide in an aqueous paste of ammonium chloride and zinc chloride electrolytes.

The active materials remain zinc and manganese dioxide in the cell that is commonly called the **alkaline battery**. This cell, which goes by the technical name of "alkaline manganese-zinc", is shown in Figure 5-3. This primary cell has evolved greatly from the dry cell's original design. It has been turned "inside out", with the zinc negative now occupying the axis of the cylindrical cell, in the form of a gelled core of powdered zinc, held within a sleeve of cation-selective separator. The active material for the positive – manganese dioxide mixed with carbon – is pressed into the space remaining between the

---

[514] Mercury, which also lessens corrosion, was used for this purpose until environmental concern led to prohibition.

[515] A modern "D size" dry cell has a capacity of 7 A h when slowly discharged to a cut-off voltage of 0.75 V. Calculate the mass of zinc utilized and estimate the total energy delivered. See Web#515.

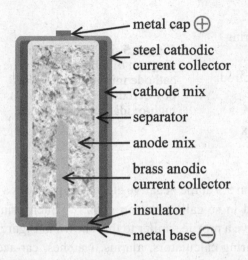

metal cap ⊕

steel cathodic
current collector

cathode mix

separator

anode mix

brass anodic
current collector

insulator

metal base ⊖

Figure 5-3 Cross-sectional diagram of an alkaline battery. The anode mix is composed of zinc powder in a gelled aqueous solution of potassium hydroxide. The separator is a cation-selective membrane. The cathode mix is made of manganese dioxide and carbon powders in the same KOH solution.

separator and the nickel-plated steel can, which also serves as the positive current collector. A concentrated aqueous potassium hydroxide[516] solution serves as electrolyte and permeates throughout the battery. The anodic and cathodic reactions are

5:9   **negative :** $Zn(s) + 2OH^-(aq) \rightarrow 2e^- + ZnO(s) + H_2O(\ell)$

and

5:10   **positive :** $MnO_2(s) + H_2O(\ell) + e^- \rightarrow MnOOH(s) + OH^-(aq)$

respectively[517]. The nominal cell voltage of the zinc-manganese-dioxide cell is 1.5 V, though this drops to about 1.2 V under moderate load. The capacity of a typical AA sized alkaline battery is cited as 2.85 A h, equivalent to a specific energy of 143 W h kg^{-1}. The performance is substantially worse, however, when the battery is called on to provide brief heavy discharges, as in a flash camera. Nevertheless, the alkaline battery represents a considerable advance on its Leclanché cell ancestors.

It is not only in the cell described above that an aqueous solution of potassium hydroxide serves as the electrolyte; several other primary batteries use it too, with zinc retained as the active negative material. Replacing the manganese dioxide as the oxidizer may be mercury(II) oxide, silver oxide, or the oxygen from the air. The reaction at the positive electrode is then one of the following:

5:11   **positives :** $\begin{cases} HgO(s) + H_2O(\ell) + 2e^- \rightarrow Hg(\ell) + 2OH^-(aq) \\ Ag_2O(s) + H_2O(\ell) + 2e^- \rightarrow 2Ag(s) + 2OH^-(aq) \\ \frac{1}{2}O_2(g) + H_2O(\ell) + 2e^- \rightarrow 2OH^-(aq) \end{cases}$

---

[516] KOH is used in preference to the cheaper NaOH because of the higher conductivity of its aqueous solutions.

[517] Construct the cell reaction and write the equation expressing the null voltage (open-circuit voltage) in terms of standard Gibbs energies and activities. See Web#517.

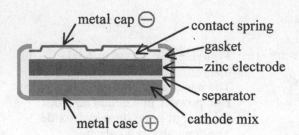

Figure 5-4 Cross-sectional diagram of a button cell. The cathode mix is composed of powdered carbon and various high-oxidation-state metals.

Each of these positives, partnered by reaction 5:9 at the negative electrode, provides a cell reaction[518] that has been, or is being, used in so-called **button cells**[519]. Such primary cells[520], which come in a variety of sizes, have a profile similar to that shown in Figure 5-4 and find many applications, such as powering calculators, alarms, watches, car-access remote controls, and liquid crystal displays. The power demands of these devices are so modest that cost ceases to be the prime consideration. Look for small holes in the casing of the zinc-air cells used to power hearing aids; these admit air to the cell but are covered by adhesive tape on a new cell to prevent atmospheric CO_2 reacting with the KOH electrolyte. The voltages of these three cell types beneficially[521] remain virtually constant throughout their useful lifetime because the chemicals participating in their reactions remain at constant activity[309]. As well, they are robust, have excellent shelf-life, an acceptable energy density for many tasks, and they perform well at low temperatures.

Zinc has served as the negative in all the primary batteries discussed above but, since the 1960s, lithium has been steadily replacing zinc in this role, with over 100 battery systems having been patented. Not only does lithium command a higher voltage [ΔE_n° for the $4Li(s) + O_2(g) \rightarrow 2Li_2O(s)$ cell reaction being 2.91 V compared with 1.66 V for the corresponding Zn reaction], but the mass of lithium required to generate one ampere hour of electricity is only 0.259 grams compared with 1.22 g for zinc[522]. These advantages are offset by the incompatibility of lithium metal with water. Lithium reacts also with most other liquids but with some of these (as with air) it forms a thin passivating layer that prevents further attack. For some organic and inorganic liquids, such as 1,2-dimethoxy-ethane $CH_3OCH_2CH_2OCH_3$ and thionyl chloride $SOCl_2$, the ionic conductivity of the layer is adequate enough for it to serve as a separator, with a salt dissolved in the liquid constituting the electrolyte solution. The poor conductivity of the passivating layers

---

[518] Write the cell reaction for each, then, via the reaction's ΔG, calculate the three null cell voltages at 25°C. See Web#518.

[519] They are called zinc-mercury-oxide, silver-zinc, and zinc-air batteries. Notice that little consistency attends the naming of batteries.

[520] Inasmuch as the oxidizer comes from outside, it could be argued that a zinc-air cell is, in part, a fuel cell.

[521] Though a downside is that the voltage does not provide information on the battery's state of charge.

[522] Confirm these masses. See Web#522.

necessitates thin electrodes of large area and most commercial **lithium primary batteries** have either the flat geometry of "coin" cells or consist of pasted lithium foil spirally wound into a "swiss roll" (or "jellyroll") configuration. Most lithium batteries have a voltage close to 3.0 V, about twice that of their zinc-based competitors.

Scant logic attends the naming of battery types. One convention when designating a cell is to cite the negative electrode first, as in zinc-air, sodium-sulfur, and lithium-manganese-dioxide. A different convention, however, applies to cells that employ metal oxide positives; for these, it is traditional to reverse the order and omit the word "oxide". Thus the technical term for the common alkaline battery is alkaline-manganese-zinc. The names of newer, lithium-containing batteries have yet to be standardized.

The reactions occurring in the **lithium-manganese-dioxide cell** are

5:12 $$\text{negative}: \quad Li(s) \rightarrow e^- + Li^+(soln)$$

and

5:13 $$\text{positive}: \quad MnO_2(s) + Li^+(soln) + e^- \rightarrow \tfrac{1}{2}Li_2Mn_2O_4(s)$$

The formula $Li_2Mn_2O_4$ represents an intercalation compound (page 97) in which lithium ions enter the MnO_2 lattice, with manganese thereby undergoing a progressive change in oxidation state. The **lithium-thionyl-chloride cell** is perhaps unique in that the $SOCl_2$ serves both as an oxidizer and as the solvent for the $LiAlCl_4$ supporting electrolyte. The reaction at the negative lithium electrode is again 5:12; the reaction at the positive is a multistep process with

5:14 $$\text{positive}: \quad 4SOCl_2(\ell) + 10e^- \rightarrow 2S(s) + S_2O_4^{2-}(soln) + 8Cl^-(soln)$$

as the overall equation. With a nominal voltage of 3.4 V and excellent power density, shelf-life, and low-temperature characteristics, it has filled military and spacecraft needs.

To conclude this brief exposition on primary batteries, we mention two specialized applications. Life vests and life rafts for support in case of disasters at sea are equipped with rescuer-attracting lamps operating from batteries that function only on submersion in the ocean; seawater serves as the electrolyte solution. In the most common design, magnesium is the negative electrode, with silver chloride as the positive. The cell reaction for this **seawater-activated battery**[523],

5:15 $$\text{cell}: \quad Mg(s) + 2AgCl(s) \rightarrow Mg^{2+}(aq) + 2Ag(s) + 2Cl^-(aq)$$

provides more than enough voltage for a 1.5 V flashlight bulb. The second application also saves lives, but in a very different context. Implantable pacemaker batteries experience only a modest drain, but must be compact, safe and totally reliable. Moreover, a lower limit to their lifetime should be long and predictable. An all-solid battery is desirable and one is provided by the **lithium-iodine battery**. The underlying cell reaction is

---

[523] What mass of each reactant must be provided for a seawater battery to power a 500 mA flashlight bulb for 12 hours? See Web#523.

5:16   **cell :** $2Li(s) + I_2(s) \rightarrow 2LiI(s)$

but the iodine is in the form of its charge-transfer adduct with poly-2-vinylpyridine, mentioned on page 9. The battery comprises a layer of this adduct on lithium foil. Lithium iodide salt forms at the junction and deleteriously increases in thickness over time. Nevertheless, the layer has sufficient conductivity to provide guaranteed performance for 8 years.

Secondary Batteries: charge, discharge, charge, discharge, charge, ...

The requirements for a secondary battery are much more stringent than those for a primary battery, so it is surprising that Planté's[524] invention of the lead-acid secondary cell in 1860 predates the Leclanché primary cell. Even more surprising is that the **lead-acid battery** remains by far the most important and widespread secondary battery to this day. The chemistry[525] and electrochemistry of this remarkable cell were described in Chapter 3. To review, the **positive**, **negative** and net **cell** reactions are[526]

5:17 $PbO_2(s) + HSO_4^-(aq) + 3H_3O^+(aq) + 2e^- \underset{\text{recharge}}{\overset{\text{discharge}}{\rightleftarrows}} PbSO_4(s) + 5H_2O(\ell)$

5:18 $Pb(s) + HSO_4^-(aq) + H_2O(\ell) \underset{\text{recharge}}{\overset{\text{discharge}}{\rightleftarrows}} 2e^- + PbSO_4(s) + H_3O^+(aq)$

and

5:19 $Pb(s) + PbO_2(s) + 2H_3O^+(aq) + 2HSO_4^-(aq) \underset{\text{recharge}}{\overset{\text{discharge}}{\rightleftarrows}} 2PbSO_4(s) + 4H_2O(\ell)$

The standard cell voltage, as calculated on page 59, is 1.925 V, but the Nernst equation

5:20 $\Delta E_n = (1.925\,V) - \dfrac{RT}{2F}\ln\left\{\dfrac{a_{H_2O}^4}{a_{H_3O^+}^2\,a_{HSO_4^-}^2}\right\}$

shows the null voltage to depend on the sulfuric acid concentration. For a fully-charged concentration of 5.2 M, the voltage is about 2.13 V at 25°C, but this declines on discharge as the ions are consumed to reach a cut-off voltage of about 1.8 V.

Though the electrochemistry remains unchanged, there have been many modifications to the lead-acid cell since Planté's time. Nowadays, the electrodes are usually prepared in

---

[524] Gaston Planté, 1834–1884, French physicist. His studies extended earlier work by Karl Wilhelm (later Sir Charles William) Siemens, 1823–1883, a German-born British inventor, and by Wilhelm Josef Sinsteden, 1803–1891, German physician and physicist.

[525] The equilibria $H_2SO_4 + 2H_2O \rightleftarrows HSO_4^- + H_3O^+ + H_2O \rightleftarrows SO_4^{2-} + 2H_3O^+$ occur but, under lead-acid cell conditions, the first dissociation is virtually complete and the second occurs to only about 1%.

[526] How much electricity, and how much energy, can a typical automotive lead-acid battery (84 A h, 12 V) supply? What mass of PbO_2 is destroyed, per cell, during this discharge? See Web#526.

the discharged state by pressing a paste of powdered lead oxides in sulfuric acid onto a grid of lead-antimony[527] alloy, which serves both as a mechanical support and as a current collector. During a "curing" process, the paste converts to an adherent porous mass of basic lead sulfates, $3PbO \cdot PbSO_4 \cdot H_2O$ and $4PbO \cdot PbSO_4$. Finally, the electrodes are "formed" by charging electrochemically in sulfuric acid solution to produce lead dioxide PbO_2 at the positive and spongy lead metal at the negative.

There are four extraneous reactions that concern designers of lead-acid batteries: corrosion of the positive electrode

5:21 positive : $\qquad Pb(s) + 6H_2O(\ell) \rightarrow 4e^- + PbO_2(s) + 4H_3O^+(aq)$

oxygen evolution from the positive electrode

5:22 positive : $\qquad 3H_2O(\ell) \rightarrow 2e^- + \frac{1}{2}O_2(g) + 2H_3O^+(aq)$

hydrogen evolution at the negative electrode

5:23 negative : $\qquad 2H_3O^+(aq) + 2e^- \rightarrow H_2(g) + 2H_2O(\ell)$

and oxygen reduction at the negative electrode

5:24 negative : $\qquad \frac{1}{2}O_2(g) + 2H_3O^+(aq) + 2e^- \rightarrow 3H_2O(\ell)$

Reactions 5:22 and 5:23 compete with the charging reactions 5:17 and 5:18 and these competitive processes take over completely when recharge is complete, leading to loss of water from the cell and necessitating periodic "topping up" of the cells with pure water. On newer "sealed" batteries[528], no topping up is needed. Moreover, whereas earlier designs had to be used upright, this is no longer essential. On charging, oxygen is formed before hydrogen is liberated. The oxygen, formed at the positive, then diffuses to the lead electrode, where it undergoes reaction 5:24, replenishing the lost water.

Of course, the major use of the lead-acid cell is for internal-combustion-engined road vehicles where, assembled into a six-cell battery, it serves two purposes: to provide the very large power demanded for starting the engine, and to provide a prolonged supply of modest current for such tasks as illuminating parking lights and powering the radio. The design of lead-acid automotive batteries thus represents a compromise between the competing demands for high specific power and high specific energy. Such batteries degrade from loss of active material from the positive plates and, especially if left discharged for long periods, by "sulfation" – the recystalization of lead sulfate into larger, electrochemically inactive, crystals.

Relieved of the need to provide high power, the so-called "deep cycle" batteries are more suitable for such applications as electric cars and long-term storage. Deep cycle lead-acid batteries are much more robust, both mechanically and electrically, largely because they employ thicker positive plates and separators of more sturdy design.

---

[527] The antimony adds stiffness to the lead; it plays no chemical role.

[528] Securely sealing the batteries would be dangerous; in fact they are provided with pressure-actuated relief valves. The apt description "valve regulated" is often used.

Lead-acid batteries are heavy, with a low specific energy[529] of about 40 W h kg^{-1}, but this is a serious disadvantage only in mobile applications[530], and the reliability, longevity, and low production costs are offsetting advantages of this enduring battery type. Lead is sometimes viewed negatively as an environmental hazard; on the other hand, it is easily recycled, much of the lead in today's batteries having had a previous life. Other disadvantages of the lead-acid battery are its reluctance to charge rapidly and its poor performance at very low temperatures[531].

There are several secondary batteries that employ nickel oxyhydroxide positives. The so-called nickel-cadmium secondary cell, often referred to as the "nicad battery"[532] is probably the best known. The plates of a **nickel-cadmium cell** are usually spirally wound in a "swiss roll" configuration and packed into a steel can, using nylon cloth as the separator and 9 M potassium hydroxide KOH aqueous electrolyte. The electrode reactions are

5:25   **positive :** $$NiOOH(s) + H_2O(\ell) + e^- \xrightleftharpoons[\text{recharge}]{\text{discharge}} Ni(OH)_2(s) + OH^-(aq)$$

and

5:26   **negative :** $$Cd(s) + 2OH^-(aq) \xrightleftharpoons[\text{recharge}]{\text{discharge}} 2e^- + Cd(OH)_2(s)$$

Notice from the cell reaction

5:27   **cell :** $$Cd(s) + 2NiOOH(s) + 2H_2O(\ell) \xrightleftharpoons[\text{recharge}]{\text{discharge}} Cd(OH)_2(s) + 2Ni(OH)_2(s)$$

that there is no net consumption of electrolyte and that only species at constant activity[533] are involved, suggesting that the cell voltage remains constant, though in practice the voltage falls from a fully-charged value of about 1.35 V to a cut-off of about 1.05 V. Even though their 1.2 V average voltage compares poorly with the 2.0 V of lead-acid competitors, nickel-cadmium batteries have a superior specific energy. Because of the toxic nature of cadmium, nickel-cadmium batteries require careful disposal.

The **nickel-hydrogen cell** is similar to the nickel-cadmium cell in its positive electrode and electrolyte, but it uses hydrogen gas in the reaction

5:28   **negative :** $$H_2(g) + 2OH^-(aq) \xrightleftharpoons[\text{recharge}]{\text{discharge}} 2e^- + 2H_2O(\ell)$$

at the negative. The difficulty of storing gaseous hydrogen inhibits the adoption of this

---

[529] The specific energy decreases with discharge rate for all batteries, but the decrease is unusually large for lead-acid types.

[530] and may even be advantageous, as in forklift trucks where they provide a convenient counterweight.

[531] largely due to the increasing viscosity of the sulfuric acid, which lowers the mobilities of the ions and hence increases cell resistance.

[532] though "NiCad" is the registered trademark of the SAFT Corporation. The nickel-cadmium cell was invented in 1899 by Waldemar Jungner, 1869 – 1924, Swedish inventor.

[533] unit activity, in fact, except for water, which is the solvent of a very concentrated solution.

battery, but it has found a niche application in space vehicles. However, by making use of solid metal hydrides, in which hydrogen is combined with metals, the difficulty of storing elemental hydrogen is overcome in **nickel-metal-hydride** secondary batteries, which use the same electrochemistry as in reactions 5:25 and 5:28. Representing the metal hydride as MH, the overall reaction becomes simply

5:29   cell : $MH(s) + NiOOH(s) \underset{\text{recharge}}{\overset{\text{discharge}}{\rightleftarrows}} M(s) + Ni(OH)_2(s)$

MH is, in fact, an alloy of several metal hydrides, designed to have a hydrogen vapor pressure sufficient to provide viable electrochemistry at all states of charge, but low enough not to endanger the integrity of the cell casing at full charge. The components of the alloy, which also serves as the current collector for the negative, are proprietarily secret but appear to include various proportions of two components: (a) metals such as Zr, Ti, La, and misch metal[534] that avidly take up hydrogen, and (b) metals such as Ni, Co, and Al, that form weak hydrides. Though sharing the same voltage characteristics, nickel-metal-hydride cells are 70% better in capacity than nickel-cadmium cells, which they seem destined to replace.

The surge in portable consumer electronics has spurred a need for a robust secondary battery that can be conveniently packaged in various sizes and shapes. This need has largely been met by the **lithium-ion cell**. Intercalation was mentioned earlier in the context of the lithium-manganese-dioxide primary cell and it is crucial in the lithium-ion cell. When a small cation intercalates a crystalline solid, it forces its way into the crystal lattice of its host. Because electroneutrality must be preserved, **intercalation**, or insertion as it is sometimes called, requires an inflow also of electrons, thereby changing the oxidation state of the host, but having little effect on the size of the crystals. The change in oxidation state is seldom by a whole number because intercalation compounds are generally non-stoichiometric[301].

In the positive of the lithium-ion cell the active material is a metal oxide, usually of cobalt[535], intercalated by lithium ions. The reaction is sometimes written

5:30   positive : $Li^+(soln) + CoO_2(s) + e^- \underset{\text{recharge}}{\overset{\text{discharge}}{\rightleftarrows}} LiCoO_2(s)$

but this is misleading in two respects. Firstly, the lithium content progressively decreases during the deintercalation that accompanies charging, so an appropriate formula would be Li_xCoO_2, where $x = 1$ only at full discharge. Secondly, though x decreases on charge, it cannot decrease as far as equation 5:30 suggests. In practical cells, the lithium content of the intercalation solid is not allowed to decrease below a mole fraction of 0.55, as otherwise irreversible changes occur. The most appropriate stoichiometric equation might therefore be

---

[534] an unresolved mixture of lanthanum and rare-earth elements.

[535] or, to save cost while retaining performance, by alloying with nickel and other metals.

5:31   **positive :** $\frac{9}{20}Li^+(soln) + Li_{11/20}CoO_2(s) + \frac{9}{20}e^- \xrightleftharpoons[\text{recharge}]{\text{discharge}} Li_1CoO_2(s)$

with the understanding that the lithium content progressively increases, $0.55 \le x \le 1$, during discharge. Equation 5:31 describes the overall reaction, from full charge to total discharge; at intermediates states of charge, the active material has the formula Li_xCoO_2, with cobalt having an effective oxidation number of $(4-x)$.

Lithium ions will intercalate graphitic carbons, too, forming an intercalation compound Li_yC with a lithium content of up to a mole fraction of one-sixth. This serves as the negative electrode in the lithium-ion battery, for which the reaction may be written

5:32   **negative :** $Li_{1/6}C(s) \xrightleftharpoons[\text{recharge}]{\text{discharge}} \frac{1}{6}e^- + Li_0C(s) + \frac{1}{6}Li^+(soln)$

with the caveat that the lithium content of the intercalation solid decreases progressively, $0 \le y \le 0.17$, during discharge[536]. The overall stoichiometry of the cell reaction, corresponding to the exchange of one electron and a single lithium Li^+ ion, may be represented

5:33   **cell :** $\frac{20}{9}Li_{11/20}CoO_2(s) + 6Li_{1/6}C(s) \xrightleftharpoons[\text{recharge}]{\text{discharge}} \frac{20}{9}Li_1CoO_2(s) + 6Li_0C(s)$

In use, the lithium ions pass from the negative to the positive electrode, just as the electrons do, though by a different route. The ions make the converse journey[537] on recharge. The lithium ions do not change their oxidation state: there is no lithium metal in this battery.

The electrolyte used in most lithium-ion cells is a solution of lithium perchlorate $LiClO_4$ or lithium hexafluoroarsenate $LiAsF_6$ in a mixture of organic carbonates, absorbed in a microporous polyolefin separator. Such an electrolyte has sufficient conductivity and, though it does slowly react with the negative electrode, the lithium carbonate so formed soon establishes a protective conductive layer.

The lithium-ion battery operates over a voltage range falling from about 4.0 V to a cut-off at 2.5 V. Its high specific energy and energy density (about 150 W h kg^{-1} and 400 W h L^{-1}), coupled with high reliability and cycle life, make this a very successful secondary battery. However, it has two main drawbacks. One is its unattractive shelf-life which causes its performance to deteriorate from the instant of its manufacture, whether or not it is used. The second is that, to maintain the battery safe and operative, its voltage must never be allowed to rise above 4.2 V or fall below 2.0 V. This means that circuitry must accompany the battery to open the circuit at low voltages and to ensure a safe but reasonably rapid recharging protocol (constant current until the voltage reaches 4.0 V; constant voltage thereafter). Because high temperatures must be avoided, measures to prevent over-heating are incorporated, as well as a pressure-relief valve. Moreover, if

---

[536] What oxidation number does carbon have when the lithium-ion cell is partly charged? See Web#536.

[537] They may be said to "rock" or "shuttle" back and forth, leading imaginative publicists to coin the names "rocking-chair battery" and "shuttlecock battery" for the lithium-ion cell.

lithium-ion cells are configured in parallel (to provide more current) or series (for more voltage), it is necessary to ensure that the malfunction of one cell does not lead to a (possibly dangerous) fault in others. The need for all this protective circuitry is the reason that lithium-ion cells are not offered for sale as such, but only as engineered units, each for a specific purpose.

The sodium-sulfur secondary battery should be placed, perhaps, in the "glorious failure" category. Sodium and sulfur are abundant and cheap materials. The electrode reactions, which proceed at the surfaces of the separator in the **sodium-sulfur cell**, are simply[538]:

5:34   **positive :** $2Na^+(sep) + 5S(\ell) + 2e^- \underset{\text{recharge}}{\overset{\text{discharge}}{\rightleftharpoons}} Na_2S_5(\ell)$

5:35   **negative :** $Na(\ell) \underset{\text{recharge}}{\overset{\text{discharge}}{\rightleftharpoons}} e^- + Na^+(sep)$

The cell operates at temperatures in the 300–400°C range, at which temperature the sodium is molten and the positive active material consists of liquid sulfur and immiscible liquid sodium pentasulfide in a porous carbon current collector. Sodium ions travel through the separator, which is a ceramic known as **beta alumina** or, specifically, β″ alumina. Its

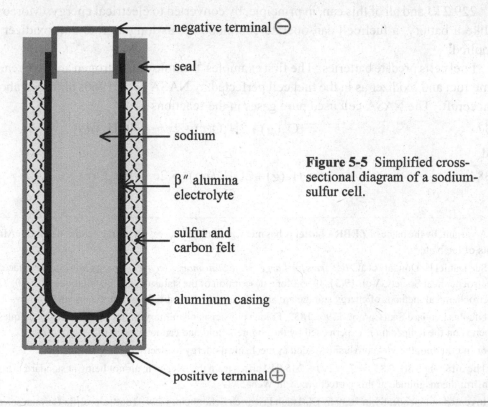

negative terminal ⊖

seal

sodium

β″ alumina electrolyte

sulfur and carbon felt

aluminum casing

positive terminal ⊕

Figure 5-5 Simplified cross-sectional diagram of a sodium-sulfur cell.

---

[538] There is a further cathodic step on prolonged discharge, but it is accompanied by a lessening of cell voltage.

composition is Al_2O_3, with about 0.16 mole fraction of sodium oxide Na_2O, and it has a healthy conductivity of about 20 S m^{-1} at 350°C. Prototype cells of the design shown in outline in Figure 5-5 have performed satisfactorily but, despite major efforts being devoted to this battery since the 1980s, it has never been successfully commercialized[539] as a small unit. Difficulties surround the integrity of the separator and of the seals between the various cell components. However, the sodium-sulfur battery has found application for large-scale stationary energy storage[540].

Fuel Cells: limitless electrical energy in principle, many problems in practice

Carnot[541] proved that when fuels are burned in a heat engine, such as the motor of a car, only a fraction of the Gibbs energy of combustion can be converted to mechanical energy; and, in practice, the conversion efficiency rarely exceeds 30%. No such limitation applies to the conversion of a fuel's chemical energy into electricity in a fuel cell[542]. For example, the energy liberated when 32.0 g (1.00 mol) of methanol is "burned" with air in a fuel cell by the reaction

5:36 $CH_3OH(\ell) + \frac{3}{2}O_2(g) \rightarrow 2H_2O(\ell) + CO_2(g)$ $\Delta G° = -235.0$ kJ mol^{-1}

is[543] 229.2 kJ and all of this can, in principle, by converted to electrical energy. Moreover, unlike a battery, a fuel cell can operate continuously, as long as fuel and oxidizer are supplied.

Fuel cells predate batteries. The first examples[544] employed hydrogen and oxygen, the same fuel and oxidizer as in the fuel cell perfected by NASA in the1960s and used aboard spacecraft. The NASA cell used pure gases in the reactions

5:37   cathode : $O_2(g) + 2H_2O(\ell) + 2e^- \rightarrow 4OH^-(aq)$

and

5:38   anode : $H_2(g) + 2OH^-(aq) \rightarrow 2e^- + 2H_2O(\ell)$

---

[539] A variant, by the name of **ZEBRA battery**, has met with limited success. Its name echoes the South African roots of the battery.

[540] See Daniel H. Doughty et al., *Batteries for Large-Scale Stationary Electrical Energy Storage*, Interface, The Electrochemical Society, Vol. 19(3), 49–53 for a description of the status of such installations in 2010. Other electrochemical methods of energy storage are also discussed in this article and in the one that follows it.

[541] Nicolas Léonard Sadi Carnot, 1796–1832, French physicist and military engineer. The fraction in question depends on the temperatures experienced by the engine's fluid and cannot exceed $(T_{hot} - T_{cold})/T_{hot}$.

[542] or in other nonthermal conversions, such as mechanical energy to electricity or vice versa.

[543] The difference of -5.8 kJ = 229.2 kJ $-$ 235.0 kJ arises from the oxygen in air not being at standard pressure. Confirm the magnitude of this correction; as in Web#543.

[544] Invented independently in 1839 by Christian Friedrich Schönbein, 1799–1868, a prolific German scientist, and contemporaneously by William Robert Grove, 1811–1896, a Welsh lawyer, judge, and amateur scientist.

with an aqueous KOH electrolyte and catalyst-loaded electrodes. An energy efficiency of 70% was claimed. For earthbound operation, it would clearly be more economical to use air instead of oxygen. However, this is not possible because the CO_2 in the air reacts with the KOH. This exemplifies one of the fundamental problems with fuel cells: because fuel and oxidizer are delivered continuously into the cell, there is a danger of impurities accumulating in the cell and ultimately impairing its operation. Ensuring that the consumables are of high purity is, of course, an added expense.

Reaction 5:36 describes the overall chemistry of the **direct-methanol fuel cell**[545], the component electrode reactions being:

5:39 **cathode :** $\quad 3O_2(air) + 12H_3O^+(aq) + 12e^- \rightarrow 18H_2O(\ell)$

5:40 **anode :** $\quad 2CH_3OH(aq) + 14H_2O(\ell) \rightarrow 12e^- + 2CO_2(g) + 12H_3O^+(aq)$

The fuel is an aqueous solution of methanol of not more than 2 molar concentration. The electrolyte is acidic and the electrodes are separated by a proton-selective membrane. Only about 50 kJ of electrical energy per mole of methanol is presently obtainable from this fuel cell, which faces two serious problems. Firstly, reaction 5:40 is slow, even when catalyzed by an expensive 50:50 alloy of platinum and ruthenium and operated at temperatures in the 90–140°C range. Secondly, methanol slowly diffuses through the membrane, interfering with the cathodic reaction.

To be economically viable as electricity generators, fuel cells must use cheap fuels and oxidizers. Air fills the latter need and the former has been met by natural gas, which is mostly methane CH_4. Fuel cells do not use methane directly; instead the methane is "reformed" by high-temperature reaction with steam

5:41 $\qquad CH_4(g) + H_2O(g) \xrightarrow[900°C]{Ni} CO_2(g) + 3H_2(g)$

The hydrogen-rich mixture from reforming could be used as a fuel in several types of fuel cell. There are problems, however. Natural gas has a significant content of sulfur-containing compounds and, without their removal, these compounds can poison the catalysts needed for the anodic hydrogen reaction. Moreover, a large concentration of carbon monoxide, also a catalyst poison, is present. Dropping the temperature after this **reforming reaction** to about 300°C favors the right-hand side of the water-gas equilibrium,

5:42 $\qquad\qquad CO(g) + H_2O(g) \rightleftarrows CO_2(g) + H_2(g)$

but also slows the interconversion rate. Subsequent purification can further reduce the CO content but the upshot is that hydrogen made by reforming always contains some catalyst poison. The relative success of the various fuel cells is largely a question of how well they can tolerate poisoning, rather than their innate electrochemistry.

A brief description of five fuel-cell designs that use air as the oxidizer, and hydrogen or other gases as fuel, follows:

---

[545] The "direct" in this name distinguishes it from other fuel cell schemes in which methanol is first decomposed and its decomposition products used as fuel.

(a) **PEM**[546] **fuel cells**. These operate at temperatures ranging from ambient to 90°C. A solid proton-conducting polymer (such as Nafion®) serves as electrolyte. Electrodes are carbon with platinum catalyst.

(b) **Alkaline fuel cells**, similar to the NASA cell, operating in the vicinity of 100°C. The electrolyte is 10 molar potassium hydroxide, with OH^- as the charge carrier. Electrodes are catalyst-loaded carbon. CO_2 must be removed from the hydrogen fuel and from the air oxidizer.

(c) **Phosphoric acid fuel cells**, operating close to 180°C. Employing proton conduction, the electrolyte is a very concentrated aqueous solution of H_3PO_4 held within a porous silicon carbide separator. Electrodes are platinum-loaded graphite.

(d) **Molten carbonate fuel cells**, operating at about 625°C. The electrolyte, through which charge is carried by the carbonate anion CO_3^{2-}, is a molten mixture of carbonate salts held within a porous matrix of $LiAlO_2$. The electrodes are porous non-precious metals, with nickel serving as the catalyst. Carbon monoxide, as well as hydrogen, can fuel this cell; methane, too, with reforming taking place within the cell.

(e) **Solid oxide fuel cells**, operating at about 850°C. Oxide ions O^{2-} carry the charge through a solid electrolyte of yttria-stabilized zirconia ZrO_2. The cathode is strontium-doped lanthanum manganite $LaMnO_2$; the nickel content of the anode serves as catalyst. Carbon monoxide is an alternative fuel.

Figure 5-6 is a schematic diagram of the operation of a **hydrogen-air fuel cell**, but it comes nowhere near portraying the actual geometry of such cells. Generally, these fuel cells employ a stack of very thin anode|electrolyte|cathode layers, with interlayer channels incorporated through which the fuel, oxidizer, and water vapor can flow. Additional tubes for coolant may be needed to maintain an optimum temperature. Careful pressure regulation is required too, to ensure that the liquid|gas interface remains within the pores of the porous electrodes.

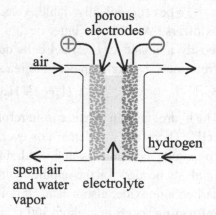

Figure 5-6 Schematic diagram of a hydrogen-air fuel cell.

Though engineering and materials problems escalate, fuel-cell efficiency improves with increasing operating temperature because the electrode reactions are then faster and because poisoning is less troublesome. A vast amount of effort has been expended in getting fuel cells to their present stage of development. Whether eventual widespread commercialization will vindicate this investment remains to be seen.

---

[546] The acronym stands either for polymer electrolyte membrane or proton-exchange membrane, both names being equally appropriate. Another name is **solid polymer electrolyte fuel cell**.

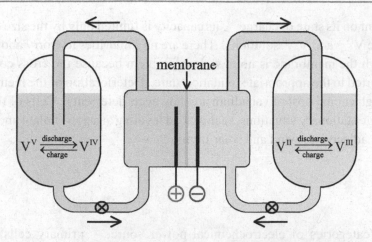

Figure 5-7 Schematic diagram of a vanadium redox battery.
The solution composition throughout each pumped loop
changes steadily during discharge and charge.

Although they have been touted as mobile power sources, and even for such miniature applications as in mobile phones, it seems likely that if fuel cells prosper, it will first be as stationary installations generating electricity at times of high demand. Indeed, such applications are already at the stage of being evaluated.

Because demand for electricity varies with the time of day, power utilities welcome satisfactory means of storing power. One electrochemical way of doing this is with a **redox battery**, **flow battery**, or **redox flow battery**, though "secondary fuel cell" might be a more descriptive name. As in a fuel cell, the reactants are fed into the cell from the outside. In this case though, the electroreactants are dissolved in the electrolyte solution, which flows into the cell. The products are also soluble in the electrolyte and are removed by the flowing stream. The reaction is reversible, permitting "secondary" operation. There are no solid phases to degenerate, giving the cell an almost indefinite lifetime. One manifestation of this principle is the **vanadium redox battery**, illustrated schematically in Figure 5-7. Each half-cell contains a carbon-felt electrode and the two are linked by a proton-conductive membrane. The cell exploits vanadium's ability to exist in four oxidation states: V^{II}, V^{III}, V^{IV}, and V^{V}, all of which are soluble in the sulfuric acid electrolyte. The reactions at the electrodes are

5:43   **positive :** $VO_2^+(aq) + 2H_3O^+(aq) + e^- \underset{\text{recharge}}{\overset{\text{discharge}}{\rightleftarrows}} VO^{2+}(aq) + 3H_2O(\ell)$

and

5:44   **negative :** $V^{2+}(aq) \underset{\text{recharge}}{\overset{\text{discharge}}{\rightleftarrows}} e^- + V^{3+}(aq)$

Though the voltage of this cell is rather low, with an open-circuit value averaging 1.4 V, and is dependent on its state of charge[547], its capacity is limited only by the size of the tanks that contain the $V^{IV,V}$ and $V^{II,III}$ solutions. There are no impurities to worry about and even leakage through the membrane is not a serious problem because the cross contaminants soon get converted to the appropriate oxidation state. Deterioration of the membrane with use, and the high capital cost of vanadium are, however, deterrents. Cells of this type are suitable for large stationary situations, such as load leveling at a power plant and temporary energy storage adjacent to wind and solar farms.

Summary

All three categories of electrochemical power source – primary cells, secondary batteries, and fuel cells – are rapidly developing fields of electrochemical technology.

Notwithstanding the hyperbole from environmentalists and battery fabricators – and, of late, from governments and car manufacturers – it is most unlikely that electrochemical power sources will ever rival the internal combustion engine in power density. Never-theless, it seems certain that environmental and economic constraints will increasingly force the replacement of many internal combustion engines by rechargeable power sources. At the time of writing, the lead-acid, nickel-metal-hydride, and lithium-ion secondary batteries are all competing for the potentially lucrative electric car market. All have disadvantages, as have other contending secondary batteries and fuel cells. Sadly, the paragon secondary battery remains elusive.

---

[547] See Web#547 for a discussion of how the open-circuit voltage of the vanadium flow cell changes with its state of charge.

6

Electrodes

Two is the minimum number of electrodes needed for an electrochemical cell although, for reasons that will be appreciated after reading Chapter 10, three are commonly employed. An electrode cannot be studied in isolation but, nevertheless, one often seeks to investigate the properties of a single electrode, rather than to study an electrochemical cell as a whole. Whenever interest is in a single electrode, that electrode is named the **working electrode**[601] (WE); the second electrode, the role of which is merely to complete the cell, is called the **reference electrode** (RE). In other circumstances, as in the cells discussed in the previous chapter, both electrodes are of equal importance and the "working" and "reference" descriptors are not then applicable.

Electrode Potentials: the reference electrode is the key

The cell voltage ΔE always means the potential of the working electrode with respect to the reference electrode, rather than vice versa. Thus, provided that a meaning can be attached to the terms on the right-hand side of the equation

6:1
$$\Delta E = E_{WE} - E_{RE}$$

then the voltage of any cell may be split into two terms, each arising from an individual electrode. But there is no obvious way to make the split, because the potential of a single electrode cannot be measured. Indeed, there are difficulties in even *defining* a single electrode potential[602]. Nevertheless, the concept of an **electrode potential** (or **half-cell potential**) is so appealing that electrochemists have agreed on ways to interpret the term "electrode potential" and so allot a quantitative value to the symbol E_{WE}, at least for cells in which the ionic conductor is an aqueous electrolyte solution.

---

[601] or sometimes, especially in the context of potentiometry (page 119), the **indicator electrode**.

[602] though a non-thermodynamic assignment can be made on the basis of a plausible model (see A.J. Bard, G. Inzelt and F. Scholtz, *Electrochemical Dictionary*, Springer, 2008, page 528). An electrode potential on this Kanevskii scale is more negative by 4.44 volts than the SHE-based electrode potential described overleaf.

Electrochemical Science and Technology: Fundamentals and Applications, First Edition. Keith B. Oldham, Jan C. Myland, Alan M. Bond.
© 2012 John Wiley & Sons, Ltd. Published 2012 by John Wiley & Sons, Ltd.

Before introducing the convention that allows working electrode potentials to be assigned, the reference electrode must be discussed[603]. The reference electrode is the one that we are less interested in. We want to be able to incorporate the RE into our cell and then forget about it. An ideal reference electrode would be one that maintains a constant potential E_{RE} whether we use it as an anode or a cathode, and irrespective of the current, if any, that we pass through it. To come close to this ideal requires, among other attributes, that there be an abundant supply of all the species involved in the reference electrode reaction and that the activities of all these species be constant. As well, a satisfactory reference electrode must resist polarization (Chapter 10) and be physically robust.

These criteria are close to being met by the **Ag | AgCl electrode**[604]. This consists of bulk silver coated by a substantial layer of silver chloride and immersed in solution rich in chloride ion, as in the left-hand electrode of Figure 6-1. The electrode reaction in aqueous media is

6:2 $$AgCl(s) + e^- \rightleftarrows Ag(s) + Cl^-(aq)$$

its rapid kinetics being one reason for the success of Ag | AgCl as a reference electrode. Another is the porous nature of the silver chloride layer, which allows solution to penetrate to the underlying metal.

The difficulty in splitting a cell voltage into the difference of two electrode potentials is akin to the conundrum encountered in Chapter 2, where it was noted that the Gibbs energies of individual ions could not be measured. In that case, the remedy was to assign one special ion, the hydronium ion H_3O^+, to have a specified G^o value. A similar convention enables individual electrode potentials to be given quantitative significance. If one particular reference electrode is chosen and its potential is assigned a value of zero, then equation 6:1 shows that any working electrode paired with this selected reference has a potential equal to the measurable cell voltage. For electrochemistry in which the ionic conductor is an aqueous solution, the chosen reference electrode is the **standard hydrogen electrode**[605] (SHE):

6:3 $$E_{SHE} = 0$$

This electrode consists of a metal[606] in contact with an aqueous solution containing hydronium ion at unit activity and saturated with gaseous H_2 at 1.000 bar pressure (unit activity), as in the right-hand electrode of Figure 6-1. The SHE reaction is

---

[603] H. Kahlert addresses reference electrodes in greater depth in: F. Scholz (Ed.), *Electroanalytical Methods*, 2nd edn, Springer, 2010, pages 291–308.

[604] and also by the **saturated calomel electrode**, an excellent reference electrode that, however, is disappearing from use on account of its hazardous mercury content. Note that $E_{SCE} = 0.2412$ V.

[605] also known as the **normal hydrogen electrode** (NHE).

[606] In principle, any metal (or, indeed, any electronic conductor) would serve, but platinum is invariably used because it catalyzes reaction 6:4. A high surface area is achieved by electrodepositing a powdery coating of platinum onto a platinum plate.

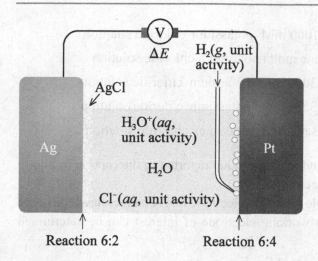

Figure 6-1 Very careful measurements of the voltage of this equilibrium cell have established that the potential of the Ag | AgCl reference electrode (in unit activity aqueous hydrochloric acid) at 25.00°C is $\Delta E = 0.22216$ V.

6:4
$$2H_3O^+(aq) + 2e^- \rightleftarrows H_2(g) + 2H_2O(\ell)$$

The choice of this reference electrode to establish a scale of electrode potentials was made for thermodynamic reasons[607], not for experimental convenience. Sadly, the standard hydrogen electrode has few of the desirable features listed above and, in fact, it is implemented only for meticulous metrological purposes.

Fortunately, there is no need to employ the impractical standard hydrogen electrode in day-to-day electrochemistry because equation 6:1 may be rewritten as

6:5
$$\Delta E = \left(E_{WE} - E_{SHE} \right) - \left(E_{RE} - E_{SHE} \right) \qquad \begin{matrix} \text{adjustment to} \\ \text{practical RE} \end{matrix}$$

and the term ($E_{RE} - E_{SHE}$) may be determined once and for all in a separate experiment. Figure 6-1 diagrams such a calibration experiment using the Ag | AgCl electrode as the substitute electrode. In practice, the Ag | AgCl electrode rarely employs hydrochloric acid [ions: $H_3O^+(aq)$, $Cl^-(aq)$]. More usually, an aqueous solution of potassium chloride[608] [ions: $K^+(aq)$, $Cl^-(aq)$] is used, either saturated with this salt, or at a variety of specified concentrations. With such an Ag | AgCl electrode, the experiment shown in Figure 6-1 could not be carried out without a separator to prevent the mixing of the two solutions. The separator would introduce a junction potential, as discussed Chapter 3. Corrections for this junction potential, or its elimination, would be needed. In such ways, the potential $E_{Ag|AgCl}$ has been accurately determined at 25°C to be[609]

---

[607] One way of looking at the convention is that it defines the activity of electrons in the standard hydrogen electrode to be unity.

[608] Sometimes sodium chloride is used, resulting in somewhat different, slightly less positive, potentials.

[609] from H. Kahlert in: F. Scholz (Ed.), *Electroanalytical Methods*, 2nd edn, Springer, 2010, page 269. These potentials have a temperature coefficient of about −0.7 mV K⁻¹.

6:6 $\qquad E_{Ag|AgCl} = \begin{cases} 0.2363 & \text{in 1000 mM potassium chloride solution} \\ 0.2223 & \text{in one molal potassium chloride solution} \\ 0.2070 & \text{in 3000 mM potassium chloride solution} \\ 0.2037 & \text{in 3500 mM potassium chloride solution} \\ 0.1970 & \text{in saturated potassium chloride solution} \end{cases}$

These values differ from each other, and from the value reported in the caption to Figure 6-1, because of differing chloride ion activities.

Once the potential E_{RE} of a suitable reference electrode has been measured, or found from the literature, the potential of a working electrode of interest can be determined directly from experiment as[610]

6:7 $$E = \Delta E + E_{RE}$$

An example of such an experiment is shown in Figure 6-2. This cell is equipped with a salt bridge (page 66) to eliminate most of the junction potential that would otherwise exist. When the current direction is as shown, the copper working electrode functions as a cathode, but with a less positive applied voltage it could become an anode. Or, on opening the switch, it would soon behave as an equilibrium cell. There are two ways of reporting the potential of this working electrode:

6:8 $\qquad E = \begin{cases} 0.018\,\text{V versus Ag}\,|\,\text{AgCl(1000 mM KCl)} \\ 0.317\,\text{V} \end{cases}$

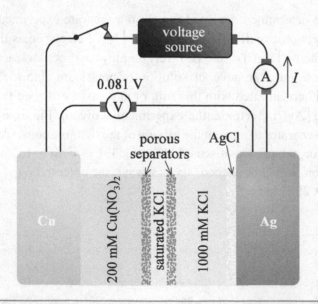

Figure 6-2 Schematic diagram of a typical electrochemical cell. The Ag|AgCl|KCl(1000 mM) electrode serves as the reference electrode and a salt bridge links the two half cells.

[610] Henceforth, the $_{WE}$ subscript will be dropped, as it becomes obvious which electrode is being discussed.

In the second option, the absence of any mention of a reference electrode implies that the measured voltage has been converted to the SHE scale. In such cases, *the electrode potential expresses what the measured cell voltage would have been had the reference electrode been a standard hydrogen electrode functioning ideally.*

Notice that the terms "galvanic" or "electrolytic" are no longer useful descriptors of such a cell as that in Figure 6-2, because such a classification depends on the (arbitrary) choice of a reference electrode.

The convention of using a hydrogen-based standard reference electrode can be applied to cells that employ water-like solvents, such as methanol CH_3OH and liquid ammonia NH_3, but it is mostly restricted to aqueous systems. In other cases, the upper option in 6:8 is adopted by specifying explicitly what the reference electrode is. When the temperature departs from 25°C, it is wise to report the reference conditions, even when the system is aqueous. One popular and effective **internal reference** system for many nonaqueous solvents is the ferrocenium | ferrocene[611] couple based on the

6:9 $$(C_5H_5)_2Fe^+(soln) + e^- \rightleftarrows (C_5H_5)_2Fe(soln)$$

equilibrium, with both species present at equal small concentrations[612].

Standard Electrode Potentials: they are related to standard Gibbs energies

As described above for aqueous systems, the potential of a working electrode is $E = \Delta E - E_{RE}$, where RE is the reference electrode. Because it equals the value that the cell voltage would have had with an ideal SHE as reference, the supposed cell reaction would have been

6:10 $$Cu^{2+}(aq) + H_2(g) + 2H_2O(\ell) \rightarrow Cu(s) + 2H_3O^+(aq)$$

for the cell diagramed in Figure 6-2.

Because of polarizations (Chapter 10), it is difficult to interpret the electrode potential of a cell through which current is flowing. However, if the switch shown in Figure 6-2 is opened and the cell is allowed to come to rest then, as we saw in Chapter 3, the cell voltage is related, through the equation

6:11 $$E_n = \Delta E_n = \frac{-\Delta G}{nF} \qquad \text{null potential}$$

to the thermodynamics of the cell reaction. Here the italic n represents the number of electrons that were cancelled out in forming the cell reaction from the electrode reactions,

---

[611] Ferrocene is bis-cyclopentadienyl iron, the Fe atom being sandwiched between the aromatic rings. The full symbol is η^5-$(C_5H_5)_2Fe$, indicating that the iron atom is symmetrically placed with respect to the 10 carbon atoms. Fc is a frequent abbreviation.

[612] How the referencing is practiced in voltammetry is explained on pages 353–354. With some solvents, the ferrocenium cation may be unstable.

while each subscript $_n$ serves as a reminder that the cell is operating under null conditions. ΔG denotes the change in Gibbs energy that would accompany an SHE-referenced cell reaction, such as that described by equation 6:10.

In the case of the cell shown in Figure 6-2, the Gibbs energy change would have been

6:12 $$\Delta G = G_{Cu} + 2G^{o}_{H_3O^+} - G_{Cu^{2+}} - G^{o}_{H_2} - 2G^{o}_{H_2O}$$

The Gibbs energies of the species participating in the SHE reference electrode reaction have been written as their standard values because unit activities are prescribed for the standard hydrogen electrode. Moreover, these three terms disappear because of the thermodynamic conventions that $G^{o}_{H_3O^+} = G^{o}_{H_2O}$ and that $G^{o}_{element} = 0$. The latter convention also enables one to set $G_{Cu} = G^{o}_{Cu} = 0$. Only one nonzero term, $G_{Cu^{2+}}$, remains. As we saw in Chapters 2 and 3, this expression can be written in terms of the standard Gibbs energy and the activity. Thereby one finds[613]

6:13 $$E_n = \frac{G_{Cu^{2+}}}{2F} = \frac{G^{o}_{Cu^{2+}}}{2F} + \frac{RT\ln\{a_{Cu^{2+}}\}}{2F} = E^o - \frac{RT}{2F}\ln\left\{\frac{1}{a_{Cu^{2+}}}\right\}$$

In 6:13, we have defined a quantity E^o, which is the value adopted by the electrode potential under null conditions, when all the species that are involved in the reaction at the working electrode are at unit activity. It is called the **standard electrode potential**[614]. Long lists of standard electrode potentials have been compiled[615]; the table[616] on page 392 is one of these. In compiling such tables, it is usual to cite the equilibrium to which the potential relates, with the electrons on the left-hand side[617], as in

6:14 $$Cu^{2+}(aq) + 2e^- \rightleftarrows Cu(s), \qquad E^o = 0.340 \text{ V}$$

As a glance at the table will confirm, some standard electrode potentials are known to very high precision, others less so. Different sources may list slightly diverse values.

With the v's being stoichiometric coefficients, let the general equation

6:15 $$v_A A + v_B B + \cdots + ne^- \rightleftarrows v_Z Z + v_Y Y + \cdots$$

describe an electrode reaction in which the ionic conductor is aqueous. Then the standard

---

[613] Explain the origin of the "2" divisor in equation 6:13 and then predict the potential of the cell in Figure 6-2 after opening the switch. The (mean ionic) activity coefficient of the copper ion in 200 mM aqueous copper(II) nitrate solution is 0.466. See Web#613.

[614] Notice that in several electrochemical contexts, the adjective "standard" implies unit activities.

[615] Beware that many of these tables still relate to an older standard atmospheric pressure of 101325 Pa. The data in this book have been corrected to the present 100000 Pa (one bar) standard. Demonstrate how this correction was made. See Web#615.

[616] Use G^o values from this table to confirm the standard potential of the $MnO_4^-(aq) + 8H_3O^+(aq)$ $+5e^- \rightleftarrows Mn^{2+}(aq) + 12H_2O(\ell)$ electrode reaction. Compare this with the E^o value from the table on page 392 and with Web#616.

[617] and standard electrode potentials are therefore sometimes known as **standard reduction potentials**.

electrode potential is

6:16 $$E^\circ = \frac{-[\nu_Z G_Z^\circ + \nu_Y G_Y^\circ + \cdots - \nu_A G_A^\circ - \nu_B G_B^\circ - \cdots]}{nF}$$ standard electrode potential

Thus it follows that a table of standard electrode potentials contains no more information than is available from a table of standard Gibbs energies, such as that on page 391. Furthermore, the information is much more compact in the latter form[618]. Nonetheless, E° values do have many uses. In fact, it is from accurate standard electrode potentials, painstakingly measured by electrochemists[619], that tabulated Gibbs energies and other thermodynamic data have been largely compiled.

Novice electrochemists sometimes make the error of trying to add or subtract standard electrode potentials. They argue, for example, that because

6:17 $$Cu^{2+}(aq) + 2e^- \rightleftarrows Cu(s) \qquad E^\circ = 0.340 \text{ V}$$

and

6:18 $$Cu^+(aq) + e^- \rightleftarrows Cu(s) \qquad E^\circ = 0.521 \text{ V}$$

then the standard potential of the electrode

6:19 $$Cu^{2+}(aq) + e^- \rightleftarrows Cu^+(aq)$$

should be the difference between these values, namely -0.181 V. There is no justification for this assumption. The E° of reaction 6:19 *can* be calculated from 6:17 and 6:18, but not by a simple subtraction[620].

It is only for aqueous and similar systems that it is unnecessary to cite the reference electrode in quoting an electrode potential. In all other circumstances, any statement about the value of an electrode potential (standard or otherwise) is meaningless without a specification of the reference electrode.

The Nernst Equation: how activities influence electrode potentials

Rarely are the activities of all the species involved in an electrode reaction constrained to equal unity[621]. More generally, the electrode potential is influenced by those activities. When nonunity activities are encountered, the electrode potential is given by

6:20 $$E = E^\circ - \frac{RT}{nF} \ln\left\{ \frac{a_Z^{\nu_Z} a_Y^{\nu_Y} \cdots}{a_A^{\nu_A} a_B^{\nu_B} \cdots} \right\}$$ Nernst equation

---

[618] because a few species can be combined into a multitude of reactions.

[619] See Web#619 for an example of how accurate standard electrode potentials are determined by careful measurement of cell voltages.

[620] Correctly calculate the E° of reaction 6:19 from 6:17 and 6:18. See Web#620.

[621] Such a rare instance would be the $Ag | AgCl$ electrode in a chloride melt.

for the general electrode reaction 6:15, and by 6:13 for the cell shown in Figure 6-2. This is the **Nernst equation**[622]; it is, perhaps, the most important equation in electrochemistry. In principle[623], the Nernst equation applies to all electrodes when there is no current. It *may* also apply when current does flow, with the caveat that the activities which appear in the equation then refer to the conditions *at the electrode surface*, and not necessarily elsewhere. When an electrode obeys the Nernst equation despite current flow, it is said to be behaving in a **nernstian** or **reversible** fashion.

Though 6:20 is the most fundamental representation of the Nernst equation, you may find it written differently. The activities of certain species may be omitted because those species are in their standard, unit activity, states, as was done in equation 6:13. The properties of logarithms permit the exponents of the activities to be manipulated, as in

$$6:21 \quad E = E^\circ - \frac{v_Z RT}{nF} \ln \left\{ \frac{a_Z a_Y^{v_Y/v_Z} \cdots}{a_A^{v_A/v_Z} a_B^{v_B/v_Z} \cdots} \right\} \quad \text{or} \quad E = E^\circ + \frac{RT}{F} \ln \left\{ \frac{a_A^{v_A/n} a_B^{v_B/n} \cdots}{a_Z^{v_A/n} a_Y^{v_Y/n} \cdots} \right\}$$

or logarithms to base 10 may be employed[624]:

$$6:22 \quad E = E^\circ - \frac{2.303 RT^\circ}{nF} \log_{10} \left\{ \frac{a_Z^{v_Z} a_Y^{v_Y} \cdots}{a_A^{v_A} a_B^{v_B} \cdots} \right\} = E^\circ - \frac{(59.159 \text{ mV})}{n} \log_{10} \left\{ \frac{a_Z^{v_Z} a_Y^{v_Y} \cdots}{a_A^{v_A} a_B^{v_B} \cdots} \right\}$$

The Nernst equation may be written in inverse form[625] as:

$$6:23 \qquad \frac{a_Z^{v_Z} a_Y^{v_Y} \cdots}{a_A^{v_A} a_B^{v_B} \cdots} = \exp \left\{ \frac{-nF}{RT} (E - E^\circ) \right\}$$

showing that activities respond exponentially to electrode potential.

When a reactant or product is a solute, the Nernst equation may be approximated by replacing the activity of the solute by a concentration ratio; for example, equation 6:13 could be written as

$$6:24 \qquad E \approx E^\circ - \frac{RT}{2F} \ln \left\{ \frac{c^\circ}{c_{Cu^{2+}}} \right\}$$

However, this can be a poor approximation in the case of an ion[626] and what is often done in such cases is to incorporate the activity coefficient into the E° term by writing, for example,

---

[622] Walther Hermann Nernst, 1864 – 1941, Prussian physical chemist and 1920 Nobel laureate.

[623] though not always in practice. When it fails it is usually because the exchange current density (page 134) of the reaction is inadequate compared with currents generated by extraneous processes.

[624] The cited millivoltage applies only at 25.00°C. Calculate the chloride ion activity coefficient in 3000 mM aqueous KCl solution at 25°C. See Web#624.

[625] Notice that either side of 6:23 expresses the equilibrium constant K of reaction 6:15, electrons being ignored.

[626] Show that the error in making this approximation is about 10 millivolts if $\gamma_{Cu^{2+}} = 0.466$. See Web#626.

$$6{:}25 \quad E = E^{o\prime} - \frac{RT}{2F}\ln\left\{\frac{c^o}{c_{Cu^{2+}}}\right\} \quad \text{with} \quad E^{o\prime} = E^o - \frac{RT}{2F}\ln\left\{\frac{1}{\gamma_{Cu^{2+}}}\right\}$$

Here $E^{o\prime}$ is named the **formal electrode potential** or the **conditional electrode potential**[627]; it may behave as a constant during certain experiments, or in a series of similar experiments. The second equation in 6:25 calls for the activity coefficient of a single ion which, as noted on page 44, cannot be measured; the mean ionic activity coefficient appropriate to the cell in question is used instead, for example $\gamma_\pm = \gamma_{Cu^{2+}}^{2/3}\gamma_{NO_3^-}^{1/3}$ in the case of the cell depicted in Figure 6-2.

Electrochemical Series: elaboration into Pourbaix diagrams

The table of electrode potentials, page 392, contains several entries in which one of the participating species is an element, generally either a metal, as for

$$6{:}26 \qquad Fe^{2+}(aq) + 2e^- \rightleftarrows Fe(s) \qquad E^o = -0.447\ V$$

or a gas, for example

$$6{:}27 \qquad Cl_2(g) + 2e^- \rightleftarrows 2Cl^-(aq) \qquad E^o = 1.3579\ V$$

The element in question may be either oxidized or reduced by the electrode reaction. In either case, the element and its partner in the reaction constitute an **electrochemical couple** or **redox couple**[628]. Thus $Cu^{2+}|Cu$ and $Cl_2|Cl^-$ are examples of such couples. The **electrochemical series** of the elements is a listing of redox couples according to the magnitude of the electrode potential. Figure 6-3 is an example with eleven couples ranked in this way. The elements with the most positive E^o values are strong oxidizing agents and come highest in the series, whereas powerful reducing elements, such as the alkali metals, have the most negative electrode potentials[629].

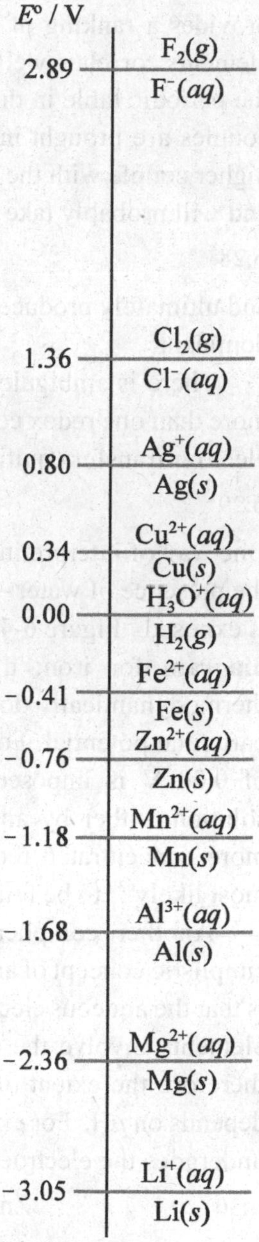

Quite apart from the quantitative data conveyed by the electrochemical series much qualitative chemical information is inherent in a listing such as that in Figure 6-3. The series

E^o / V	
2.89	$F_2(g)$ / $F^-(aq)$
1.36	$Cl_2(g)$ / $Cl^-(aq)$
0.80	$Ag^+(aq)$ / $Ag(s)$
0.34	$Cu^{2+}(aq)$ / $Cu(s)$
0.00	$H_3O^+(aq)$ / $H_2(g)$
−0.41	$Fe^{2+}(aq)$ / $Fe(s)$
−0.76	$Zn^{2+}(aq)$ / $Zn(s)$
−1.18	$Mn^{2+}(aq)$ / $Mn(s)$
−1.68	$Al^{3+}(aq)$ / $Al(s)$
−2.36	$Mg^{2+}(aq)$ / $Mg(s)$
−3.05	$Li^+(aq)$ / $Li(s)$

Figure 6-3 An electrochemical series.

---

[627] often erroneously called "the standard electrode potential". See Web#720 for elaboration.

[628] "Redox" is a contraction of "reduction-oxidation" and indicates that the two members are related by electron transfer.

[629] Ironically, these metals are said by inorganic chemists to be the most electropositive elements.

provides a ranking of elements by oxidizing power. Many chemical properties of the elements correlate well with their position in the electrochemical series, often surpassing the periodic table in this respect. If the species participating in any two electrochemical couples are brought into contact, then the reaction of the more oxidized member of the higher couple with the reduced partner of the lower couple is thermodynamically favored and will probably take place. For example, Ag^+ will react with Zn:

6:28
$$2Ag^+(aq) + Zn(s) \rightleftarrows 2Ag(s) + Zn^{2+}(aq)$$

and ultimately produce an equilibrium in which the two other partners of each couple are dominant.

There is ambiguity in the ranking of some elements because the element may form more than one redox couple. For example, in addition to reaction 6:26, iron enters into an electron-transfer equilibrium with another ion:

6:29
$$Fe^{3+}(aq) + 3e^- \rightleftarrows Fe(s) \qquad E^\circ = -0.037 \text{ V}$$

One way of interpreting the chemistry of iron is that the form adopted by this element in the presence of water – $Fe^{3+}(aq)$, $Fe^{2+}(aq)$, or $Fe(s)$ – depends on the potential to which it is exposed. Figure 6-4 provides a **dominance diagram** for iron: it reveals which is the thermodynamically dominant species for any particular potential. For example, if a potential of 0.00 V is imposed on a iron-containing solution, either by an electrode or by some more concentrated redox couple, the iron is most likely[630] to be found in the $Fe^{2+}(aq)$ state.

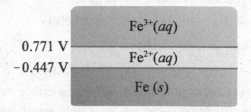

Figure 6-4 A dominance diagram for the element iron. The horizontal lines correspond to the standard potentials of the electrode reactions:
$$Fe^{3+}(aq) + e^- \rightleftarrows Fe^{2+}(aq)$$
and
$$Fe^{2+}(aq) + 2e^- \rightleftarrows Fe(s)$$

Another complication undermining the simplistic concept of an electrochemical series is that the aqueous electrode reactions of many elements involve the ions H_3O^+ or OH^- and therefore the extent of the electrode reaction depends on pH. For example, the element zinc undergoes the electrochemical reaction

6:30
$$Zn(OH)_2(s) + 2H_3O^+(aq) + 2e^- \rightleftarrows Zn(s) + 4H_2O(\ell)$$

and, because the activities of solvents and pure solids are unity, the corresponding Nernst equation is[631]

---

[630] The "most likely" caveat reflects the fact that factors other than the thermodynamic stability of the aqueous ions affect dominance. The presence of counterions may play a role by complexing or precipitating one of the ions as does the slowness of many feasible reactions.

[631] Validate the second equality in 6:31. Compare with Web#631.

6:31
$$E = E^\circ - \frac{RT^\circ}{2F} \ln\left\{ \frac{1}{a_{H_3O^+}^2} \right\} = E^\circ - (59.16 \text{ mV})\, pH$$

The dominance is affected, in such cases, by both the electrode potential and the pH. This consideration led **Pourbaix**[632] to design the diagrams that now bear his name.

Figure 6-4 is a one-dimensional dominance diagram, showing the species that have dominant thermodynamic stability at each value of the electrode potential. **Pourbaix diagrams**[633] are two dimensional and show how dominance is affected by both potential and pH. Figure 6-5 is the Pourbaix diagram for zinc and shows the regions in which four distinct species are dominant. Horizontal lines in such diagrams represent simple redox couples, involving an electron transfer only. Diagonal lines represent processes in which both electrons and protons are exchanged, as in the equilibria 6:30 and

6:32
$$ZnO_2^{2-}(aq) + 2H_2O(\ell) + 2e^- \rightleftarrows Zn(s) + 4OH^-(aq)$$

Vertical lines represent purely chemical processes[634]; those in Figure 6-5 correspond to the equilibria

6:33
$$Zn(OH)_2(s) + 2H_3O^+(aq) \rightleftarrows Zn^{2+}(aq) + 4H_2O(\ell)$$

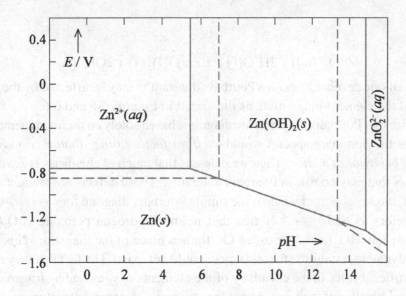

Figure 6-5 Pourbaix diagram for the element zinc. Full and dashed lines respectively represent ionic activities of unity and 0.001.

---

[632] Marcel Pourbaix, 1904 – 1998; born in Russia but it was in Belgium that he researched electrochemical thermodynamics.

[633] also known as **potential-pH diagrams**.

[634] Use Figure 6-5 to estimate the solubility product of $Zn(OH)_2$; see Web#634.

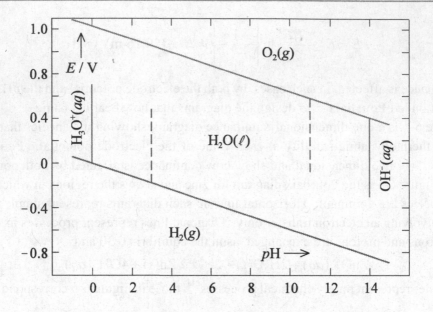

Figure 6-6 Combined Pourbaix diagrams for oxygen and hydrogen, showing the zone of thermodynamic stability of water. Full and dashed lines respectively represent ionic activities of unity and 0.001.

and

6:34
$$ZnO_2^{2-}(aq) + 2H_2O(\ell) \rightleftarrows Zn(OH)_2(s) + 2OH^-(aq)$$

The location of one or more lines on a Pourbaix diagram[635] may be affected by the activity of the ions of the element in question, as illustrated in Figures 6-5 and 6-6.

Recognize that Pourbaix diagrams are diagrams based solely on thermodynamics: they identify what the dominant species would be *if no factors other than thermodynamics governed the behavior of matter*. Their weakness is that much of chemistry is governed by kinetic factors that play no role in Pourbaix diagrams. These defects are brought out well in Figure 6-6, the upper half of which is the simple Pourbaix diagram for oxygen, while the lower half refers to hydrogen. Notice that neither hydrogen peroxide $H_2O_2(\ell)$, the hydroperoxide ion $HO_2^-(aq)$, nor ozone O_3, finds a place in the diagram. This reflects their thermodynamic instability: these species should not exist! The fact that they do exist, and play significant roles in the chemistry of the elements oxygen and hydrogen, demonstrates the slow kinetics of their decomposition reactions. A second defect revealed by the diagram is that it leads one to believe that, at any pH, water is stable only over a potential

---

[635] From the data below, construct a Pourbaix diagram for cadmium showing the zones of dominance of the species $Cd(s)$, $Cd^{2+}(aq)$ and $Cd(OH)_2(s)$. Compare your answer to Web#635.
$Cd^{2+}(aq) + 2e^- \rightleftarrows Cd(s)$, $E^\circ = -0.403$ V; $Cd(OH)_2(s) + 2e^- \rightleftarrows Cd(s) + 2OH^-(aq)$, $E^\circ = -0.825$ V; $Cd(OH)_2(s) + 2H_3O^+(aq) \rightleftarrows Cd^{2+}(aq) + 4H_2O(\ell)$, $K = 5.01 \times 10^{13}$

"window" of width 1.229 V. This is untrue! Again, slow kinetics takes credit for widening the actual window of water's stability almost twofold, thereby opening up a fruitful field of aqueous electrochemistry that would not have existed had thermodynamics held sway.

Usefully, Pourbaix diagrams show what is feasible, but they cannot be relied upon to predict what will actually occur. Their impact has been especially pronounced in the field of corrosion (Chapter 11).

Working Electrodes: constructed from many materials in many shapes and sizes

Working electrodes[636] are ones at which processes of interest occur. They may be divided into two broad classes. Sometimes the working electrode is an **active electrode**; that is, it is involved as a reactant or product in the electrode reaction. Otherwise, when the working electrode acts purely as a source or sink of electrons, not being stoichiometrically involved in the electrode reaction, it is called an **inert electrode**. The adjective "inert" can be misleading; a reaction at one inert electrode may proceed at quite a different rate than at an electrode made of another "inert" material. In extreme cases, a reaction may proceed rapidly on one inert electrode but not perceptively on another electrode material[637], or it may lead to a different product[638]. This is because electronic conductors often have catalytic properties in addition to their roles as electron sources or sinks.

Materials from which inert working electrodes are typically made include noble metals such as platinum or gold, and various forms of carbon, such as graphite, glassy carbon, carbon nanotubes, and even boron-doped diamonds. Of course, a requirement is that the material not undergo any chemical or electrochemical reaction with the reactant or product of the reaction of interest, or with any other component of the ionic conductor. Or at least, if they are removed by part of the reaction mechanism, they must be reformed by another part.

Glass, covered with a conducting layer of indium-doped tin oxide SnO_2, provides an inert working electrode that is optically transparent, valuably permitting electrochemistry and spectrophotometry to be carried out simultaneously. The surface of **modified electrodes** has been changed by the creation of an adherent layer by some appropriate chemical, physical or electrochemical process. The motives for such modification are numerous but they frequently have the aim of facilitating some specific electrode reaction. To be electrochemically useful, the layer must be able to transport electricity; but, in

---

[636] For a more extensive discussion, see Š. Komorsky-Lovrić in: F. Scholz (Ed.), *Electroanalytical Methods*, 2nd edn, Springer, 2010, pages 273–288.

[637] For example, in basic solution, the hexacyanoiron(III) ion $Fe(CN)_6^{3-}(aq)$ is readily reducible at a nickel, but not at an aluminum, electrode. For a second example, refer to pages 78–79.

[638] Gentle oxidation of water produces oxygen at some electrodes, hydrogen peroxide at others.

addition to familiar electronic and ionic conduction, **electron hopping** may occur through certain sufficiently thin modifying layers. Often enzymes are attached, so as to catalyze an electrode reaction that otherwise would not occur. The glucose sensor described on pages 175–177 provides an example.

The behavior of a large electrode depends on its area, but very little on its shape. For the small electrodes used in voltammetry, however, the shape of an inert working electrode is more important than might be expected, for reasons discussed in Chapter 12. Electrodes for voltammetric purposes are often of cylindrical shape, shrouded by an insulator, so that the surface presented to the ionic conductor is that of an **inlaid disk**, as illustrated in cross section in Figure 6-7a. Such electrodes are rather easy to construct and to clean by polishing. However, the behavior of a disk electrode depends awkwardly on its size[639], as discussed in Chapter 12, and therefore the current is often difficult to predict or interpret. Working electrodes of a hemispherical or spherical shape are far better from a predictive standpoint though, with one exception, they are difficult to make and clean. The exception is the **mercury drop electrode** which is simple to make and, because it is easily renewed, does not need cleaning. Solution can easily

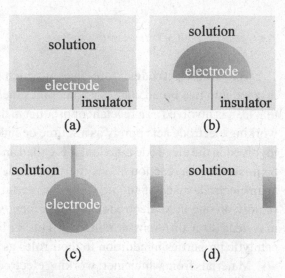

Figure 6-7 Cross-sectional diagrams of four common electrode shapes: (a) an inlaid disk electrode, (b) a hemispherical electrode, (c) a hanging drop electrode and (d) a tubular band electrode. The figures have been scaled to represent configurations with identical electrode|solution interfacial areas.

be changed with electrodes of the shape shown cross sectionally in Figure 6-7d, known as a **tubular band electrode**. Such electrodes can be conveniently constructed by drilling a hole through an insulator | conductor | insulator sandwich.

The majority of working electrodes are static and there is no relative motion between the electrode and the ionic conductor. With a liquid conductor, however, current can be greatly enhanced by stirring, by bubbling gas, or by sonication[640]. The effect of such forced convection is difficult to predict quantitatively. With careful cell design, however, reproducible convection can be achieved. Foremost in this regard is the **rotating disk**

---

[639] because the disk has a flat section and an edge, from both of which currents arise. The edge effect is more pronounced for small disks.

[640] that is, exposure to high-frequency sound.

electrode, described in more detail on pages 248–252. Somewhat similar results are achieved by electrodes embedded in the walls of pipes through which electrolyte solutions flow, as in the tubular band electrode. The **dropping mercury electrode** also benefits from controlled convection. Some features of this classic electrode are described on page 260. Much of our present day electrochemical knowledge rests on experiments conducted in the last century with mercury drop electrodes.

The measurement of cell voltages in the absence of current flow, known as **potentiometry**, is used to furnish thermodynamic information and to measure ionic activities. Devices used for the latter application employ a working electrode in conjunction with an exterior reference electrode[641]. A working electrode used in this way to measure the activity of a specific ion is known as an **ion-specific electrode** or, more realistically, an **ion-selective electrode** (ISE).

What is usually measured in analytical potentiometry is the ratio of the two activities of the target ion. It should be emphasized that ion-selective electrodes respond to activities *not concentrations*. Sometimes the purpose of the investigation is well served by an activity measurement but, more often, it is a concentration that is sought. Fortunately, calibration or standardization procedures, such as those discussed later in this section, enable accurate concentration information to be gathered by potentiometry.

According to the Nernst law, an electrode of pure copper metal develops a null potential

$$6{:}35 \qquad E_n = E^\circ - \frac{RT}{2F}\ln\left\{\frac{a_{Cu}}{a_{Cu^{2+}}}\right\} = E^\circ + \frac{RT}{2F}\ln\left\{a_{Cu^{2+}}\right\}$$

when placed in a solution containing $Cu^{2+}(aq)$ ions. Therefore, paired with a reference electrode, it may be used to measure copper ion activities. In an obsolescent terminology, such an electrode is described as being an **electrode of the first kind** and it could be considered a rudimentary ion-selective electrode. An ideal ion-selective electrode responds in a nernstian fashion to the target ion – Cu^{2+} in this case – but is totally unaffected by other ions. In practice, most electrodes fail to live up to this ideal, because species other than the target interfere. For example, silver ions interfere with the copper electrode because they react chemically with the electrode:

$$6{:}36 \qquad Cu(s) + 2Ag^+(aq) \rightarrow Cu^{2+}(aq) + 2Ag(s)$$

and play havoc with the analysis.

A silver electrode coated with insoluble silver chloride exhibits a null potential of

$$6{:}37 \qquad E_n = E^\circ - \frac{RT}{F}\ln\left\{\frac{a_{Ag}a_{Cl^-}}{a_{AgCl}}\right\} = E^\circ - \frac{RT}{F}\ln\left\{a_{Cl^-}\right\}$$

---

[641] The "exterior" descriptor distinguishes it from the "interior" reference electrode (Figure 6-8). Sometimes the exterior reference electrode is incorporated, with the ISE, into a single unit called a **combination electrode**.

and this so-called **electrode of the second kind**[642] therefore has a nernstian response to chloride ions. It is rather unselective, however, because the presence in solution of any anion, such as $Br^-(aq)$ or $CN^-(aq)$ that forms a salt more insoluble than AgCl, will react and interfere, as will $Ag^+(aq)$. The $Ag \mid Ag_2S$ electrode has fewer interferences, because Ag_2S is less soluble, and does serve as a satisfactory[643] ISE for the $S^{2-}(aq)$ target ion[644].

Many successful ion-selective electrodes[645] employ a discriminating membrane and have the general geometry shown in Figure 6-8. The principle used by such an ISE in measuring the activity of a target ion in a test solution is straightforward. The end of a chamber is sealed by a membrane permeable to the target ion, say ion i and, ideally, it is permeable *only* to that ion. This membrane separates two aqueous solutions, each containing the $i^{z_i+}(aq)$ ion. One is the test solution, into which the ISE is dipped: this solution contains the target ion with an unknown activity a_i^{outer}. The other is the solution inside the chamber, with a fixed activity a_i^{inner} of ion i. The target ions establish equilibrium across the membrane, creating an electrical potential difference between the two solutions. Equation 2:25 is obeyed in the form[646]

6:38
$$\phi^{inner} - \phi^{outer} = \frac{RT}{z_i F} \ln \left\{ \frac{a_i^{outer}}{a_i^{inner}} \right\}$$
membrane potential difference

The membrane has transformed a disparity in activities of the target ion between the outer and inner solutions into a difference in the electrical potentials of those solutions. This

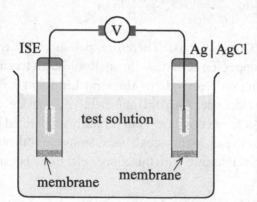

Figure 6-8 Potentiometry with an ion-selective membrane electrode. The electrode inside the ISE, known as an "interior reference electrode", could also be an $Ag \mid AgCl$ electrode.

---

[642] These have two solid (and hence unit-activity) phases, in contrast to the single phase of electrodes of the first kind. There are also **electrodes of the third kind**, which have three.

[643] though the $Hg^{2+}(aq)$ ion interferes seriously.

[644] From G^o data, calculate the standard potential of this electrode. Write an expression for its null potential and draw a logarithmic graph showing how the null potential depends on sulfide ion activity. SeeWeb#644.

[645] Though they are universally called "electrodes", the name "half-cell" would be more appropriate.

[646] Notice that the Henderson equation, 3:24, gives rise to equation 6:38 by setting either of the mobility terms to zero, as befits a semipermeable membrane.

so-called **membrane potential difference** or **Donnan potential difference**[647] cannot be measured directly, so we must rely on a pair of electrodes to transduce the potential difference between two solutions into a potential difference between two metallic conductors, as in the arrangement shown in Figure 6-8. One electrode is placed inside the ISE chamber, as illustrated, a second in the test solution. The two electrodes[648] need not be identical, though they could be. The figure shows an Ag|AgCl reference electrode used for each purpose. The voltmeter measures

$$6:39 \qquad \Delta E_n = E_{RE}^{internal} + \phi^{inner} - \phi^{outer} - E_{RE}^{external} = E_{constant} + \frac{RT}{z_i F}\ln\left\{a_i^{outer}\right\}$$

equation 6:38 having been incorporated. The $E_{constant}$ includes any difference between the two electrode potentials and a term arising from the invariant activity of the target ion in the inner solution.

The **fluoride-ion sensor** provides a typical example of an ion-selective membrane electrode. The membrane is a crystal of europium fluoride-doped lanthanum fluoride which, as described on page 10, conducts by F^- ion migration. A transfer equilibrium, or **Donnan equilibrium**, is established (sometimes slowly) between fluoride ions in the test solution and those within the ISE chamber. This chamber holds a mixed solution of sodium fluoride and potassium chloride, the latter to supply chloride ions for the Ag|AgCl interior reference electrode. To measure an unknown fluoride ion concentration in the test solution, one would normally first calibrate the ISE with at least two solutions[649] of known c_{F^-} and then interpolate logarithmically to find the unknown concentration, as in Figure 6-9 overleaf. Equation 6:39 predicts that the cell voltage will change by[650] $2.3026RT/z_i F$ (equal to -59.159 mV at $25.00°C$ for $z_i = -1$) for a tenfold increase in the activity of the target ion. As the graph illustrates, the slope of an experimental graph of cell voltage versus the logarithm of concentration is often of a somewhat smaller magnitude[651].

The antibiotic **nonactin**[652] selectively complexes the potassium ion, thereby "disguising" the hydrophillic nature of the K^+ ion and enabling it to cross hydrophobic cell walls that would otherwise be barriers. The resulting "potassium leak" disrupts bacterial metabolism and so confers bactericidal properties on nonactin. A **potassium-ion sensor**

---

[647] Federick George Donnan, 1870–1956, British chemist. His concern was largely with biological membranes.

[648] The inner electrode is often called the "indicator electrode"; the outer electrode is the reference electrode. In a sense, they are both reference electrodes.

[649] These should closely match the test solution, other than in their fluoride ion concentrations. The hydroxide ion OH^- is the only significant interferant, so the test solution is adjusted to about pH 5 by a reagent that also imposes a high and reproducible ionic strength.

[650] The numerical factor arises because, for any x, $\ln\{x\} = \ln\{10\}\log_{10}\{x\} = 2.3026\log_{10}\{x\}$.

[651] Can you explain why? See Web#651.

[652] A cyclic organic compound with eight oxygen atoms which octahedrally surround a space that closely matches the size of the K^+ ion. Compounds with this, or similar, structures are know as **crown ethers**.

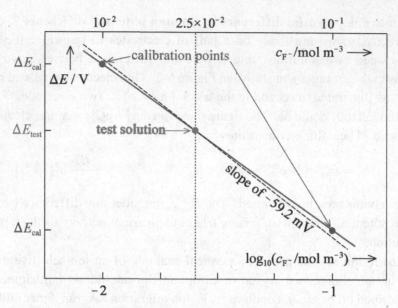

Figure 6-9 Calibration procedure for an ion-selective electrode.

uses nonactin dissolved in an organic solvent held in the pores of a thin porous disk of poly(vinyl chloride) to provide a potassium-ion-selective membrane for ISE use[653].

The oldest, and still the most widely used, ion-selective electrode is the **glass electrode**. The simple design of this ISE is illustrated in Figure 6-10. It consists of a chamber holding an interior $Ag\,|\,AgCl$ reference electrode, with a thin-walled ($50\text{--}100\ \mu m$ thick) glass membrane separating the internal solution (usually $100\ mM$ hydrochloric acid HCl) from the test solution[654]. Though the mechanism by which it operates is probably that described elsewhere[655], the membrane behaves as if it were selectively permeable to hydronium ions. The cell voltage obeys equation 6:39 with $z_i = z_{H_3O^+} = 1$, so that

6:40 $$\Delta E = E_{constant} + \frac{RT}{F}\ln\left\{a_{H_3O^+}^{outer}\right\}$$

Recall that pH is defined as the negative of the decadic logarithm of the hydronium ion activity, and therefore

6:41 $$\Delta E = E_{constant} - \frac{2.3026RT}{F}pH = E_{constant} - E_{slope}pH$$

---

[653] Using **linear regression** or otherwise, determine the potassium ion concentration of a solution that gives a cell voltage of $-97\ mV$ with such a sensor. Calibrating solutions of 0.20, 0.60 and 2.00 mM potassium sulfate give voltages of -133, -104 and $-78\ mV$. See Web#653.

[654] Some commercial pH sensors incorporate the "exterior" reference electrode in the same unit as the glass electrode. Others incorporate a temperature sensor to take account automatically of the T in equation 6:41.

[655] See Web#655.

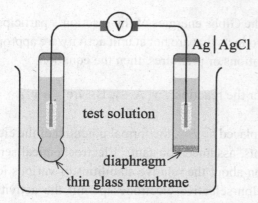

Ag|AgCl

test solution

diaphragm

thin glass membrane

Figure 6-10 A glass electrode in use measuring the pH of a test solution. Special voltmeters are needed to cope with the high impedance of the glass electrode. The electrode inside the ISE could also be an Ag|AgCl electrode.

where E_{slope} is theoretically 59.159 mV at 25°C, but in practice is somewhat less than this. The glass electrode is usually used to measure *p*H in conjunction with a **pH meter**. This is an electronic voltmeter[656], specially designed to cope with the high resistance of the glass electrode and to output $(E_{\text{constant}} - \Delta E)/E_{\text{slope}}$ instead of the cell voltage ΔE itself. The *p*H meter is first "standardized" by using two buffer solutions of known *p*H; this procedure establishes appropriate values of E_{constant} and E_{slope}. Once standardized in this way, the *p*H meter provides a direct reading of the *p*H of a test solution. Unfortunately Na^+ ions do transit the membrane to some small extent and constitute a substantial interference when H_3O^+ ions are scarce, that is for test solutions of high *p*H.

Summary

By contriving one of the electrodes to remain stable despite current flow, a cell voltage ΔE may be expressed as the difference between the potential E of the working electrode and the constant potential E_{RE} of the reference electrode, so that

6:42 $$\Delta E = E - E_{\text{RE}}$$

In aqueous electrochemistry E_{RE}, and hence E, is referenced to E_{SHE}. The standard electrode potential $E°$ is the null potential of the working electrode when all the reactants and products of the working electrode reaction are in their standard states. The Nernst equation gives the open-circuit potential of an electrode in terms of the activities of the electrode reaction participants:

6:43 $$E_n = E° - \frac{RT}{nF} \ln\left\{ \frac{a_Z^{v_Z} a_Y^{v_Y}}{a_A^{v_A} a_B^{v_B}} \right\}$$ for the reaction $$v_A A + v_B B + ne^- \rightarrow v_Z Z + v_Y Y$$ Nernst equation

The standard potential $E°$ provides access to thermodynamic data about the reaction at the

---

[656] invented by the American, Kenneth H. Goode, 1902–1967, when he was a 19-year old undergraduate at the University of Chicago.

electrode, and permits calculation of the Gibbs energies of the reaction's participants. If the activities of all the reactants and products that are not at unit activity are appropriately replaced in equation 6:43 by concentrations or pressures, then the equation

6:44 $\qquad E_n = E^{o\prime} - \dfrac{RT}{nF} \ln\left\{ \dfrac{c_Z^{v_Z} c_Y^{v_Y}}{c_A^{v_A} c_B^{v_B}} \right\}$ for the reaction $\quad v_A A + v_B B + ne^- \rightarrow v_Z Z + v_Y Y$

applies approximately. E^{o} has been replaced by $E^{o\prime}$, the formal potential of the electrode, which incorporates activity coefficients, assumed constant. Electrochemical series and Pourbaix diagrams portray information about the relative stabilities of various ions and compounds of particular elements. Ion-selective electrodes permit the activities and concentrations of target ions to be determined potentiometrically.

7

Electrode Reactions

On page 46, the study of chemical reactions was said to be the search for answers to four questions. The same questions recur in studying *electro*chemical reactions. Answering each of these questions leads into a different realm of physical electrochemistry:

What electrode reactions occurs? $\Rightarrow$ Electrochemical stoichiometry

Why does the reaction occur? $\Rightarrow$ Electrochemical thermodynamics

How fast does the reaction occur? $\Rightarrow$ Electrochemical kinetics

How does the reaction occur? $\Rightarrow$ Mechanisms of electrode reactions

This chapter is largely concerned with the third and fourth questions, which are interconnected because studies of the kinetics of an electrode reaction provide information on its mechanism. The first and second questions were addressed in previous chapters, but initially these "what?" and "why?" questions will be revisited briefly.

Faraday's Law: necessities for an electrode reaction

Inasmuch as it is the basis of all electrochemistry, how have we managed to arrive at Chapter 7 without even mentioning **Faraday's law**? In truth, this law, which states that *the amount (number of moles) of any substance produced or consumed in an electrode reaction is proportional to the quantity of charge passed*, is implicit in the equation of every electrode reaction that is written in this book. By quantitatively linking electrons to molecules and ions, each of these balanced stoichiometric equations demonstrates the proportionality of the amount of electricity to the amount of chemical change. Nowadays this equivalence seems almost self-evident, but it was not when Michael Faraday[151] put electrochemistry on a quantitative footing in 1832 by enunciating the law that now bears his name.

In Chapter 13 we shall find that an electric current may sometimes flow transiently through a cell *without* an accompanying chemical reaction, as a consequence of the capacitive property of electrodes. Such currents are described as **nonfaradaic**. In contrast, currents that are obedient to Faraday's law are said to be **faradaic**.

Coulometry is conceptually the simplest electroanalytical technique[701], being based directly on Faraday's law. If Q is the amount of electricity needed to oxidize or reduce *all* of species i from an electrolyte solution or other ionic conductor, then

7:1
$$Q = -nFVc_i / v_i$$

where n and v_i are respectively the number of electrons[702] and the number of molecules or ions of the target species i in the electrode reaction. V is the cell volume. Usually, the electrolysis is carried out with the cell voltage held constant at a value chosen to preclude any competitive electrode reaction. The solution is stirred, or the cell is of minute size[703], as otherwise the electrolysis time would be unacceptably long. The current is monitored[704] and integrated, permitting the original concentration of the target to be found from

7:2
$$c_i = \frac{v_i Q}{nFV} = \frac{v_i}{nFV} \int_0^{t_\infty} I \, dt$$

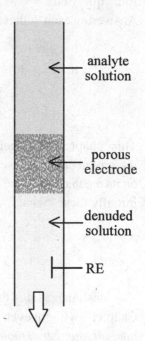

where t_∞ is the time by which the current has fallen to an insignificant or background value. Even with vigorous stirring, coulometry is a slow method. A faster variant is **flowing coulometry**, in which the analyte solution flows steadily through a long narrow tubular working electrode, or a porous metal working electrode[705] as illustrated in Figure 7-1. Because the flow rate $\dot{V}$ (m³ s⁻¹) is chosen to be small enough that the emergent solution is completely denuded of the target species i, the entrant concentration c_i is calculable as $v_i I / nF\dot{V}$ from the *steady* current I.

analyte solution

porous electrode

denuded solution

The general stoichiometric equation for an electrode reaction is

7:3
$$v_A A + v_B B + \cdots + ne^- \rightleftarrows v_Z Z + v_Y Y + \cdots$$

RE

with the understanding that n is negative for an anodic reaction and positive when the reaction is cathodic. By using the $\rightleftarrows$ symbol in equation 7:3, we indicate that the discussion relates to the left-to-right progress of the reaction, while reminding the

Figure 7-1 Flowing coulometry.

---

[701] Electroanalytical chemistry is the use of electrochemistry to measure concentrations. Related to coulometry is the technique of **coulometric titration**, described at Web#701, as is **amperometric end-point detection**, another electrochemical aid to titrimetry.

[702] The signs in equations 7:1 and 7:2 apply because a positive I corresponds to a negative n in equation 7:3.

[703] See Web#703 for a brief description of **thin-layer coulometry** and a worked problem based thereon.

[704] The current falls exponentially during coulometry. Why? Derive an expression showing, in terms of the initial current, how the current changes with time. See Web#704.

[705] Such an electrode is often, though erroneously, said to be "three-dimensional".

reader that the reverse process is also proceeding to a lesser extent. Reaction 7:3 embodies Faraday's law inasmuch as[706]

7:4
$$\begin{cases} \text{quantity of charge passed} = Q \\ \text{amount of Z produced} = n_Z^{\text{final}} - n_Z^{\text{initial}} = \Delta n_Z = -v_Z Q / nF \propto Q \\ \text{amount of A consumed} = n_A^{\text{initial}} - n_A^{\text{final}} = -\Delta n_A = -v_A Q / nF \propto Q \end{cases}$$

with similar proportionalities for other reactants and products[707]. Each subscripted n in equations 7:4 represents an amount (numbers of moles). In the cathodic reduction of oxygen in aqueous solution,

7:5
$$O_2(g) + 2H_2O(\ell) + 4e^- \rightleftharpoons 4OH^-(aq)$$

for example, the passage of $4 \times (-96485)$ coulombs of charge produces four moles of hydroxide ion and consumes one mole of oxygen, as well as two moles of water. Such relationships were often used in Chapters 4 and 5 to interrelate charge and chemical amounts.

There is no single "correct" way to write the equation of an electrode reaction. For instance, equation 7:5 may be rewritten as

7:6
$$\tfrac{1}{2}O_2(g) + H_2O(\ell) \rightleftharpoons 2OH^-(aq) - 2e^-$$

or in innumerable other ways, without affecting the stoichiometry. The electrode potentials of reactions 7:5 and 7:6 are identical; the Nernst equations for each are equivalent. With a change in the sign of their stoichiometric number, we may even move species, as well as electrons, from one side of the equation to the other, as in

7:7
$$\tfrac{1}{2}O_2(g) + H_2O(\ell) - OH^-(aq) + 2e^- \rightleftharpoons OH^-(aq)$$

and, as we shall see later, it may sometimes be advantageous to do so. Of course, the requirement for balancing atoms and charges by adjusting the stoichiometric coefficients (the subscripted v's) must always be met in the equations for all reactions, chemical or electrochemical.

Provided that the standard Gibbs energies of the reactants and products of an electrode reaction are known, its stoichiometric equation permits the standard potential of the electrode to be determined as

7:8
$$E^\circ = \frac{-\Delta G^\circ}{nF} = \frac{-1}{nF}\left[v_Z G_Z^\circ + v_Y G_Y^\circ + \cdots - v_A G_A^\circ - v_B G_B^\circ - \cdots \right]$$

Moreover, with this **standard electrode potential** known, the null electrode potential can be calculated for any pertinent activities of the reactants and products, through the Nernst

---

[706] The signs in 7:4 are appropriate if charge passing anodically is regarded as positive. This matches the convention that anodic current at a working electrode is reckoned as positive.

[707] Calculate the charge that must be passed to produce one tonne (10^3 kg) of pure copper by the electrorefining process summarized in reactions 4:4 and 4:5. Check at Web#707.

equation,

7:9
$$E_n = E^\circ - \frac{RT}{nF} \ln\left\{ \frac{a_Z^{v_Z} a_Y^{v_Y} \cdots}{a_A^{v_A} a_B^{v_B} \cdots} \right\} \qquad \text{null potential}$$

The **null potential** is that adopted by an electrode when no current flows.

Unlike the standard potential, the null potential of an electrode is not a constant. It depends on the prevailing activities of the reactants and products at the electrode surface and may change temporarily during an experiment as the corresponding surface concentrations alter as a result of current flow. Rarely is the change permanent in laboratory experiments, because seldom is the change in the amount of any species significant in comparison with the total content of that species within the cell. Of course, coulometric experiments are exceptions to that generalization and another exception occurs often in voltammetry (Chapters 12, 15, and 16), in which an electrode reaction is frequently studied in the initial absence of at least one of the products. In such experiments, the initial null potential is theoretically either $-\infty$ for an oxidation, or $+\infty$ for a reduction. In practice when product is absent, the resting potential will be irreproducible or determined by some other redox couple, frequently arising from impurities[708].

Obviously, the fact that one can write down a stoichiometric equation, and even perform pertinent thermodynamic calculations, does not mean that the electrode reaction will actually occur. There follows a list of some requirements for an electrode reaction of interest to be sustained:

(a) *There must be an adequate supply of reactants* to the electrode interface. Some reactants may be abundantly present at the electrode, as when a reactant is the electrode material itself, or the solvent of an ionic conductor, or present as a layer on the electrode. In those cases, supply is not a constraint. However, for reactants that are present as solutes, supply to the electrode is possible only through the motion of the dissolved species. Such motion is known as **transport** and is the subject of Chapter 8. Reaction and transport destroy the original uniformity of concentration, making it necessary to distinguish between the *bulk* concentration of a species and its concentration at the electrode *surface*. In this book we use alternative

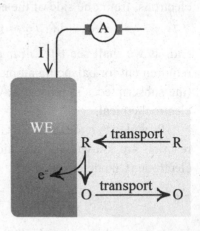

[708] Estimate the null potential for the reaction $Fe(CN)_6^{3-}(aq) + e^- \rightarrow Fe(CN)_6^{4-}(aq)$ when the solution contains, in addition to supporting electrolyte, 1.00 mM $K_3Fe(CN)_6$ and either (a) 0.1 mM $K_4Fe(CN)_6$ (b) a mere trace (say 10^{-6} mM) of $K_4Fe(CN)_6$ present as impurity or (c) no $K_4Fe(CN)_6$ whatsoever. See Web#708.

superscripts, as in c^b and c^s to differentiate between the two sites[709].

(b) *There must be somewhere for the products to go* without blocking the access of reactants to the electrode. If the product is a solute, it leaves by transport away from the electrode. If the product is a gas, it is produced initially as a solute, but once the solubility of the gas in the solvent is exceeded, troublesome bubbles may form[710]. If the product is a solid, it must be porous. Cases arise in which a solid product, though nonporous, is semiconducting and itself supports a continuing reaction. An example is in the electro-polymerization of pyrrole C_4H_4N to form polypyrrole (page 9). There are many examples of porous products; the silver chloride produced by the anodic oxidation of silver metal in a chloride medium was mentioned in the context of reaction 6:2. But there are numerous counterexamples, too; the anodic oxidation of aluminum in an aqueous medium produces an impervious layer of Al_2O_3, so the reaction ceases almost immediately.

(c) *The electrode reaction must have sufficiently fast kinetics*. The topic of electrode kinetics occupies most of the remainder of this chapter.

(d) *There must be an adequately conducting pathway* through the circuit formed by the ionic conductor, the two electrodes and the associated wiring. The ionic conductor is usually the least conductive circuit element. With solutions in water or other solvents, it is standard practice to add **supporting electrolyte**[711] to improve the solution's conductivity[712]. A supporting electrolyte is usually a salt[713], and enough[714] should be added that the concentration of the supporting ions far exceeds the concentration of the **electroactive ions** (those that are reactants or products). Though it ranks first in importance, the increase in conductivity is not the only beneficial effect of excess supporting ions. The other benefits, and some drawbacks, are discussed on pages 199 and 200. Of course, no supporting electrolyte is needed if the ionic conductor is an ionic liquid or a solid electrolyte.

(e) If the study is to be meaningful, *no other electrode reaction should occur to a significant extent* in the vicinity of the null potential of the reaction in question. This includes a prohibition on the supporting electrolyte or the solvent undergoing anodic oxidation or cathodic reduction at nearby potentials. If an extreme potential is applied, the solvent will inevitably undergo oxidation or reduction, leading to current-voltage curves

---

[709] Commonly an asterisk, as in c^* is used to indicate a bulk concentration.

[710] though bubbles are sometimes advantageous. In aluminum electrowinning (page 71), bubbles from the cathode provide valuable agitation, aiding transport. Bubbles are the key to electroflotation (page 184).

[711] Also known as **base electrolyte**, **indifferent electrolyte**, or **inert electrolyte**. The last two of these names reflect an essential property of a supporting electrolyte: its ions should not undergo any reaction.

[712] On the basis of conductivity alone, which of the following would make the best supporting electrolyte solution: (a) 800 mM KCl, (b) 450 mM Na_2SO_4 or (c) 500 mM $Mg(NO_3)_2$? See Web#712.

[713] Potassium chloride is a favorite in aqueous solutions, such salts as tetraalkylammonium hexafluoro-phosphates are preferred in nonaqueous solvents because of their higher solubility.

[714] see the discussion following equation 8:61.

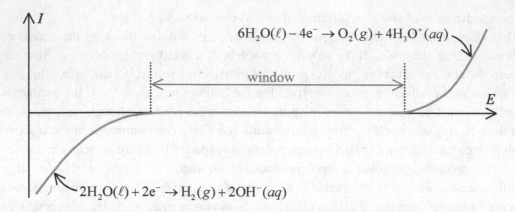

Figure 7-2 An example of the current-voltage curve resulting from a totally polarized working electrode, in this case a gold electrode in an aqueous potassium nitrate solution. The solvent-oxidation and solvent-reduction reactions occurring at extreme potentials are shown. Between them is the "window" within which other electrode reactions may be investigated.

of the typical shape shown in Figure 7-2 for an oxygen-free[715] aqueous solution of potassium nitrate. In the region where the current is virtually zero, the working electrode is said to be **totally polarized**. The figure demonstrates that it is only within a **potential window**[716] that an electrode reaction may be studied without interference from solvent electrochemistry.

Kinetics of a Simple Electron Transfer: the Butler-Volmer equation

Most electrode reactions take place through complicated mechanisms involving a number of sequential steps. Such mechanisms are addressed in a later section. Here we consider the simplest oxidation reaction in which a single electron is transferred to the electronic conductor from a dissolved species R

7:10 $$R(soln) \rightleftarrows e^- + O(soln)$$

thereby creating O, a second dissolved species. R and O represent the reduced and oxidized partners of a one-electron redox couple; either or both of them will be an ion. Examples of reactions that probably occur in such an elementary one-step process are

---

[715] Freedom from oxygen is needed because this dissolved gas can be cathodically reduced. Usually, prior to the experiment, an inert gas (nitrogen or argon) is bubbled through a liquid electrolyte to displace oxygen.

[716] The location of the potential limits of the window depend markedly, not only on the solvent, but also on the nature of the electrode. For the system illustrated in Figure 7-2, the anodic limit is about 1.0 V for a platinum electrode, but about 0.0 V for mercury.

7:11
$$Cr(CN)_6^{4-}(aq) \rightleftarrows e^- + Cr(CN)_6^{3-}(aq)$$

and the oxidation of ferrocene[611] to the ferrocenium cation

7:12
$$(C_5H_5)_2Fe(soln) \rightleftarrows e^- + (C_5H_5)_2Fe^+(soln)$$

In choosing to address an electrode reaction in which the stoichiometric coefficients and the electron number are all unity, and in which there is a single reactant and a single product, we are, of course, choosing the simplest instance. The choice is not merely for convenience, however, because we believe that electrode reactions that overall are more complicated than 7:10, nevertheless proceed by steps that are often no more complicated than 7:10.

Our concept of chemical equilibrium is of equality between forward and backward reaction rates. When these opposing rates are unequal, their difference results in a **net reaction rate** v_{net}. Applied to electrode reaction 7:10, the net rate is the difference between the rates of the forward (oxidation) and backward (reduction) processes:

7:13
$$v_{net}(E) = v_{ox}(E) - v_{rd}(E)$$

An "(E)" has been placed following each term to emphasize that the value of each rate depends on the electrode potential E or, to put it in another way, on the activity of electrons in the electrode. The unit in which each rate is measured is mol m^{-2} s^{-1}, because this is a heterogeneous reaction (page 47). The net rate, which alone is directly measurable[717], is also the rate at which electrons are being generated, and therefore is proportional, through Faraday's constant, to the current density i. The expression

7:14
$$\frac{i(E)}{F} = v_{net}(E) = v_{ox}(E) - v_{rd}(E)$$

conforms to the convention that *anodic current at a working electrode is positive*.

The rate $v_{ox}(E)$ of the oxidative process in 7:10 will be proportional to the activity of species R at the electrode surface, or equivalently to the product of R's activity coefficient and a concentration ratio:

7:15
$$v_{ox}(E) = k_{ox}(E)a_R^s = k_{ox}(E)\gamma_R c_R^s / c^o = k'_{ox}(E)c_R^s$$

The final step in 7:15 has incorporated the activity coefficient γ_R and the standard concentration c^o into the rate constant k_{ox} thereby creating a new **potential-dependent rate constant**[718] k'_{ox} with units of m s^{-1}. A similar treatment of the reductive rate leads to $v_{rd}(E) = k'_{rd}(E)c_O^s$. When these two rate expressions are substituted into equation 7:14, the important result

---

[717] usually. Isotopic exchange experiments can measure the individual rates.

[718] "potential-dependent rate *constant*" appears oxymoronic but it is the name generally used, though **rate coefficient** is also encountered. Even at constant potential, k' is not truly constant because it incorporates an activity coefficient. In solutions wherein the supporting-electrolyte-bolstered ionic strength is large, γ (and hence $\bar{k}'$) will vary little, at a constant potential, during an experiment.

7:16
$$\frac{i(E)}{F} = k'_{ox}(E)c^s_R - k'_{rd}(E)c^s_O$$

emerges[719]. Equation 7:16 is equally valid whether the oxidative rate exceeds its reductive counterpart or vice versa. That is, the equation applies equally to the reduction reaction

7:17
$$O(soln) + e^- \rightleftharpoons R(soln)$$

as it does to the oxidation 7:10. Result 7:16 shows how the current density depends on the concentrations, at the electrode surface, of the members R and O of the redox couple, in terms of two potential-dependent rate constants. Our goal in this section is to elucidate the form of those potential dependences.

Firstly, however, let us examine the form adopted by equation 7:16 after a prolonged null condition. Then the current density will be zero, the two surface concentrations will have returned to their bulk values, and the potential will have acquired its null value. Hence

7:18
$$0 = k'_{ox}(E_n)c^b_R - k'_{rd}(E_n)c^b_O$$

Moreover, in the absence of current, the Nernst equation 6:44

7:19
$$E_n = E^{o\prime} - \frac{RT}{F}\ln\left\{\frac{c^b_R}{c^b_O}\right\}$$

will apply, and it may be combined with 7:18 into

7:20
$$E_n = E^{o\prime} - \frac{RT}{F}\ln\left\{\frac{k'_{rd}(E_n)}{k'_{ox}(E_n)}\right\}$$

By suitably adjusting the bulk concentration ratio, *any* potential may be made a null potential, and therefore equation 7:20 must hold whatever the potential; that is:

7:21
$$E = E^{o\prime} - \frac{RT}{F}\ln\left\{\frac{k'_{rd}(E)}{k'_{ox}(E)}\right\}$$

One sees that at the formal potential $E^{o\prime}$ the two rate constants are equal; this common value is given the symbol $k^{o\prime}$ and the name **formal rate constant**[720]: $k^{o\prime} = k'_{rd}(E^{o\prime}) = k'_{ox}(E^{o\prime})$.

Next, after dropping the (E) suffixes, equation 7:21 may be rearranged into

7:22
$$\frac{RT}{F}\ln\left\{\frac{k^{o\prime}}{k'_{rd}}\right\} + \frac{RT}{F}\ln\left\{\frac{k'_{ox}}{k^{o\prime}}\right\} = E - E^{o\prime}$$

and differentiated with respect to E. This gives

---

[719] In other texts you will find this equation written without the primes. Confirm, as in Web#719, that equation 7:16 is dimensionally consistent.

[720] though it is often inappropriately called the "standard rate constant" and symbolized k^o or k_s. See Web#720 for a fuller description of $k^{o\prime}$ and how it differs from k^o.

7:23
$$\frac{RT}{F}\frac{d}{dE}\ln\left\{\frac{k^{o\prime}}{k'_{rd}}\right\} + \frac{RT}{F}\frac{d}{dE}\ln\left\{\frac{k'_{ox}}{k^{o\prime}}\right\} = 1$$

The first left-hand term in 7:23 is named the **reductive transfer coefficient** and represented by the symbol[721] α. The second left-hand term is the **oxidative transfer coefficient**, which the equation shows to equal $1-\alpha$. Thus

7:24
$$\alpha = \frac{RT}{F}\frac{d}{dE}\ln\left\{\frac{k^{o\prime}}{k'_{rd}}\right\} \qquad \text{and} \qquad 1-\alpha = \frac{RT}{F}\frac{d}{dE}\ln\left\{\frac{k'_{ox}}{k^{o\prime}}\right\}$$

whence on integration[722]

7:25
$$k'_{rd} = k^{o\prime}\exp\left\{\frac{-\alpha F}{RT}(E-E^{o\prime})\right\} \qquad \text{and} \qquad k'_{ox} = k^{o\prime}\exp\left\{\frac{(1-\alpha)F}{RT}(E-E^{o\prime})\right\}$$

Alternative names for the dimensionless α and $1-\alpha$ quantities are **symmetry factors** and **charge-transfer coefficients**. Though they are commonly derived by dubious arguments based on vague energy profiles, it is seen here that their appearance is a consequence of purely mathematical arguments. Those arguments throw no light on whether or not α is a constant. As used in the Butler-Volmer equation described below, the transfer coefficients are usually treated as empirical constants that sum to unity, with experimental values typically clustering in the range between 0.3 and 0.7, and that is their status in this book. The **Marcus-Hush theory**[723] predicts that the transfer coefficient will have a constant value close to 0.5 at potentials near $E^{o\prime}$ but that α may become potential dependent at more remote potentials. Though the theory has proved of value in special cases, our inability to study electron-transfer reactions over wide ranges of potential makes further discussion of the distinction between the Butler-Volmer and Marcus-Hush treatments generally rather moot.

With the transfer coefficients regarded as constants, equations 7:25 show that the oxidative rate constant increases exponentially with potential, whereas the reductive rate constant declines exponentially with potential. Figure 7-3 overleaf illustrates this behavior graphically for two values of α.

Substitution from 7:25 into equation 7:16 leads to the result[724]

7:26
$$\frac{I}{A} = i = Fk^{o\prime}\left[-c_O^s\exp\left\{\frac{-\alpha F}{RT}\left(E-E^{o\prime}\right)\right\} + c_R^s\exp\left\{\frac{(1-\alpha)F}{RT}\left(E-E^{o\prime}\right)\right\}\right]$$

---

[721] The symbol β is often used in the literature, α being reserved for what we denote as α_{rd}.

[722] Carry out the required integration to derive 7:25, or see Web#722.

[723] Rudolph A. Marcus, 1923– , Canadian-American theoretical chemist, Nobel laureate 1992; Noel S. Hush, 1925– , Australian theoretical chemist. See Web#723 for more information on the Marcus-Hush theory.

[724] A parallel derivation can produce an equation similar to 7:26 with activities replacing concentrations and without primes. Though preferable to a purist, such a relationship is hardly ever used in practice.

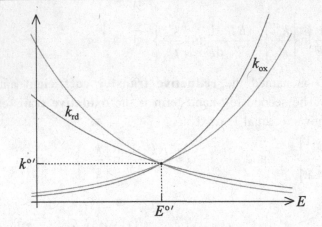

Figure 7-3 The reductive and oxidative rate constants depend exponentially on electrode potential and share a common value at the formal potential. The reductive transfer coefficients $\alpha = 0.35$ and $\alpha = 0.50$ were chosen for the **red** and **green** curves respectively.

This is the usual way in which the important **Butler-Volmer equation**[725] is formulated, but an alternative[726] is

$$7{:}27 \qquad i = i_n \left[-\frac{c_O^s}{c_O^b}\exp\left\{\frac{-\alpha F}{RT}(E - E_n)\right\} + \frac{c_R^s}{c_R^b}\exp\left\{\frac{(1-\alpha)F}{RT}(E - E_n)\right\} \right]$$

where i_n, known as the **exchange current density**[727], equals $Fk^{o\prime}(c_R^b)^\alpha (c_O^b)^{1-\alpha}$. Both versions describe how the current density i depends on the electrode potential E and on the two surface concentrations c_O^s and c_R^s. The Butler-Volmer equation reduces to different versions of the Nernst equation under two circumstances: when the current density is zero, and when the formal rate constant is very large[728].

7:26 is an equation of some complexity. There are thirteen distinct quantities whose symbols are either present in the equation or are closely associated with it. At this juncture it may be useful to review the meanings and status of these quantities in a typical electrochemical study. The table opposite provides such an appraisal. Generally one of the variables shown in **blue** is imposed, the other **blue** variable being monitored. Values of the two variables shown in **green** are inferred from the electrical observations during the experiment, because there is no electrochemical way of measuring them directly[729]. There are exceptions depending on the motive for the study, but frequently the object of the experiment is to determine the value of one or all of the quantities shown in **red**.

[725] John Alfred Valentine Butler, 1899 – 1977, English electrochemist; Max Volmer, 1885 – 1965, German physical chemist. In addition to these scientists, the Hungarian electrochemist Tibor Erdey-Grúz, 1902 – 1976, should also receive credit.

[726] Derive equation 7:27 from 7:26 or see Web#726.

[727] In 1000 mM NaCl, using a platinum electrode of area 9 mm², at 25°C, measurements on the reaction $Fe(CN)_6^{3-}(aq) + e^- \rightleftarrows Fe(CN)_6^{4-}(aq)$ reveal $k^{o\prime}$ and α values of 1.0×10^{-4} m s^{-1} and 0.5. Show that the exchange current density, when each hexacyanoiron ion has a concentration of 2.0 mM, is 19 A m². Web# 727.

[728] Demonstrate the truth of this statement or consult Web#728.

[729] With special instrumentation, they may be measurable optically in certain cases.

Symbols	Meanings	Status during most experiments
F, R	Faraday's constant, gas constant	physical constants, known and invariant
T, A, c_R^b, c_O^b	temperature, electrode area, bulk concentrations of R and O	experimental parameters, known[730] and constant
$E^{o'}, k^{o'}, \alpha$	formal potential, formal rate constant, transfer coefficient	unknown parameters, usually treated as constants but subject to mild variation[731]
I, E	current, electrode potential	variables, either controlled or measured
c_R^s, c_O^s	concentrations of R and O at the electrode surface	variables, unknown and not measurable electrochemically

The term $\exp\{-F(E - E^{o'})/RT\}$ may be thought of as the activity a_{e^-} of electrons in an electrode at potential E, compared with their activity at the formal potential $E^{o'}$. From this viewpoint, the Butler-Volmer equation may be written

7:28
$$\frac{i}{F} = v_{net} = k^{o'} c_R^s a_{e^-}^{\alpha-1} - k^{o'} c_O^s a_{e^-}^{\alpha}$$

Note that α and $1-\alpha$ appear as powers and thereby play roles akin to fractional stoichiometric coefficients. Equation 7:28 correctly indicates that increasing the activity of electrons has two effects: it accelerates the reductive process and it retards the oxidative process. If the electrode potential is made more negative, the transfer coefficient α is the fraction of the increased electron activity devoted to increasing the reductive rate, while the complementary $1-\alpha$ fraction serves to decrease the oxidative rate. The Butler-Volmer equation, in the 7:28 version, closely resembles the equations describing the kinetics of ordinary chemical reactions, as in Chapter 2. And, just as stoichiokinetic equations usefully portray chemical reactions, so the representation

7:29
$$R + (\alpha - 1)e^- \rightleftharpoons \alpha e^- + O$$

serves to describe both the stoichiometry and the kinetics of the simple $R \rightleftharpoons e^- + O$ oxidation. The "shape" of the rate law 7:28 is exactly as expected from the stoichiokinetic equation 7:29. The only surprise, perhaps, is that the forward and reverse rate constants are equal; this is a consequence of the electron activities being referenced to the formal potential. Of course, apart from a change to $\rightleftharpoons$ arrows, rate law 7:28 and stoichiokinetic

---

[730] Occasionally, voltammetry (or polarography[1305]) is employed as a method of chemical analysis, the bulk concentration of O or R being unknown and sought. Voltammetry has also been used to measure electrode areas.

[731] With ionic strength in particular, because of its effect on activity coefficients. Also with temperature.

equation 7:29 are equally applicable to the cathodic reduction of O.

A word of warning about nomenclature! Thus far, we have been analyzing an electrode reaction occurring in a single elementary step. Such reactions are rare. In the next section, attention will be directed to reactions that involve a sequence of elementary steps, electrochemical and chemical. During our analysis of such mechanisms, it will transpire that the current generally obeys an equation of the same "shape" as 7:26 (or 7:27 or 7:28) but with modified terms. Confusingly, the name "Butler-Volmer equation" is also given to such expressions. Moreover, the terms that replace the α and $1-\alpha$ are also commonly referred to as "transfer coefficients". Yet again, the phrase "exchange current density" is used in the context of both one- or multistep reactions. In this book, we shall endeavor to reduce the risk of confusion by including the adjective "composite" in names that relate to multistep mechanisms and by attaching subscripts to α to denote composite transfer coefficients.

A further word of warning. Except in potentiometry or in carefully designed steady-state experiments (Chapter 12), time is an important factor in the study of electrode reactions but, in the interest of introducing concepts gradually, its role has been sidestepped in this chapter. All of the factors colored in blue or green in the table on page 135 are liable to change with time. There are three ways in which time may enter the picture:

(a) The experimenter may decide to impose a time-dependent signal (such as a voltage ramp, or an a.c. current) on the cell.

(b) Transport laws involve time, causing concentration profiles to evolve.

(c) Chemical and electrochemical reactions cause changes in concentrations with time. These matters will be taken up in subsequent chapters.

Elsewhere you may find equations similar to 7:26 or 7:27 but written with an n incorporated, so as to cater to a concerted transfer of several electrons

7:30 $$R(soln) \rightleftarrows ne^- + O(soln)$$

However, it is our position that there is no evidence to suggest that even a reaction as stoichiometrically simple as

7:31 $$Tl^+(aq) \rightleftarrows 2e^- + Tl^{3+}(aq)$$

does, in fact, occur by concerted electron transfer. Multiple electron transfers can be explained by successive events as in the following section.

For a simple **surface-confined reaction**

7:32 $$R(ads) \rightleftarrows e^- + O(ads)$$

the formalism of this chapter applies. The only changes are that the electrode reaction rate constants have the s^{-1} unit, rather than $m\ s^{-1}$, and that volumetric concentrations at the electrode surface c^s are replaced by areal surface concentrations, Γ, with mol m^{-2} units. On page 269 the voltammetry of surface-confined reactions is briefly discussed. Such reactions, which often have a bearing on biochemical processes, are of increasing importance in electrochemistry.

Multistep Electrode Reactions: studying kinetics to elucidate mechanisms

It is from studying the kinetics of an electrode reaction that evidence concerning the reaction's mechanism can most easily be gleaned. The kinetics is summarized by the **rate law** that mathematically describes how the current density is influenced by concentrations and by the electrode potential; for example, equation 7:26 is the rate law for the simple electron transfers described in the prior section. From multiple experiments at various potentials and with various concentrations of reactants and products, the experimental rate law may be determined. As for the simple case discussed in the previous section, this composite rate law will generally consist of two[732] rate terms: a positive oxidative moiety and a negative reductive moiety[733], both usually potential dependent

7:33
$$\frac{i}{F} = v_{net} = v_{ox} - v_{rd}$$

Though they are not independently measurable electrochemically, we again view the current density to be composed of two components: a positive oxidative current density i_{ox} and a negative reductive current density i_{rd}

7:34
$$i = Fv_{net} = Fv_{rd} - Fv_{ox} = i_{ox} + i_{rd}$$

Figure 7-4 is a diagram showing how these partial current densities combine to form a typical graph of current density versus potential.

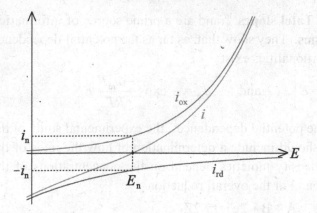

Figure 7-4 The **oxidative current density** and the **reductive current density** sum to give the measured **net current density**. At the null potential E_n the two partial current densities are identical in magnitude but opposite in sign.

As the diagram implies, at potentials remote[734] from the null potential the current density is dominated by one or other of the partial currents and this fact opens up the

---

[732] though not uncommonly, because of experimental constraints, only one can be established reliably.

[733] Note that we assume the mechanism for the reduction mechanism to be the converse of that for the oxidation reaction. One can *conceive* of mechanisms in which this is not so but a thermodynamic argument, known as the principle of **microscopic reversibility**, proves otherwise. That is not to say, however, that different mechanisms may not apply *at different potentials*.

[734] As little as 120 mV from E_n is often adequate. Confirm this for the simple reaction 7:10, or see Web#734.

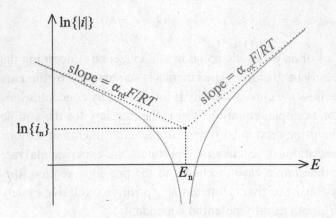

Figure 7-5 The Tafel plot. The straight line portions occur where $i \approx i_{rd}$ and where $i \approx i_{ox}$. The graph is curved close to the null potential E_n, where neither approximation holds. The construction lines show how the null potential and the composite exchange current density i_n may be measured.

opportunity of deciphering the experimental rate law by measuring i_{rd} and i_{ox} separately. This procedure is associated with the name of **Tafel**[735]. Because electrode reaction rates vary exponentially with potential, the Tafel approach involves plotting the logarithm of the absolute value $|i|$ of the current density versus potential. This produces graphs resembling Figure 7-5, with linear portions. The slopes of those linear segments yield the **composite transfer coefficients**

$$7:35 \qquad \alpha_{ox} = \frac{RT}{F} \frac{\partial}{\partial E} \ln\{i\}\Big|_{i \approx i_{ox}} \qquad \text{and} \qquad \alpha_{rd} = \frac{-RT}{F} \frac{\partial}{\partial E} \ln\{-i\}\Big|_{i \approx i_{rd}}$$

The slopes themselves are called **Tafel slopes**[736] and are a prime source of information about electrode reaction mechanisms. They show that, as far as the potential-dependence is concerned, the following proportionalities exist:

$$7:36 \qquad i_{ox} \propto \exp\left\{\frac{\alpha_{ox}F}{RT}E\right\} \qquad \text{and} \qquad i_{rd} \propto -\exp\left\{\frac{-\alpha_{rd}F}{RT}E\right\} .$$

In addition to investigating the potential dependences, the experimental study of the kinetics of an electrode reaction should include a determination of how the rates of the reductive and oxidative current density moieties depend on the concentrations of the species participating in the reaction. For the overall reduction

$$7:37 \qquad\qquad\qquad\qquad A + B + 2e^- \; \rightleftarrows 2Z$$

to take an example, one would expect i_{rd} to reflect the concentration of A at the electrode surface. But this dependence is not necessarily a proportionality to c_A^s itself; i_{rd} could well be proportional to c_A^s raised to some power other than unity, perhaps to $(c_A^s)^2$, $(c_A^s)^{1/2}$,

[735] Julius Tafel, 1862–1919, a Swiss-German organic chemist who discovered the $\log\{|i|\}$ versus E relationship empirically.

[736] Often in the literature this term refers not to $d(\ln\{|i|\})/dE$, but to $dE/d(\log_{10}\{|i|\})$ and equals $2.303RT/(\alpha F)$, or about 59 mV divided by the composite transfer coefficient at 25°C. What is the anodic Tafel slope for the one-step $O - e^- \; \rightleftarrows R$ reaction? Check at Web#736.

$(c_A^s)^0$, or even $(c_A^s)^{-1}$. In principle[737], the appropriate power can be found from the operation

7:38
$$\frac{\partial \ln\{-i_{rd}\}}{\partial \ln\{c_A^s\}} = \Omega_{A,rd}$$

which yields the **order** $\Omega_{A,rd}$ of the reductive partial current density with respect to reactant A. The surface concentrations of B and Z likely play roles in determining the i_{rd} partial current, too. Similar considerations apply to the concentration dependences of i_{ox}. Hence the kinetic study should include a hunt for the various reductive and oxidative kinetic orders; that is, the numerical values of the exponents represented by the Ω's in the proportionalities

7:39
$$i_{ox} \propto (c_A^s)^{\Omega_{A,ox}} (c_B^s)^{\Omega_{B,ox}} (c_Z^s)^{\Omega_{Z,ox}} \quad \text{and} \quad -i_{rd} \propto (c_A^s)^{\Omega_{A,rd}} (c_B^s)^{\Omega_{B,rd}} (c_Z^s)^{\Omega_{Z,rd}}$$

Determining the experimental rate law for the electrode reaction 7:37 thus devolves into finding eight dimensionless numbers – the six Ω's and the two α's – in the experimental reductive and oxidative rate laws:

7:40
$$\begin{cases} i_{ox} \propto (c_A^s)^{\Omega_{A,ox}} (c_B^s)^{\Omega_{B,ox}} (c_Z^s)^{\Omega_{Z,ox}} \exp\left\{\dfrac{\alpha_{ox} F}{RT} E\right\} & \text{general} \\[2ex] i_{rd} \propto -(c_A^s)^{\Omega_{A,rd}} (c_B^s)^{\Omega_{B,rd}} (c_Z^s)^{\Omega_{Z,rd}} \exp\left\{\dfrac{-\alpha_{rd} F}{RT} E\right\} & \text{law} \end{cases}$$

A requirement is that setting i_{ox} and i_{rd} equal each other in magnitude, and thereby invoking the null potential, in 7:40, produces a result that is compatible with the Nernst equation.

Having determined the experimental rate laws, how can they be used to investigate the mechanism? Figure 7-6 will help to explain the process. Relying on chemical savvy,

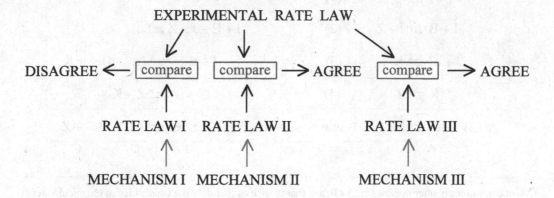

Figure 7-6 A chart illustrating how the determination of a rate law is used to rule out certain putative mechanisms. Here Mechanism I is unacceptable, but II and III remain possibilities.

---

[737] but less straightforwardly in practice. The experimenter has direct control over *bulk* concentrations only.

possible mechanisms are envisaged and the rate law appropriate to each is deduced. As the figure illustrates, each putative mechanism (three in the chart) is analyzed to find the rate law which that mechanism predicts. Each mechanistic rate law is then compared with the experimental rate law. Disagreement rules out that mechanism; agreement shows that the mechanism is *possibly* correct. No mechanism may be proved unequivocally, though persuasive circumstantial evidence can accumulate.

The electrode reaction 7:10 that was addressed in the prior section had the simplest possible mechanism: it occurred in a single step. Steps in a multistep mechanism may be chemical or electrochemical. Each step is itself a reaction; it has its own rate constants and, if it is an electrochemical step, its own formal potential and transfer coefficients. As well as involving reactants and products, these steps usually involve **intermediates**[738], ephemeral species that are created in one step and consumed in another. As discussed for chemical reactions in Chapter 2, it is nearly always possible to arrange the steps in a logical sequence such that an intermediate created in one step is consumed in a step later in the list. Conceivably, an intermediate might be transported away from the electrode or be involved in some side reaction, but no such events are considered here.

Continuing with the contrived $A + B + 2e^- \rightleftarrows 2Z$ example of an overall reaction, one possible four-step mechanism – involving three intermediates I, J and K – is that listed on the left-hand side of the table below. Of course the mechanistic steps, of which two are chemical and two electrochemical, are required to add up to the overall stoichiometry. In numbering the steps, we have "hatted" the third, to indicate that this is the "slow step"[739]

Mechanistic equations	Step	Modified equations
$A + e^- \rightleftarrows I$	(1)	$A - I + e^- \rightleftarrows$
$I + B \rightleftarrows J + Z$	(2)	$I + B - J - Z \rightleftarrows$
$J + e^- \rightleftarrows K$	$(\hat{3})$	$J + \alpha_3 e^- \rightleftarrows (\alpha_3 - 1)e^- + K$
$K \rightleftarrows Z$	(4)	$\rightleftarrows Z - K$
$A + B + 2e^- \rightleftarrows 2Z$	sum	$A + B - Z + (1 + \alpha_3)e^- \rightleftarrows (\alpha_3 - 1)e^- + Z$

---

[738] Not uncommonly intermediates are **radicals**; that is species that defy the usual rules of chemical valency. Unlike most organic species, be they ionic or neutral, organic radicals have an *odd* number of electrons, the unpaired electron being indicated by a superscript dot in such formulas as $H_3C^\bullet$ or $HO^\bullet$. **Radical anions** have both an unpaired electron and a negative charge.

[739] It might be expected that the designation "slow step" would mean the step proceeding at the slowest rate. In kinetics terminology, however, it describes the step with the smallest rate constant, that is, the one that has the potential to be slowest. When locked into a mechanism, each step generally proceeds at the *same* rate.

or rate-determining step (page 48). All the other steps are fast in comparison[740] and can therefore[741] be treated as at equilibrium. How can one construct the overall rate law from the mechanism? That is, how are the procedures represented by red arrows in Figure 7-6 actually performed? Elaborate algebra may be applied on a case-by-case basis[742] but a construct operating through the stoichiokinetic principle (pages 49–50), works well in most cases[743]. It involves eight straightforward rules that will be illustrated by reaction 7:37 and the table shown on the facing page:

(a) Write the equations of all pre-slow-step reactions, that is Steps (1) and (2) in the example, as equilibria, by transferring all species from the right-hand side to the left, introducing negative signs on transfer.

(b) Write the equation of the rate-determining step, here Step (3), in the normal way. If this is an electrochemical step, split the electron in the manner of equation 7:29.

(c) Write the equations of all post-slow-step reactions, just Step (4) in the example, by transferring species to the right-hand side, introducing negative signs on transfer.

(d) Leaving the rate-determining step unchanged, multiply or divide other equations, if necessary (it is unnecessary in the example) by a small integer (seldom other than 2), so that all intermediates disappear on implementing rule (e).

(e) Add the modified equations to generate the stoichiokinetic equation. Combine and cancel terms on addition, but do *not* combine terms on one side of the equation with those on the other. The stoichiokinetic equation provides the key to formulating the rate law.

(f) The terms A+B−Z on the left-hand side of the stoichiokinetic equation, show that the reductive moiety of the reaction proceeds at a rate proportional to the concentrations of A and B, and inversely proportional to the concentration of Z. That is, $v_{rd} \propto c_A c_B / c_Z$. Similarly the right-hand terms show that $v_{ox} \propto c_Z$.

(g) The stoichiometric coefficients of the electrons in the stoichiokinetic equation reveal the potential dependences of the reductive and oxidative rates. Thus $v_{rd} \propto \exp\{-(1+\alpha_3)FE/RT\}$ and $v_{ox} \propto \exp\{-(\alpha_3-1)FE/RT\}$.

(h) From the proportionalities, an expression for the current density, which equals $F(v_{ox} - v_{rd})$, may finally be written down as

7:41
$$\frac{i}{F} = k'_{ox} c_Z^s \exp\left\{\frac{(1-\alpha_3)F}{RT} E\right\} - k'_{rd} \frac{c_A^s c_B^s}{c_Z^s} \exp\left\{\frac{-(1+\alpha_3)F}{RT} E\right\}$$

by supplying appropriate rate constants. According to this mechanism, the **composite transfer coefficients** α_{ox} and α_{rd}, will equal $1-\alpha_3$ and $1+\alpha_3$ respectively and are thereby

---

[740] Of course it is possible for two steps to be comparably slow. However, considering that the timescale for reactions ranges from nanoseconds to centuries, this is unlikely.

[741] For a justification of this approximation, see Web#741.

[742] The algebra appropriate to the mechanism proposed for reaction 7:37 will be found at Web#742.

[743] Employ the stoichiokinetic construct method to elucidate the rate laws appropriate to the three mechanisms, detailed on page 78, governing hydrogen evolution from aqueous solutions on different metals. See Web#743.

expected to have numerical values close to 0.5 and 1.5 respectively[744]. Note the unexpected retarding effect of Z on the reductive moiety. Note also that 7:41 meets the requirement[745], when $i = 0$, of being compatible with the Nernst equation for reaction 7:37. The k'_{ox} and k'_{rd} terms are composites of chemical and/or formal rate constants, formal potentials, and other constants but, because they are of no mechanistic utility, we shall not analyze them further.

As an interesting real example, consider the cathodic reduction of 1-bromonaphthalene, $C_{10}H_7Br$ in an inert organic solvent. Representing the naphthyl $C_{10}H_7$ core by $N\ell$, the stoichiometric equation of the reaction

7:42
$$N\ell\,Br(soln) + 2e^- \rightleftarrows N\ell^-(soln) + Br^-(soln)$$

is simple but the experimental rate law turns out to be surprisingly elaborate. A three-step mechanism that can explain the experimental facts is given in the left-hand column of the following table. It involves a free radical and a radical anion[738] as intermediates. Rule (d) must be implemented carefully in this case. The rate law predicted by the mechanism is

7:43
$$\frac{i}{F} = k'_{ox}\sqrt{c^s_{N\ell^-}\,c^s_{Br^-}\,c^s_{N\ell Br}}\,\exp\left\{\frac{(1-\alpha_1)F}{RT}E\right\} - k'_{rd}c^s_{N\ell Br}\exp\left\{\frac{-\alpha_1 F}{RT}E\right\}$$

with the conclusion that α_{ox} and α_{rd} sum to 1, and not the 2 that might have been expected from the stoichiometry of reaction 7:42.

Mechanistic equations	Step	Modified equations
$N\ell\,Br + e^- \rightleftarrows N\ell\,Br^{\bullet-}$	$(\hat{1})$	$N\ell\,Br + \alpha_1 e^- \rightleftarrows (\alpha_1 - 1)e^- + N\ell\,Br^{\bullet-}$
$\frac{1}{2}N\ell\,Br^{\bullet-} \rightleftarrows \frac{1}{2}N\ell^{\bullet} + \frac{1}{2}Br^-$	(2)	$\rightleftarrows -\frac{1}{2}N\ell\,Br^{\bullet-} + \frac{1}{2}N\ell^{\bullet} + \frac{1}{2}Br^-$
$\frac{1}{2}N\ell\,Br^{\bullet-} + \frac{1}{2}N\ell^{\bullet} \rightleftarrows \frac{1}{2}N\ell\,Br + \frac{1}{2}N\ell^-$	(3)	$\rightleftarrows -\frac{1}{2}N\ell\,Br^{\bullet-} - \frac{1}{2}N\ell^{\bullet} + \frac{1}{2}N\ell\,Br + \frac{1}{2}N\ell^-$
$\frac{1}{2}N\ell\,Br + e^- \rightleftarrows \frac{1}{2}N\ell^- + \frac{1}{2}Br^-$	sum	$N\ell Br + \alpha_1 e^- \rightleftarrows (\alpha_1 - 1)e^- + \frac{1}{2}N\ell^- + \frac{1}{2}Br^- + \frac{1}{2}N\ell\,Br$

Because the molecules are held together by electron-*pair* bonds, cathodic reductions and anodic oxidations of organic compounds usually involve *even* numbers of electrons. The mechanism shown on the left in the next table has found widespread application to two-electron oxidations

7:45
$$R \rightleftarrows 2e^- + O$$

---

[744] At 25°C, the Tafel slopes $dE/d(\log 10\{|i|\})$ will be close to 120 mV and 40 mV respectively.

[745] Confirm this, or see Web#745.

Mechanistic equations	Step	Modified equations
$R - e^- \rightleftarrows I$	(1)	$R - e^- - I \rightleftarrows$
$I \rightleftarrows J$	($\hat{2}$)	$I \rightleftarrows J$
$J - e^- \rightleftarrows O$	(3)	$\rightleftarrows e^- + O - J$
$R - 2e^- \rightleftarrows O$	sum	$R - e^- \rightleftarrows e^- + O$

Two intermediates (perhaps radical cations) appear in the mechanism, in which the interconversion of two intermediates is the slow step. The stoichiokinetic method leads to the simple rate law

7:46
$$\frac{i}{F} = k'_{ox} c_R^s \exp\left\{\frac{F}{RT}E\right\} - k'_{rd} c_O^s \exp\left\{\frac{-F}{RT}E\right\}$$

with both composite transfer coefficients equaling unity. However, were Step (1) to be rate determining, $\alpha_{ox} \approx 0.5$ and $\alpha_{rd} \approx 1.5$, the converse being the case if Step (3) determines the rate.

A notation that is commonly adopted in categorizing complex electrode mechanisms is composed of the letters E and C, standing for "electrochemical" and "chemical". Thus the symbol EC$\hat{E}$C describes the mechanism of which 7:41 is the rate law, the "hat" showing which of the four steps is rate determining. Likewise E$\hat{C}$E describes the mechanism to which 7:46 relates.

As a final example, consider the CC$\hat{E}$ mechanism listed on the left below

Mechanistic equations	Step	Modified equations
$I_3^-(aq) \rightleftarrows I_2(ads) + I^-(aq)$	(1)	$\frac{1}{2}I_3^-(aq) - \frac{1}{2}I_2(ads) - \frac{1}{2}I^-(aq) \rightleftarrows$
$I_2(ads) \rightleftarrows 2I(ads)$	(2)	$\frac{1}{2}I_2(ads) - I(ads) \rightleftarrows$
$I(ads) + e^- \rightleftarrows I^-(aq)$	($\hat{3}$)	$I(ads) + \alpha_3 e^- \rightleftarrows I^-(aq)$ $-(1-\alpha_3)e^-$
$I_3^-(aq) + 2e^- \rightleftarrows 3I^-(aq)$	sum	$\frac{1}{2}I_3^-(aq) - \frac{1}{2}I^-(aq) + \alpha_3 e^- \rightleftarrows I^-(aq)$ $-(1-\alpha_3)e^-$

for the reduction of the triiodide ion I_3^- in aqueous solution to the iodide ion I^-. The predicted rate law is

7:47
$$\frac{i}{F} = k'_{ox}c^s_{I^-} \exp\left\{\frac{(1-\alpha_3)F}{RT}E\right\} - k'_{rd}\sqrt{\frac{c^s_{I_3^-}}{c^s_{I^-}}} \exp\left\{\frac{-\alpha_3 F}{RT}E\right\}$$

which leads to an expectation that the composite transfer coefficients are related by $\alpha_{ox} = 1 - \alpha_3$ and $\alpha_{rd} = \alpha_3$ to the transfer coefficients of Step (3) and will therefore sum to 1, rather than the 2 that would have been expected from a concerted electron transfer. The experimental values of $\alpha_{ox} = 0.78$ and $\alpha_{rd} = 0.20$, do indeed sum to close to unity, even though their individual values are abnormally far from the usual $0.3 - 0.7$ range. Moreover, the predicted concentration dependences match experiment. Evidently, the mechanism predicts the experimental rate and is likely to be correct.

Because of its simplicity, we shall use the simple $R(soln) \rightleftarrows e^- + O(soln)$ reaction as the generic representation of an electrode reaction in much of this book, notwithstanding our belief that it is a poor descriptor of most electrochemical reactions. It does, however, provide a framework that can be adapted to suit more realistic electrode mechanisms.

Summary

Though the feasibility and direction of an electrode reaction are governed by the applied voltage and the Gibbs energy change accompanying the cell reaction, thermodynamics provides no information on the rate of the reaction. The stoichiometry of an electrode reaction satisfies Faraday's law, but the equation for the reaction may be written in several ways, only one of which also conveys information correctly about the kinetics of the reaction. Most electrode reactions occur by multistep mechanisms, but the simplest electrode reactions involve the transfer of a single electron between the electrode and an electroactive solute and take place in a single bidirectional step. Both the forward and backward directions of this process are influenced by the electrode potential, as conveyed by the Butler-Volmer equation

7:48
$$\frac{i}{F} = k'_{ox}c^s_R - k'_{rd}c^s_O = k^{o\prime}\left[c^s_R \exp\left\{\frac{(1-\alpha)F}{RT}(E-E^{o\prime})\right\} - c^s_O \exp\left\{\frac{-\alpha F}{RT}(E-E^{o\prime})\right\}\right]$$

which applies to the $R(soln) - e^- \rightleftarrows O(soln)$ or $O(soln) + e^- \rightleftarrows R(soln)$ reaction. The rates of multistep electrode reactions are governed by the mechanism of the process, which is not known a priori and can never be established with absolute certainty. The "slowest" step in the mechanism, which may or may not involve electrons, is crucial in determining the reaction rate. Information about the mechanism can be gleaned from how the current depends on potential and on the surface concentrations of all the nonunity-activity species that participate in the reaction. This information is summarized by the values of α_{ox} and α_{rd} (which sum to an integer) and by the stoichiokinetic equation, which must match the stoichiometry of the overall reaction, as well as the experimental orders.

8

Transport

In most electrochemical cells, reactants must travel to reach the electrode and/or products must be transported from the electrode. To proceed in our study of electrochemistry, we now need to look in some detail at the various ways in which such transport occurs and the laws it obeys. Though transport occurs in gases[801], and even in certain solids, it is the transport of molecular and ionic solutes in liquids, and especially in aqueous solutions, that will be our primary concern. This is sometimes called "mass transport" to distinguish it from other kinds of transport, such as heat transport or momentum transport, but for simplicity and because the transport of a solute does not necessarily imply the transport of mass, we shall use the unqualified term "transport".

Flux Density: solutes in motion obey conservation laws

Transport is the motion of a solute though space. The **flux density**[802] of a species i (an ion or a molecule), quantifies the motion; it is measured in the *SI* unit of mol m^{-2} s^{-1} and is symbolized j_i in this book. Flux density can be defined by

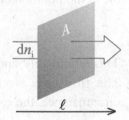

8:1
$$j_i = \frac{1}{A}\frac{dn_i}{dt}$$
definition of flux density

in terms of the number dn_i of moles of i crossing a surface of area A in a time interval dt. The area in question is perpendicular to the direction of motion. Flux density is a vector quantity, but in electrochemical usage it is always measured in the direction in which the transport occurs – that is, along the **flux lines** – being positive or negative in accordance with the choice of coordinate direction. Illustrated above is

---

[801] and more rapidly than in liquids. For example, Na$^+$ migrates 6500 times faster in nitrogen than in water and O$_2$ diffuses in air 8000 times faster than in water.

[802] The simpler term "flux" is also used, but this sometimes means the quantity with units mol s^{-1}, which we call the total flux and denote by J. The distinction between J and j, mirrors that between I and i.

Electrochemical Science and Technology: Fundamentals and Applications, First Edition. Keith B. Oldham, Jan C. Myland, Alan M. Bond.

transport for which j_i is positive. In general, the flux density depends on position and time, and we will frequently use a notation such as $j_i(\ell,t)$ to indicate these dependences. Flux density equals the product of the concentration c_i of a solute species i and its average velocity $\bar{v}_i$ in the direction of motion:

8:2
$$j_i = \bar{v}_i c_i$$

The flux densities of all charged species contribute to the current density through the relationship

8:3
$$i = F \sum_i z_i j_i$$

where each z is a charge number[803].

   If transport of a species is occurring solely along parallel flux lines in the x direction, as in Figure 8-1, then the amount of that species in any narrow wafer of volume V lying between x and $x + dx$ will change from its initial value $Vc(x,t)$ in response to flux densities in and out of the wafer. In fact

8:4
$$\begin{pmatrix} \text{new} \\ \text{amount} \end{pmatrix} = \begin{pmatrix} \text{old} \\ \text{amount} \end{pmatrix} + \begin{pmatrix} \text{influx} \\ \text{from left} \end{pmatrix} - \begin{pmatrix} \text{efflux} \\ \text{to right} \end{pmatrix}$$

or symbolically, with A denoting the surface areas across which transport occurs

8:5 $$Vc(x,t+dt) = Vc(x,t) + Aj(x,t)dt - Aj(x+dx,t)dt$$

Such transport is described as planar[804] because the equiconcentration surfaces are planes. Because $V = Adx$, equation 8:5 may be rearranged into

8:6
$$\frac{c(x,t+dt)-c(x,t)}{dt} = \frac{j(x,t)-j(x+dx,t)}{dx}$$

or, in the notation of the partial differential calculus,

8:7
$$\frac{\partial c}{\partial t} = -\frac{\partial j}{\partial x} \qquad \text{planar transport}$$

This remarkably simple equation is known as the **conservation law for planar transport**. The flux lines are not parallel in Figure 8-2 and, for transport in this more general circumstance, the conservation equation takes the more elaborate form

8:8
$$\frac{\partial c}{\partial t} = -\frac{\partial j}{\partial \ell} - j\frac{d}{d\ell}\ln\{A(\ell)\}$$

because account must be taken of the change in area of the **equiconcentration surfaces**[805]

---

[803] Confirm that each of equations 8:1, 8:2, and 8:3 is dimensionally consistent. See Web#803.

[804] It is sometimes described as "linear transport" because the flux lines are straight lines, but that descriptor does not distinguish it from spherical transport, in which the flux lines are also straight lines.

[805] Equiconcentration surfaces are orthogonal (at right angles, normal, perpendicular) to the flux lines. Note the similarity to equipotential surfaces (page 197); in fact the two surfaces usually coincide in electrochemistry. Equiconcentration surfaces are planes in the case of Figure 8-1 and portions of spheres for Figure 8-3.

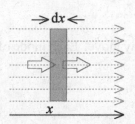

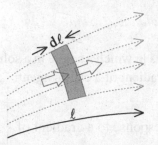

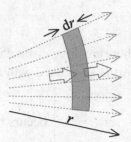

Figure 8-1 When transport occurs along parallel flux lines, the conservation equation takes the simple form given in equation 8:7.

Figure 8-2 When flux lines are not parallel, the conservation law takes the form given in equation 8:8.

Figure 8-3 When flux lines radiate from a point, equiconcentration surfaces are spheres, or portions thereof, and the conservation equation is as reported in 8:9.

with the coordinate ℓ. Equation 8:8, the general **conservation law**, is derived elsewhere[806]. As an example, consider the case illustrated in Figure 8-3, in which the flux lines radiate in three dimensions from a single point. Then the equiconcentration surfaces are spherical or hemispherical and their areas increase with the square of the radius r, so that $d\ln\{A(r)\}/dr$ equals[807] $2/r$. Therefore, in this case, equation 8:8 becomes

$$8:9 \qquad \frac{\partial c}{\partial t} = -\frac{\partial j}{\partial r} - \frac{2j}{r} \qquad \text{spherical transport}$$

This is the **conservation law for spherical transport**.

The conservation law requires amendment if the species in question is being created or removed within the transport medium. For instance, if the homogeneous chemical reaction

$$8:10 \qquad A(soln) + B(soln) \underset{\overset{\vec{k}}{\longleftarrow}}{\overset{\vec{k}}{\longrightarrow}} Z(soln)$$

takes place in a solution through which the planar transport of species A is occurring, then the conservation law 8:7 needs supplementation by two terms:

$$8:11 \qquad \frac{\partial c_A}{\partial t} = -\frac{\partial j_A}{\partial x} - \vec{k}'c_A c_B + \bar{k}'c_Z$$

where $\vec{k}'$ and $\bar{k}'$ are concentration-based rate constants (respectively bimolecular and unimolecular – see Chapter 2)[808].

---

[806] The general conservation law is derived in Web#806. Equations 8:7, 8:8, and 8:9 are referred to as **continuity equations**.

[807] Confirm this or see Web#807.

[808] Confirm that the four terms in 8:11 have the same *SI* unit. See Web#808.

Three Transport Modes: migration, diffusion and convection

There are three distinct ways in which a solute in a solution may move. Each mode of transport is associated with a gradient of a causal agent:

migration ⎫
diffusion ⎬ occurs in response to a gradient of ⎰ electrical potential
convection ⎭   ⎱ activity or concentration
                                                  pressure

With one exception, the motion is from a higher to a lower value of the property in question. It is the migration of anions that is the exception: they migrate towards higher (more positive) electrical potentials.

The force that causes migration is provided by an electric field and energy is dissipated by the motion, appearing as heat. Similarly, a hydrostatic force usually[809] instigates convection. There is no comparable force that can be identified with diffusion, which is driven by the entropy increase that accompanies the passage of solute into a more dilute region of the solution. Stated differently but equivalently, diffusion arises from the **brownian motion**[810] of the solute molecules in a liquid; this jostling inevitably leads to more molecules leaving a concentrated region for a more dilute adjacent region than are recouped by the reverse journey.

A common characteristic of both migration and diffusion is that, in these transport modes, the solute moves through the solvent, which remains more-or-less static. In convection, the solute moves by being entrained with the solvent in a moving solution. Thus convection is very different from the other modes in being closely dependent on the geometry of the vessel which contains the liquid. Convection is often used, particularly in electrosynthesis, as a rough-and-ready means of accelerating transport. Only with favorable cell geometries can convection be modeled with help from the laws of hydrodynamics. In most other circumstances, convection is an unwelcome complication that investigative electrochemists try to avoid. In this context, we distinguish between **forced convection**, deliberately caused by stirring, pumping or sonicating[640] the solution, and **natural convection**, the unwanted motion of the solution caused by vibrations or density gradients.

Three distinct sections of this chapter address each of the three modes separately. Rarely in electrochemistry, however, is one mode of transport operative in isolation. Even forced convection, which is a very efficient transport mode, nevertheless calls upon the other modes to transfer solutes to an electrode. Moreover, migration can induce diffusion and vice versa. Indeed, migration and diffusion may be viewed as two manifestations of the same phenomenon. Recall that energy is expended in migration, but that entropy is

---

[809] Less commonly, electroosmotic force (Chapter 13) may be the cause of convection.

[810] The erratic motion of molecules in a liquid, inferred from the random motion that they impart to much larger suspended particles, as first observed microscopically by Robert Brown, 1773 – 1858, Scottish botanist.

increased in diffusion. Both transport modes thus lead to a decrease in Gibbs energy and the tendency towards this decrease can be viewed as the cause of the transport. Following this line of reasoning leads to relationships that will be discussed on page 159. For the present, however, we shall regard the three transport modes as distinct and operating singly.

That migration affects only ionic solutes, whereas all solutes are prone to diffusion and convection, is one important distinction among the transport modes. Another is a consequence of Newton's[105] laws of motion. Motion cannot start instantaneously: the moving body must be accelerated up to its terminal velocity. The time required by this acceleration period depends on the mass being accelerated. The consequential delay is so trivial[811] for migration that it can be ignored in electrochemistry: as soon as the switch is thrown, electrons and ions jump into steady motion almost instantaneously. This is not the case with natural convection, however. Large quantities of solution must be set in motion, and the inertia is considerable. This is a fortunate happenstance in voltammetry (Chapters 12, 15, and 16), for it gives the experimenter as long as 100 seconds before there is a need to worry about the effect of density gradients.

Migration: ions moving in response to an electric field

The term **mobility** was introduced on page 19 where it was defined[812] as the ratio of the average velocity of an ion to the electric field that elicits its motion.

8:12
$$u_i = \frac{\overline{v}_i}{X}$$
definition of mobility

The migratory flux density of an ion is given by an equation that combines this definition with equation 8:2 into

8:13
$$j_i^{mig} = u_i c_i X = -u_i c_i \frac{d\phi}{d\ell}$$

Mobilities change somewhat with concentration and those listed in the second column of the table overleaf are for infinite dilution, as indicated by the u° notation. Notice that the mobilities of most common inorganic ions in water are in the vicinity of $10^{-7}\,\mathrm{m^2\,V^{-1}\,s^{-1}}$ in magnitude. This may appear to be slow, for it implies that an ion moves only 9 mm in a day under a field of one volt per meter. On a molecular scale, however, the speed is respectable: this migrating ion rushes past 300 water molecules in one second.

Two features become clear on inspection of the tabulated mobilities. Firstly, contrary to the expectation that doubly charged ions might migrate twice as fast as singly charged

---

[811] though the effect can be significant for large ions at high a.c. frequencies.

[812] The alternative definitions $\overline{v}_i / |X|$, $\overline{v}_i /(z_i X)$, and $\overline{v}_i /(z_i Q_0 X)$ of "mobility" are also in use, so take care in comparing information from diverse sources.

ions, they do in fact migrate more slowly. Secondly, and quite unexpectedly, the mobilities of the $H_3O^+(aq)$ and the $OH^-(aq)$ ions are well out of line with the others. Before analyzing these experimental values in greater detail, let us consider how mobilities might be predicted. If we imagine the ions as small spheres of radius R, moving through a viscous medium such as water under the action of a force f, they should behave similarly to small metallic spheres falling through the medium under the force of gravity. Those falling spheres closely obey **Stokes' law**[813] in acquiring a terminal velocity of

i	$\dfrac{10^9 \times u_i^\circ}{m^2 V^{-1} s^{-1}}$	R_i^{stokes} pm	R_i^{cryst} pm
$Li^+(aq)$	40.1	239	68
$Na^+(aq)$	51.9	183	97
$K^+(aq)$	76.2	125	133
$Rb^+(aq)$	80.6	118	147
$Cs^+(aq)$	80.0	119	167
$NH_4^+(aq)$	76.2	125	143
$H_3O^+(aq)$	362.2	26	–
$Mg^{2+}(aq)$	54.9	347	66
$Ca^{2+}(aq)$	61.6	309	99
$Sr^{2+}(aq)$	62.5	304	112
$Ba^{2+}(aq)$	65.9	289	153
$Zn^{2+}(aq)$	54.7	348	74
$F^-(aq)$	−57.4	166	133
$Cl^-(aq)$	−79.1	120	181
$Br^-(aq)$	−80.9	118	196
$I^-(aq)$	−79.6	119	220
$OH^-(aq)$	−205.2	46	–

$$8{:}14 \qquad v = \frac{f}{6\pi\eta R} \qquad \text{Stokes' law}$$

where η is the viscosity of the medium (8.937×10^{-4} kg m^{-1} s^{-1} for pure water at 298 K). Because the force f_i experienced by a migrating ion i has a magnitude of $z_i Q_0 X$, we expect a migratory velocity of

$$8{:}15 \qquad v_i = \frac{f_i}{6\pi\eta R_i} = \frac{z_i Q_0 X}{6\pi\eta R_i}$$

and this prediction may be recast into an equation that allows the radius of the migrating ion to be calculated. Thus, making use of definition 8:12,

$$8{:}16 \qquad R_i = \frac{z_i Q_0 X}{6\pi\eta v_i} = \frac{z_i Q_0}{6\pi\eta u_i^\circ}$$

Values calculated by this Stokes' law formula from mobilities, listed in the **second column** above, are given in the **third column** and compared, in the **fourth column**, with ionic radii calculated from crystallography[814]. For the largest singly charged ions, the agreement is not bad, considering the diversity of the measurement techniques. One very evident paradox is that proceeding through the sequence of alkali metal cations $Li^+ \rightarrow Na^+ \rightarrow K^+ \rightarrow Rb^+$, or

---

[813] Sir George Gabriel Stokes, 1819–1903, Irish-born mathematician and physicist. According to Stokes' law, how fast will a mercury droplet of volume 0.050 mm^3 fall through water after it has acquired a constant speed? See Web#813.

[814] from the density of salt crystals of known packing, generally assuming anion-cation contact.

the alkaline earth sequence $Mg^{2+} \to Ca^{2+} \to Sr^{2+} \to Ba^{2+}$, during which the crystallographic radii steadily increase, we see that the Stokes' radii steadily *decrease*! A similar, though less pronounced, effect is observed for the halide anion sequence $F^- \to Cl^- \to Br^- \to I^-$. This, and the doubly-versus-singly charged anomaly noted earlier, is undoubtedly due to the **hydration** of the ions (page 41), being particularly intense for small and multiply charged ions. When they migrate, small ions do not travel alone, but are accompanied by a coterie of water molecules[815]. With the hydration effect taken into account, the crystallographic radius exceeds the Stokes-based radius somewhat. This may be due to the moving ion finding it easier to penetrate the rather open structure of water than does a sphere whose size vastly exceeds that of the water molecules.

The anomalies of the hydronium and hydroxide ions are explained by the **Grotthuss mechanism**[816], illustrated in Figure 8-4. In addition to moving as intact particles through the water, these ions are able to simulate motion by a chemical change of identity. In other solvents, the hydronium and hydroxide ions migrate with typical mobilities.

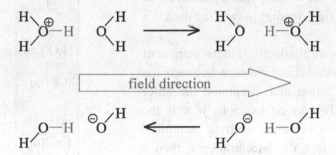

Figure 8-4 The Grotthuss mechanism. The exchange of a proton H^+ between an ion and a water molecule mimics true migration and inflates the mobility.

It was Kohlrausch[817] who observed that the increase with concentration in the conductivity of a solution of a strong[818] electrolyte was not exactly linear. Interpreted in terms of mobilities, his finding was that the ionic mobilities decline as the square-root of the electrolyte concentration. The presence of foreign ions also slows the migration of a

---

[815] Probably 3 for $H_3O^+(aq)$, 4 each for $F^-(aq)$ and $Zn^{2+}(aq)$ and 6 for $Mg^{2+}(aq)$, but different methods of assessing these hydration numbers give somewhat different values.

[816] Christian Johann Dietrich Grotthuss (he later went by the name Theodor Grotthuss), 1785–1822, Lithuanian electrochemist and photochemist. Predating ions, his explanation was based on the reversible dissociation of H_2O molecules into atoms.

[817] Friedrich Wilhelm Georg Kohlrausch, 1840–1910, German physical chemist.

[818] The **conductivity of solutions of weak electrolytes** (page 41) follows a different pattern, because the equilibrium that exists in these cases is displaced. See Web#818. Interpreting this behavior led Friedrich Wilhelm Ostwald (1853–1932, Latvian physical chemist, 1909 Nobel laureate for his work on catalysis and kinetics) and Arrhenius[232] towards our current view of ionic solutions.

particular ion i, and it turns out that the mobility decline follows the square root of the ionic strength (page 41) through a constant Υ:

8:17
$$u_i = u_i^\circ - \Upsilon\sqrt{\mu}$$

An explanation of this behavior and a theoretical basis for Υ comes from the work of Onsager[819] and his colleagues. Their mathematical treatment, which is based on the same central-ion-plus-ionic-cloud model as the Debye-Hückel theory, is elaborate[820] and will not be presented here. The mobility decline is attributed to two effects, both of which contribute to Υ, but with opposite signs. The so-called **electrophoretic effect** arises from the electrostatic drag that the moving central ion experiences from the ionic atmosphere trying to move in the opposite direction. The **relaxation effect**, associated with the names of Debye[241] and Falkenhagen[821], is caused by the moving central ion no longer being centrally located within its ionic cloud; this asymmetry leads to an additional electric field, tending to *increase* the ion's velocity.

A straightforward conductivity measurement of a salt solution reveals the difference between the mobilities of the cation and the anion, the latter being negative. To take an example, if κ is the conductivity of a calcium nitrate solution, $Ca^{2+}(aq) + 2NO_3^-(aq)$, of concentration c, then it follows from equation 1:30 that

8:18
$$u_{Ca^{2+}} - u_{NO_3^-} = \frac{\kappa}{2Fc}$$

Finding individual mobilities is less easy, but the method described below can accomplish this. Once one single mobility has been measured, all others become accessible through conductivity measurements and equations such as 8:18.

Figure 8-5 illustrates the simple apparatus used in the **moving boundary method** of measuring mobilities. In the diagrammed example, a solution

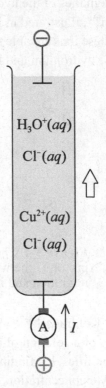

$H_3O^+(aq)$

$Cl^-(aq)$

$Cu^{2+}(aq)$

$Cl^-(aq)$

A I

Figure 8-5 The motion of the junction between two solutions of known conductivity is measured in the moving boundary method.

[819] Lars Onsager, 1903–1976, Norwegian-American physical chemist; Nobel laureate 1968 for his discovery of reciprocity relationships in thermodynamics.

[820] For a binary electrolyte comprised of cation i and anion j, it predicts that $\pi N_A \eta \sqrt{18RT\varepsilon}\, \Upsilon / z_i F^2 = 1 + z_j F \eta u_i^\circ h^2 / [2RT\varepsilon(1+h)]$ where $h^2 = z_i z_j (u_i^\circ - u_j^\circ)/[(z_i - z_j)(z_i u_i^\circ + z_j u_j^\circ)]$.

[821] Hans Falkenhagen, 1895–1971, German physical chemist, member of the productive European scientific team that included Debye, Hückel, and Onsager.

of hydrochloric acid overlies a solution of copper(II) chloride in a long tube of known cross-sectional area A. The current I flowing between the two electrodes ($Ag\,|\,AgCl$ could be used) is measured, the applied d.c. voltage being large enough that the motion of the boundary can be conveniently observed. The cations $Cu^{2+}(aq)$ and $H_3O^+(aq)$ slowly migrate upwards. The $Cl^-(aq)$ ions, which alone carry current across the boundary, migrate downwards. Because the hydronium ion has a greater mobility than the copper(II) ion, and because the denser solution occupies the lower segment of the tube, the boundary remains sharp. No significant concentration changes occur in the two solutions, so the upward velocity of the boundary must match the average velocities of the copper and hydronium ions:

8:19
$$v_{boundary}\begin{cases} = \overline{v}_{Cu^{2+}} = X^L u_{Cu^{2+}} = (i/\kappa^L)u_{Cu^{2+}} = I u_{Cu^{2+}}/(A\kappa^L) \\ = \overline{v}_{H_3O^+} = X^U u_{H_3O^+} = (i/\kappa^U)u_{H_3O^+} = I u_{H_3O^+}/(A\kappa^U) \end{cases}$$

where the ^{L} and ^{U} superscripts relate to the lower and upper segments of the tube. Though the analysis of the moving boundary method is usually presented in terms of transport numbers[153], the equations above show that a determination of the velocity of the boundary, together with other easily measured quantities, can lead directly to the mobilities of both cations.

The moving boundary experiment is an example of **electrophoresis**[822], a family of methods in which charged particles migrate under an electric field. The particles may be simple ions or much larger charged particles, frequently ones of biochemical or biomedical interest. In a typical electrophoresis experiment, a solution containing several charged particles is injected near one end of a tube, along which a strong electric field exists.

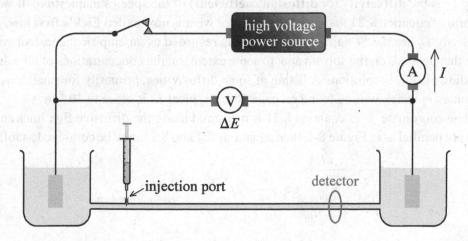

Figure 8-6 Typical apparatus for electrophoresis.

---

[822] For his pioneering work in electrophoresis the Swedish chemist Arne Wilhelm Kaurin Tiselius, 1901–1971, earned a Nobel prize in 1948.

Because they have different mobilities, the particles travel at differing speeds along the tube and arrive at different times at the detector, which is positioned near the other end. When the polarity is as shown in Figure 8-6, the apparatus is designed to investigate positively charged particles. Many detection methods exist, and most allow a quantitative assessment of the amount of each species passing the detector, so permitting both the identification of the particles present in the sample (from its residence time between injection and detection) and measurement of its concentration (from the detector output for a known injection volume). The carrier electrolyte solution is usually a buffer and added resolution is often possible by judicious selection of the pH. A popular version is **capillary electrophoresis**, in which the tube is narrow and through which solution travels by **electroosmosis** (Chapter 14), but other versions employ gels or porous paper as the electrophoretic medium.

Diffusion: Fick's two important laws

Diffusion occurs in response to a gradient of activity and the diffusive flux is proportional to that gradient.

8:20
$$j_i^{dif} \propto \frac{\partial a_i}{\partial \ell}$$

However, as in chemical kinetics (page 51) and for a similar reason, diffusion is almost always treated as a response to a gradient of *concentration* and is considered to obey

8:21
$$j_i^{dif} = -D_i \frac{\partial c_i}{\partial \ell} \qquad \text{Fick's first law}$$

where D_i is the **diffusivity** (or **diffusion coefficient**) of the species in question. It was in the form of equation 8:21 that Fick[823] enunciated what is now called **Fick's first law**. The diffusivity D_i has the *SI* unit[824] of $m^2\ s^{-1}$ and is regarded as an empirical constant with a value that depends on the solvent and, to some extent, on the concentrations of all solutes, including i itself, in solution. A listing of some diffusivities, primarily for small ions and molecules in water, will be found on page 388; a typical D_i is close to $10^{-9}\ m^2\ s^{-1}$.

The coordinate ℓ in equation 8:21 is measured along the diffusive flux lines and, if these are parallel as in Figure 8-1, then equations 8:7 and 8:21 may be combined as follows

8:22
$$\left.\begin{aligned} \frac{\partial c}{\partial t} &= -\frac{\partial j}{\partial x} \\ j &= -D\frac{\partial c}{\partial x} \end{aligned}\right\} \Rightarrow \frac{\partial c}{\partial t} = D\frac{\partial^2 c}{\partial x^2} \qquad \begin{aligned} &\text{Fick's second law} \\ &\text{planar transport} \end{aligned}$$

---

[823] Adolf Eugen Fick, 1829–1901, German physiologist who studied diffusion through membranes.

[824] Note that this is the same *SI* unit as for permeability (page 80). In a sense, diffusivity is the permeability of an unconfined solvent.

The result is **Fick's second law for planar transport**. The corresponding law for spherical transport can be written in two equivalent ways:

8:23
$$\frac{\partial c}{\partial t} = D\frac{\partial^2 c}{\partial r^2} + \frac{2D}{r}\frac{\partial c}{\partial r} = \frac{D}{r^2}\frac{\partial}{\partial r}\left\{r^2\frac{\partial c}{\partial r}\right\}$$
Fick's second law
spherical transport

It arises from combining conservation law 8:9 with Fick's first law[825].

As we shall find in later chapters, electrochemists often need to solve Fick's second law. The solution describes $c_i(\ell,t)$ explicitly; that is, it specifies what the concentration of the diffusing species is at all points of interest along the ℓ spatial coordinate[826] at all positive times t. As with any differential equation, Fick's second law cannot be solved as it stands. Each problem requires three[827] pertinent **boundary conditions** for its solution.

As an illustration of the solution to Fick's second law, we now consider its role in what could be considered as one of the prototypical experiments in electrochemistry. It is illustrated in Figure 8-7. Initially the solution contains an uncharged electrooxidizable species R at a uniform concentration c_R^b. Thus, with $c_R(x,t)$ denoting the concentration of R at any point x in solution at time t, then

8:24
$$c_R(x>0,0)=c_R^b$$

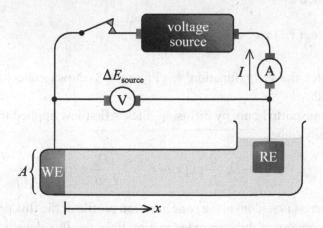

Figure 8-7 In the potential-leap experiment, the WE is suddenly brought to a potential large enough to denude the electrode surface of the electroreactant.

---

[825] Derive **Fick's second law for cylindrical diffusion** as it would apply to a cell in which the electrode is a long rod surrounded by the ionic conductor. See Web#825.

[826] ℓ would be x in cases of planar transport, r for spherical transport, or some other appropriate coordinate in other cases. Sometimes more than one spatial coordinate is required.

[827] Why are *three* boundary conditions needed? Fick's second law involves two differentiations in space and one in time. So, to remove these derivatives, we must integrate three times: twice in space, once in time. Each requires an "integration constant", which is what a boundary condition provides. Hence three boundary conditions are needed: two in space and one in time.

The solution will likely contain excess supporting electrolyte (page 129) too, but this is immaterial to the present discussion. At time $t = 0$, the switch is closed and the working electrode is suddenly brought to, and maintained at, a potential sufficiently positive to remove R totally from the electrode surface; that is, for this **potential-leap experiment**

8:25
$$c_R(0, t > 0) = 0$$

Because R is uncharged and the solution is quiescent, diffusion is the only transport mode operative for R, so Fick's laws govern transport. Clearly the flux lines will be parallel for the cell geometry as illustrated, and therefore the form adopted by Fick's second law is 8:22; that is

8:26
$$\frac{\partial}{\partial t} c_R(x, t) = D_R \frac{\partial^2}{\partial x^2} c_R(x, t)$$

Of the three boundary conditions needed to solve this equation, two are present in 8:24 and 8:25. The third is provided by the observation that, far from the electrode, the concentration will remain unchanged, which may be stated mathematically as

8:27
$$c_R(x \to \infty, t) = c_R^b$$

A derivation[828] using Laplace transformation demonstrates that the four equations[829] 8:24–8:27 are satisfied by

8:28
$$c_R(x, t) = c_R^b \, \text{erf} \left\{ \frac{x}{\sqrt{4 D_R t}} \right\} \qquad \begin{array}{l} \text{potential leap} \\ \text{concentration profile} \end{array}$$

where erf{ } denotes the error function[828]. Figure 8-8 shows concentration profiles illustrating this result.

Because R is transported only by diffusion, Fick's first law applied to equation 8:28 shows the flux density to be

8:29
$$j_R(x, t) = -D_R \frac{\partial}{\partial x} c_R(x, t) = -c_R^b \sqrt{\frac{D_R}{\pi t}} \exp \left\{ \frac{-x^2}{4 D_R t} \right\}$$

Electrochemical interest is seldom in the concentration profile or the flux profile, but in the current. The stoichiometry of the electrode reaction links the flux density of the reactant to the current. If the overall electrode reaction is

8:30
$$v_R R(soln) \to ne^- + v_O O(soln)$$

where each v is a stoichiometric coefficient (often unity and then omitted from the equation), then the flux densities of R and O at the electrode surface are related to the rate of creation of electrons, that is, to the current density by

---

[828] Not only equation 8:28, but also the equations 8:29 and 8:32, are derived in Web#828. This web document opens with a definition of the error function and reports some of its properties.

[829] Confirm that each of these four equations is, indeed, satisfied by formula 8:28. See Web#829.

[830] Beyond about 100 s, the potential-leap experiment is in danger of being affected by natural convection.

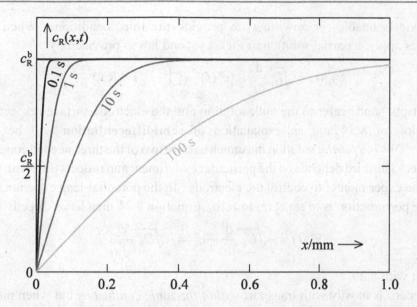

Figure 8-8 Concentration profiles resulting from the potential-leap experiment for a diffusivity $D_R = 1.00 \times 10^{-9}$ m^2 s^{-1}. Note that, even after times as long[830] as 100 s, the concentration diminution is confined to a layer of only about one millimeter thickness, easily validating condition 8:27.

8:31
$$\frac{j_O(0,t)}{\nu_R} = \frac{-j_R(0,t)}{\nu_O} = \frac{-j_e(0,t)}{n} = \frac{i(t)}{nF} = \frac{I(t)}{nAF}$$

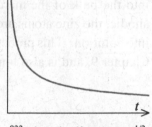

Combining this equation with the $x = 0$ version of 8:29 yields

8:32
$$I(t) = \frac{n}{\nu_R} A F c_R^b \sqrt{\frac{D_R}{\pi t}} \qquad \text{Cottrell equation}$$

This is the **Cottrell equation**[831]: it describes the current resulting from a potential-leap experiment when planar diffusive transport applies. It describes a current, initially infinite[832], that declines as $t^{-1/2}$.

It frequently happens in electrochemistry that one seeks to solve Fick's second law for planar transport, equation 8:22, under circumstance in which two of the three boundary conditions are

8:33
$$\left. \begin{array}{r} c_i(x,0) \\ c_i(x \to \infty, t) \end{array} \right\} = c_i^b$$

---

[831] Frederick Gardner Cottrell, 1877–1948, American electrochemist, inventor of the electrostatic precipitator for removing polluting particles from gas streams, and founder of the philanthropic *Research Corporation*.

[832] in principle. In practice there will always be a modicum of ohmic and kinetic polarization to temper the initial absence of transport polarization. See Chapter 10.

but where one is unable (or unwilling) to provide the third condition. When these circumstances apply, a partial solution to Fick's second law is provided by

8:34
$$j_i^s(t) = \sqrt{D_i}\,\frac{d^{1/2}}{dt^{1/2}}\left\{c_i^s(t) - c_i^b\right\} \qquad i = R, O$$

The superscripts [b] and [s] refer to the bulk solution and the electrode surface respectively. The derivation of 8:34 and an explanation of **semidifferentiation** will be found elsewhere[833]. This is a *partial* solution inasmuch as only two of the three needed conditions have been met. The third depends on the particular experiment and reflects the perturbation applied by the experimenter to control the electrode. In the potential-leap experiment, for example, the perturbation is to set $c_R^s(t)$ to zero. Equation 8:34 then leads directly to

8:35
$$j_R^s(t) = -\sqrt{D_R}\,\frac{d^{1/2}}{dt^{1/2}}c_R^b = -\sqrt{D_R}\,\frac{c_R^b}{\sqrt{\pi t}}$$

in agreement with 8:29.

Our concern is mostly with transport *within the ionic conductor*, but when mercury electrodes are used, electrochemists must also treat the diffusion of metals within the working electrode. The process of dissolving in mercury, which many metals do, is known as **amalgamation**. An example occurs when zinc ions reduce at a mercury electrode to form zinc amalgam:

8:36
$$Zn^{2+}(aq) + 2e^-(Hg) \rightleftarrows Zn(amal)$$

The amalgamated zinc is formed at the surface of the mercury electrode and then diffuses[834] into the bulk of the mercury electrode. If the mercury electrode is subsequently made anodic, the zinc atoms are transported in the opposite direction and may be oxidized back into solution. This process lies at the heart of stripping analysis, a procedure discussed in Chapter 9, and is also important in polarography[1305].

Diffusion and Migration: they may cooperate or oppose

The analysis, given in the previous section, of the potential-leap experiment is correct as far as it goes, but it is incomplete. The oxidation of the electroneutral R produces a cationic product O subject to both diffusion and migration. Moreover, the current has to flow through the solution as a whole and, unless ions are present initially, the solution is

---

[833] For more information on the partial solution and on the fractional calculus, see Web#833 and Web#1242.

[834] Combine Stokes' law with the Nernst-Einstein law to produce a relationship between the diffusivity of a solute and its radius. This is the **Stokes-Einstein law**. Use it, as in Web#834, to estimate the diffusivity D_{Zn}^{amal} of zinc atoms in an amalgam. The radii of Zn and Hg atoms are respectively 153 pm and 176 pm; the viscosity of mercury is 1.526×10^{-3} kg m^{-1} s^{-1}. Comment on the discrepancy between your estimate and the measured diffusivity value of 1.89×10^{-9} m^2 s^{-1}.

not an ionic conductor and the passage of current would be impossible. Being neutral, R cannot carry any current. To provide conductivity, ions *must* be present initially, though not necessarily in excess. Although the cation O is transported by both migration and diffusion, the latter transport mode will dominate if supporting ions are at a high concentration compared to O because their presence diminishes the electric field. Nevertheless, even then, migration of O will be important close to the electrode because *this ion carries 100% of the current at the WE*, however concentrated the supporting ions might be. More detailed analysis of transport close to an electrode will be presented later.

Migration of electroactive species occurs from the onset of any experiment when supporting ions are not in overwhelming excess. But worse, the two or more ionic concentrations then proceed to vary in both time and space, affecting the local field, which in turn affects the migratory fluxes. Prediction of the outcome of an experiment becomes a nightmare. You can see why it is that electrochemists prefer, whenever possible, to work with excess supporting electrolyte. Then, the effects of migration can be lessened to an extent that they may legitimately be ignored.

One would expect that factors – primarily the viscosity of the solution – that hinder migration would also hinder diffusion, and this is true. Mention was made earlier of the viewpoint that migration and diffusion are just different aspects of a more generalized transport, both arising from a gradient of a thermodynamic quantity known as the **electrochemical potential**. A précis of how this approach leads to the important **Nernst-Einstein equation**[622,835]

8:37 $$z_i FD_i = RTu_i$$ Nernst-Einstein equation

will be found elsewhere[836], but here we shall derive this result by recourse to an experimental argument.

A constant current I is passed through the cell illustrated overleaf in Figure 8-9. Originally the two chambers contained the same solution of a metal salt. The cations $M^{z_M}(soln)$ move from left to right through the tube, initially by migration alone but, as the salt concentration on the left builds up as a result of the dissolution of the left-hand electrode, also by diffusion. We focus attention, though, on the anions $A^{z_A}(soln)$; their leftwards migration is opposed by a rightwards diffusion. Eventually[837], the concentrations will cease to change. In that eventual condition the anions experience a **diffusive** flux density and a **migratory** flux density that must sum to zero; that is:

8:38 $$0 = j_A^{dif} + j_A^{mig} = -D_A \frac{dc_A}{dx} + -u_A c_A \frac{d\phi}{dx}$$

This equation may rearranged and integrated along the length of the tube

---

[835] Albert Einstein, 1879–1955, German-American physicist and the quintessential scientist in lay eyes; Nobel laureate, 1921, for his work on the quantum mechanical photoelectric effect.

[836] See Web#836 for a terse derivation of the equation 8:37 by a thermodynamic argument.

[837] after a *very* long time. This is a "thought experiment" only.

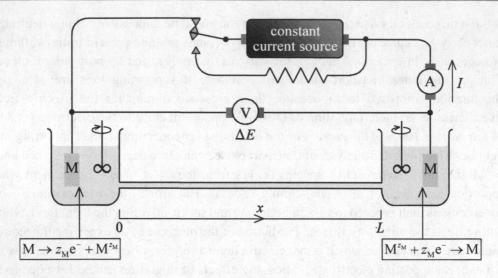

Figure 8-9 A narrow tube interconnects two electrode chambers that are gently stirred to ensure uniform composition in each chamber.

8:39
$$0 = D_A \int_{c_A^L}^{c_A^R} \frac{dc_A}{c_A} + u_A \int_{\phi^L}^{\phi^R} d\phi = D_A \ln\left\{\frac{c_A^R}{c_A^L}\right\} + u_A \left(\phi^R - \phi^L\right)$$

where the ^{R} and ^{L} superscripts refer to conditions in the right- and left-hand chambers. The fact that diffusion is more accurately described as activity driven than concentration driven permits the replacement of the concentration ratio in 8:39 by a_A^R / a_A^L. Exponentiation and rearrangement then lead to

8:40
$$a_A^R \exp\left\{\frac{u_A}{D_A}\phi^R\right\} = a_A^L \exp\left\{\frac{u_A}{D_A}\phi^L\right\}$$

in which the diffusive and migratory components are intermingled.

Here is a good place at which to draw a phenomenological distinction between the terms "steady state" and "equilibrium state". A **steady state** implies that no changes are occurring with time in any property (concentration, flux density, etc) of interest. An **equilibrium state** is a stronger condition: not only is nothing changing with time, but there are no net motions either. In the eventual state arrived at by the Figure 8-9 experiment, the metal cations M^{z_M} are in a steady state, but the anions A^{z_A} are in an equilibrium state. Being at equilibrium, the anions obey equation 2:26, the Boltzmann relation. For both that equation and equation 8:40 to be satisfied simultaneously requires that

8:41
$$\frac{u_A}{D_A} = \frac{z_A F}{RT}$$

an instance of the Nernst-Einstein[838] law 8:37. The steady-state behavior of the cations in the experiment of Figure 8-9 can also be analyzed; this analysis is carried out elsewhere[839].

With the proportionality between the diffusivity and the mobility of a solute established, it is possible to write a combined transport equation as

8:42
$$j_i = -D_i \left[\frac{\partial c_i}{\partial x} + \frac{F}{RT} z_i c_i \frac{\partial \phi}{\partial x} \right]$$
Nernst-Planck equation

which, for planar transport, is the **Nernst-Planck equation**[622,840]. This is the equation that replaces Fick's first law when diffusive and migratory modes of transport come into play. The replacement for Fick's second law,

8:43
$$\frac{\partial c_i}{\partial t} = D_i \left[\frac{\partial^2 c_i}{\partial x'^2} + \frac{F}{RT} z_i \frac{\partial c_i}{\partial x} \frac{\partial \phi}{\partial x} + \frac{F}{RT} z_i c_i \frac{\partial^2 \phi}{\partial x^2} \right]$$

is so elaborate that time-dependent problems involving both diffusion and migration have been solved analytically only in very special circumstances.

Convection: transport controlled by hydrodynamics

Equation 8:13 shows that flux density for migratory transport is proportional to the potential gradient; equation 8:21 shows that flux density for diffusive transport is proportional to the concentration gradient. Thus we might expect that the flux density for convective transport would be proportional to the pressure gradient[841] and this is indeed true, at least with some provisos. The simplest vessel for containing a moving liquid is a tube with a uniform circular cross section of radius R. Solutes entrained in the flowing liquid within such a tube share their velocity with the liquid and have an average flux density given, in terms of the **viscosity** η, by

8:44
$$\overline{j_i}^{conv} = \overline{v} c_i = -\frac{R^2 c_i}{8\eta} \frac{dp}{dx}$$

which derives from **Poiseuille's law**[842]. It is necessary to specify the *average* flux density because, as illustrated in Figure 8-10 overleaf, the solution velocity v and hence j_i^{conv} varies

---

[838] Use the Nernst-Einstein law to calculate the diffusivity of the $Zn^{2+}(aq)$ ion at 25°C from the data on page 150. Check with Web#838. Compare your result with the value 0.638×10^{-9} m^2s^{-1} measured voltammetrically in a 100 mM KNO_3 solution and suggest a reason for any discrepancy.

[839] See Web#839, where the expressions for the current through, and the potential across, the Figure 8-9 cell are derived.

[840] Max Karl Ernst Ludwig Planck, 1858–1947, German physicist, Nobel laureate 1918.

[841] Convection is *usually* caused by a pressure gradient, but not in electroosmotic flow (page 299)

[842] or the Hagen-Poiseuille law; Gotthilf Heinrich Ludwig Hagen, 1797–1884, German hydraulic engineer; Jean Louise Marie Poiseuille, 1799–1869, French physician. The law's derivation is included in Web#842.

Figure 8-10 Poiseuille flow through a tube. The velocity profile, given by $v(r) = 2\dot{V}[R^2 - r^2]/\pi R^4$, where $\dot{V}$ is the flow rate (m^3 s^{-1}), is shown in cross section.

quadratically with the radius, being twice the average in the center of the tube and falling to zero at the wall. This *lack of motion at the wall* is an important and general feature of hydrodynamic flow. The type of motion described by Poiseuille's law is called **laminar flow**. At high flow rates it is replaced by a chaotic hydrodynamic regime known as **turbulent flow**; such a regime is seldom encountered in investigative electrochemistry but is favored in electrosynthesis.

Convection within a tube is electrochemically important in capillary electrophoresis, in electrokinetic phenomena (pages 298–302), and in flowing coulometry, as illustrated on page 126. Another form of flowing coulometry has been practiced in which the tube itself is the working electrode. In that case, the lack of convective motion at the tube wall means that slow diffusion is invoked to capture solute electrochemically, so that long metal tubes of very narrow bore are needed for thorough removal[843]. Shorter tube electrodes, however, can be used to *sample* the flowing solution, as with the tubular band electrode illustrated on page 118.

The most important electrochemical device that employs forced convection is the **rotating disk electrode**, consisting of a disk (typical area 10 mm^2) of an electronic conductor (typically Pt or glassy carbon) embedded in the center of, and coplanar with, a larger disk of insulator (often teflon®). The composite disk forms the end of a rod, which is mounted vertically and rotated at high speed (the **angular velocity** ω is typically about 100 rad s^{-1}) while immersed in an electrolyte solution. Solution is thrown sideways by the spinning disk and replaced from below. The flow lines followed by the moving solution are illustrated in Figure 8-11; the coordinate system and velocity components useful for describing this flow are shown in Figure 8-12. In the presence of excess supporting electrolyte, the flux density in the x-direction is provided by convection and diffusion

8:45 $$j_i = j_i^{conv} + j_i^{diff} = (v_x c_i) + \left(-D_i \frac{\partial}{\partial x} c_i\right)$$

though, because the convective velocity is zero at the electrode surface, *only* diffusion is operative there. This equation, combined with the conservation requirement 8:7, leads to

[843] See Web#843. To ensure 99.9% capture, the tube must be at least $0.584\, \dot{V}/D$ long.

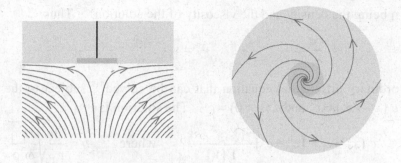

Figure 8-11 Geometry of the rotating disk electrode, showing also the flow lines followed by the convecting solution.

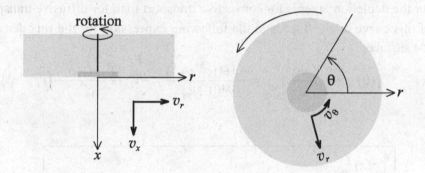

Figure 8-12 Coordinates useful in describing the behavior of the solution adjacent to a rotating disk electrode.

8:46
$$\frac{\partial c}{\partial t} = -v_x \frac{\partial c_i}{\partial x} + D_i \frac{\partial^2 c_i}{\partial x^2}$$

This is the equation that replaces Fick's second law when **diffusion** is augmented by the **convection** provided by the rotating electrode. Although modulation of the applied cell voltage, the rotation speed, and even the temperature has occasionally been applied to the rotating disk electrode, for the most part it is operated under steady conditions. In this circumstance, the left-hand member of 8:46 is zero. Hydrodynamic theory[844] can predict the motion of the moving liquid as its velocity components in the x, r, and θ directions. Close to the electrode surface, the prediction is that v_x is independent of r and θ, being approximately equal to $-v_K x^2 \sqrt{\omega^3 \rho / \eta}$. Here v_K is the von Kármán[845] number 0.51023,

---

[844] A synopsis is presented in Web#844.

[845] Hungarian Szöllöskislaki Kármán Tódor, 1881–1963, who took the name Theodore von Kármán on emigrating to the United States, where he had a distinguished career as an aeronautical engineer.

with ρ and η being the density and the viscosity of the solution[846]. Thus,

8:47
$$0 = \frac{\partial c}{\partial t} = -v_x \frac{\partial c_i}{\partial x} + D_i \frac{\partial^2 c_i}{\partial x^2} = v_K \sqrt{\frac{\omega^3 \rho}{\eta}} x^2 \frac{dc_i}{dx} + D_i \frac{d^2 c_i}{dx^2}$$

This is an ordinary differential equation that can be solved[847] subject to the boundary conditions $c_i(x \to \infty) = c_i^b$ and $c_i(x = 0) = c_i^s$. The solution is

8:48
$$c_i(x) = c_i^s + \left(c_i^b - c_i^s\right) \frac{\gamma\{\tfrac{1}{3}, x^3/b\}}{\Gamma\{\tfrac{1}{3}\}} \qquad \text{where} \qquad b = \frac{3}{v_K} \sqrt{\frac{D_i^2 \eta}{\omega^3 \rho}}$$

where $\gamma\{\tfrac{1}{3}, \}$ and $\Gamma\{\tfrac{1}{3}\}$ are respectively an incomplete and the complete gamma function of order one-third[848]. Figure 8-13 shows a graph of the concentration profile predicted by this equation. Notice from the scales of the abscissas in Figures 8-8 and 8-13 how much narrower the depletion layer is for convective transport than for diffusive transport. The slope of this curve at $x = 0$ leads to the following expression for the flux density at the electrode surface:

8:49
$$j_i(0) = -D_i \left(\frac{dc_i}{dx}\right)_{x=0} = \frac{-3D_i(c_i^b - c_i^s)}{b^{1/3}\Gamma\{\tfrac{1}{3}\}} = -v_L \frac{D_i^{2/3}\omega^{1/2}\rho^{1/6}}{\eta^{1/6}}\left(c_i^b - c_i^s\right)$$

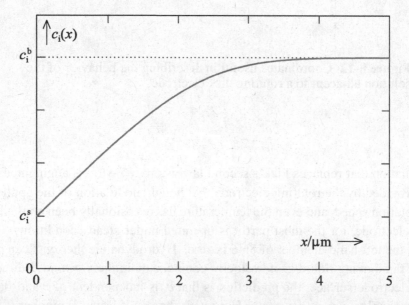

Figure 8-13 The concentration profile at a rotating disk electrode according to equation 8:48. The graph is correctly scaled for the typical value $b = (18\ \mu m)^3$.

---

[846] The ratio η/ρ is known as the **kinematic viscosity**; it equals $9.13 \times 10^{-7}\ m^2\ s^{-1}$ for water at T°.

[847] See Web#847 for details of the solution and a pedestrian derivation of equation 8:49, the Levich equation.

[848] See page 378 for definitions of these functions and Web#848 for further information.

where v_L is the Levich number, 0.62046. It was Levich[849] who recognized the applicability of the rotating disk to electrochemistry and consequently the equation describing the flux at its surface, or the current (see equation 12:31), is the **Levich equation**.

Fluxes at Electrodes and in the Bulk: transport coefficients

Let us resume our discussion of the generalized n-electron reaction 8:30, in which an R species is oxidized to an O product:

8:50 $$v_R R^{z_R}(soln) \rightarrow n e^- + v_O O^{z_O}(soln)$$

No restriction is placed on the charges of the two species, other than the stoichiometric constraint

8:51 $$v_O z_O - v_R z_R = n$$

We also consider there to be other ions, and perhaps uncharged solutes, also present in the quiescent solution, though not necessarily in excess. Concern in this section will be with the relationships that link the local flux densities and concentrations of the solute species to each other and to the current density i. We shall not delve deeply into any time dependences that the variables might have, focusing instead on instantaneous values of the j's and the c's. No particular cell geometry is assumed, but an ℓ coordinate, that measures distance outwards along flux lines from the working electrode surface, is used to express spatial dependence.

First we present the following four relationships that hold globally:

8:52 $$\sum_i z_i c_i(\ell) = 0$$

8:53 $$\sum_i z_i j_i(\ell) = \frac{i(\ell)}{F} = \frac{I}{FA(\ell)}$$

8:54 $$\sum_i \frac{z_i j_i(\ell)}{D_i} = \frac{2F}{RT} X(\ell)\mu(\ell)$$

8:55 $$\sum_i \frac{j_i(\ell)}{D_i} = -\frac{d}{d\ell} \sum_i c_i(\ell)$$

These are written using a sigma notation that implies summation over all the solutes, ionic or molecular, though uncharged species play no role in the first three summations. The first relationship is an expression of electroneutrality; the validity of this may be questioned in the immediate vicinity of the electrode interface, but it is certainly valid elsewhere. The conservation of charge is expressed by the second relationship; this also serves as a reminder that, although the same current passes through all regions of the cell, the current

---

[849] Veniamin Grigorievich Levich, 1917 – 1987, Russian physical chemist.

density is spatially variable, except when the "trough" geometry applies. The third equation of the quartet involves the local **ionic strength** (page 41). Equations 8:54 and 8:55 both derive from the Nernst-Planck relationship[850]. Making an **equidiffusivity approximation** – this means assuming that all solutes share the same diffusivity D – equation 8:54 may be combined with 8:53 into the useful expression $X(\ell) \approx [RTi(\ell)]/[2F^2D\mu(\ell)]$ for the local field. If one again invokes the equidiffusivity approximation, equation 8:55 demonstrates that the total solute content collectively satisfies Fick's first law of diffusion, notwithstanding the occurrence of migration. The elaborate equations in this paragraph apply to the electrolyte solution throughout the cell. Fortunately, simpler equations hold in the bulk of the solution, and at the working electrode itself.

In the bulk of the solution (a region denoted by a superscript ^{b}), concentrations have not had time to change from their original values. Each solute has its original uniform bulk concentration c_i^b. The conductivity κ^b is uniform and the field is given by Ohm's law

8:56
$$X^b(\ell) = i^b(\ell)\kappa^b = \frac{I\kappa^b}{A(\ell)}$$

There are no concentration gradients, so no diffusion occurs and each ion migrates according to

8:57
$$j_i^b(\ell) = \frac{FX^b(\ell)}{RT}D_i z_i c_i^b$$

There are no fluxes of neutral species but all the ions share in the carrying of current in proportion to their individual contributions to the $\sum D_i z_i^2 c_i^b$ sum.

At the electrode surface, denoted by ^{s}, the flux densities of electropassive solutes – those that are *not* involved in the electrode reaction – are necessarily zero, whereas the flux densities of those solutes that do participate in reaction 8:50 are related through the stoichiometry to the flux density of electrons and thus to the current density. One finds:

8:58
$$j_i^s = \begin{cases} 0 & i \neq O, R \\ \nu_O i^s / nF & i = O \\ -\nu_R i^s / nF & i = R \end{cases}$$

and this assignment is consistent with equation 8:53. The equation

8:59
$$i^s = \frac{2F^2 D\mu^s}{RT}X^s$$

derives[851] from the global relationships 8:53 and 8:54 by making the equidiffusivity assumption $D_R = D_O = D$, this time restricted to the electroactive species. We are now in

---

[850] Carry out the derivations or see Web#850

[851] Derive 8:59 from 8:53 and 8:54 as in Web#851.

a position to quantify the effectiveness of supporting electrolyte in nullifying the migration of electroactive species. The **migratory component** of the flux density of reactant R or product O is found from the corresponding term in the Nernst-Planck equation

$$8:60 \qquad j_i^s = (j_i^{\text{dif}})^s + (j_i^{\text{mig}})^s = -D_i \left(\frac{d}{d\ell} c_i \right)^s + \frac{FD_i}{RT} z_i c_i^s X^s \qquad i = \text{R,O}$$

and therefore the migratory current carried jointly by R and O at the electrode surface is

$$8:61 \qquad \left(i^{\text{mig}} \right)^s = z_R F (j_R^{\text{mig}})^s + z_O F (j_O^{\text{mig}})^s = \frac{F^2}{RT} (D_R z_R^2 c_R^s + D_O z_O^2 c_O^s) X^s$$

If we now divide this result by equation 8:59, once more invoking restricted equidiffusivity, we arrive at the conclusion that the fraction of the total current passed at the electrode by the migration of O and R is $(z_R^2 c_R^s + z_O^2 c_O^s)/2\mu^s$, or simply the *fractional contribution of the electroactive ions to the ionic strength*. Though the rigor of the conclusion is weakened by our having had to assume equidiffusivity, it can be concluded, for example, that if the electroactive species contribute 1% of the ionic strength, then 99% of the current at the electrode will be transported by diffusion.

It has been demonstrated that, when supporting electrolyte is present, transport *at the electrode is primarily by the diffusion of the electroactive species, whereas transport in the bulk is primarily by migration of the supporting ions*. The how and why of this changeover between the two regimes can be puzzling, and is often ignored in the literature because the algebra required to model the events is complicated. Here a rather simple, but still typical, example[852] has been selected to illustrate the transition. The chosen reaction is one in which a neutral electroreactant R is oxidized to a singly charged cationic product O, which is initially absent. Supporting electrolyte, not necessarily in excess, supplies the cation C and the anion A, both singly charged. An example would be the oxidation of ferrocene, reaction 7:12, in a solution with an electrolyte yielding a singly charged cation and anion. The results of the analysis are presented in Figures 8-14 and 8-15 overleaf, which show the concentration and flux density profiles at a particular instant in time. Observe that, despite the pronounced concentration gradients of C and A at the electrode, there is no flux there because the contributions of diffusion and migration are equal and opposite.

The bulk solution and the electrode surface are the two termini for transport and attention in electrochemistry often focuses on how conditions at one terminus relate to those at the other. For example, equation 8:49 for the rotating disk electrode may be rearranged to

$$8:62 \qquad \frac{j_i^s}{c_i^s - c_i^b} = \frac{v_L D_i^{3/2} \omega^{1/2} \rho^{1/6}}{\eta^{1/6}} = constant$$

---

[852] Web#852 contains details of the experiment, its modeling (which employs a combination of analytical and simulative techniques), and the derivation of the graphs in Figures 8-14 and 8-15.

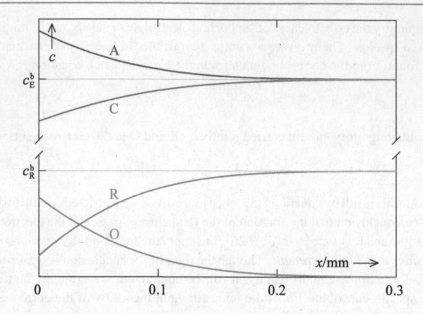

Figure 8-14 Concentration profiles during the experiment analyzed in Web#852. The neutral species **R** and the cation **O** are involved in the electrode reaction R(*soln*) $\rightleftarrows$ e⁻ + O(*soln*), while **C** and **A** are the cation and anion of the supporting electrolyte.

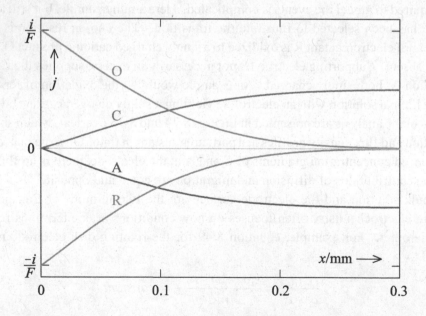

Figure 8-15 Flux density profiles corresponding to the concentrations shown in Figure 8-14. Notice the transition from the transport being largely that of the electroactive species close to the electrode to being predominantly that of the supporting ions in the bulk. See Web#852 for details.

The *constant* depends on the diffusivity of i, but its other components are independent of the electrochemistry involved. This proportionality of the flux density at the electrode surface to the surface-minus-bulk concentration difference turns out to exist for very many electrochemical techniques. The proportionality constant is symbolized m_i

8:63 $$\frac{j_i^s}{c_i^s - c_i^b} = m_i \qquad \text{definition of transport coefficient}$$

and is called a **transport coefficient**[853]. The generality of relation 8:63 will prove useful when, in Chapter 10, we consider the contribution of transport to electrode polarization. Though a transport coefficient may be assigned for most techniques, it is not a universal concept; in particular the linear relation between j^s and c^s is destroyed if migration participates in the transport of an electroactive species. The table below lists some instances of transport coefficients. The transport coefficient has the *SI* unit of m s^{-1}; it is

Experimental technique	Mode	Transport coefficient m_i	See
rotating disk electrode	conv + dif	$v_L D_i^{3/2} \omega^{1/2} \rho^{1/6} / \eta^{1/6}$	page 248
hemispherical microelectrode	diffusion	D_i / r_{hemi}	footnote 854
disk microelectrode	diffusion	$4D_i / (\pi r_{disk})$	footnote 855
thin-layer cell at steady state	diffusion	$2D_i / L$	footnote 856
Nernst transport layer	conv + dif	D_i / δ	footnote 857
macroelectrode	diffusion	$\sqrt{D_i / (\pi t)}$	footnote 855
growing spherical electrode	conv + dif	$\sqrt{7D_i / (3\pi t)}$	footnote 1305

[853] Or, less appropriately, **transport constant** or **mass-transfer coefficient** or **mass-transport coefficient**. Avoid confusion with the similarly named transport number[153] and transfer coefficient (page 133).

[854] Pages 234–245 address hemispherical microelectrodes and the voltammetry they support. See Web#854 for a discussion of the transport coefficient at such electrodes.

[855] Pages 245–248 address disk microelectrodes and the voltammetry they support. Macroelectrodes are planar electrodes large enough that their edges are of no consequence. See Web#855 for a discussion of the transport coefficient applicable to each of these electrodes.

[856] The **cell** to which this transport coefficient relates has its two electrodes separated by a very narrow distance L, with the two electrode reactions being mutual converses, such as Cu│Cu$^{2+}(aq)$│Cu. See Web#856 for further details on this, and other, kinds of **thin-layer cell**.

[857] The **Nernst transport layer**, is an empirical concept in which the potential profile is treated as two straight lines: one, adjoining the electrode, with a slope of $(c_i^b - c_i^s)/\delta$, and a second, extending over $\delta \le x < \infty$, with zero slope. The concept often satisfactorily explains transport behavior when mild convection assists diffusion. The same concept, under the name **diffusion layer**, is sometimes used as an approximation in diffusion-only transport.

a velocity that links flux density of that species at the electrode surface to its concentration there. In some techniques, as in the example of equation 8:62, the transport coefficient is unchanging; such techniques are characterized as steady-state techniques (Chapter 12). Some instances in which diffusion is supplemented by convection occur in the table, which lists both steady-state and transient techniques. With the latter techniques (Chapters 15 and 16) time is a factor in the transport coefficient. Here t is the time from the onset of the experiment and its presence as a negative power implies that transport fails to keep up with the electrode reaction. Each species involved in the electrode reaction has its own transport coefficient, though the differences between them are slight, arising from minor disparities in diffusivities. Though the coefficients may be valid for other experiments, we have in mind the response to a potential step[858] when excess supporting electrolyte is present.

Summary

The three transport modes – migration, diffusion and convection – have distinct origins and obey different transport laws

8:64 $$j_i = j_i^{mig} + j_i^{dif} + j_i^{conv} = -u_i c_i \frac{d\phi}{d\ell} + -D_i \frac{dc_i}{d\ell} + v_{soln} c_i$$

though migration and diffusion are interrelated. Being an efficient means of transport, convection is often used in preparative electrochemistry, but only in special circumstances – notably with the rotating disk electrode – is it feasible to predict the voltammetric consequences of convection quantitatively. Migration transports only ions, but is wholly responsible for carrying current through electrochemical cells other than in the vicinity of the working and reference electrodes. When supporting electrolyte is present in excess, as it usually is in investigative electrochemistry, diffusion is the transport mode that controls the current. Diffusion obeys Fick's two laws, the second of which must be solved to predict the outcome of a voltammetric experiment. The Cottrell equation arises from the simplest of those solutions.

---

[858] A potential-step experiment is one in which the potential of a previously inactive electrode is suddenly and permanently changed, resulting in the concentration of electroactive species at the surface changing suddenly and permanently from c_i^b to c_i^s. The potential-leap experiment (page 156) is the special case in which $c_i^s = 0$ and therefore $j_i^s = m_i c_i^b$.

9

Green Electrochemistry

By this title, we include not only ways in which electrochemistry can help mankind's environment, but also how it can assist humans directly through medical intervention. As well, we describe a few of the many applications of electrochemistry to biology.

Sensors for Pollution Control: keeping watch on contaminant levels

The term **pollution** is used to describe a harmful manmade change in the concentration of some species in a particular medium (air, water, soil, ecosystem, foodstuff, etc). Occasionally it is a concentration decrease that is worrisome, as when industrial effluents decrease the oxygen content of a river, or when vital nutrients are lost during food processing. But more usually it is an increase in some toxic species that is of concern. The object of remedial action is not to eliminate the pollutant entirely, for not only is that an unattainable goal, but it may even be counterproductive[901]. The aim is to bring the contaminant concentration to within acceptable bounds.

Obviously one needs to be able first to detect, and then to quantify, the presence of pollutants. Those are the tasks of the analytical chemist. The number of analytical techniques is legion and here our intent is only to give a sampling of some of the ways in which electrochemistry can assist the analytical chemist. Recognize that, in the present context, the precision of the measurement is not usually of great importance: a pollutant may be toxic at concentrations in excess of one part per million (ppm[902]), but whether its concentration is 3 ppm or 4 ppm is of little concern.

Chemical analysis can be expensive, especially if one needs to analyze many samples frequently. Devices that automatically provide measurements that can be converted to concentrations are therefore invaluable. Such devices, which may operate continuously or intermittently, are called **sensors**. Many sensors operate on electrochemical principles.

---

[901] Copper is toxic to aquatic life, for example, but small quantities are essential to many organisms.

[902] By convention, ppm means parts per million by volume in the case of gases; by mass otherwise.

Electrochemical Science and Technology: Fundamentals and Applications, First Edition. Keith B. Oldham, Jan C. Myland, Alan M. Bond.
© 2012 John Wiley & Sons, Ltd. Published 2012 by John Wiley & Sons, Ltd.

The presence of electrolytes in fresh water can be detected and quantified by a **conductivity sensor**, a simple instrument that uses a.c. to measures the conductance of a water sample in a standardized cell to find the conductivity of the water (pages 15 and 304). Such sensors are used to detect and assess the intrusion of salts into rivers, lakes, and agricultural land. Less clearly understood is the change in conductivity experienced by thin films of zinc and tin oxides, especially if they are lightly doped, on exposure to certain gases. For example, an undoped film of ZnO deposited on Al_2O_3 (synthetic sapphire) responds to arsine AsH_3 at concentrations as low as 0.015 ppm, while an SnO_2 film doped with aluminum will recognize traces of formaldehyde HCHO vapor. Because quantitation is poorly established, these devices are better described as **conductivity detectors** than sensors.

Measurement of pH by the glass electrode was described on page 122. **pH sensors** are one of today's most ubiquitous electrochemical sensors. They find applications in monitoring acid rain, in regulating the acidity of foods and drinking water, in biomedical assays, and elsewhere. You will find pH meters in quality control laboratories worldwide, and they monitor chemical flows in almost every production facility that handles aqueous solutions.

The glass electrode has been adapted to sense gases. The **ammonia sensor**, for example, employs an ammonium chloride solution, $NH_4^+(aq) + Cl^-(aq)$, in contact with a glass electrode. Ammonia in the gas stream upsets the

9:1 $\qquad NH_3(g) \rightleftarrows NH_3(aq) \qquad\qquad NH_3(aq) + H_3O^+(aq) \rightleftarrows NH_4^+(aq) + H_2O(\ell)$

equilibria and the resulting pH change will reflect the ammonia concentration in the gas stream. A **carbon dioxide sensor** operates similarly; by shifting the

9:2 $\quad CO_2(g) \rightleftarrows CO_2(aq) \qquad CO_2(g \text{ or } aq) + 2H_2O(\ell) \rightleftarrows HCO_3^-(aq) + H_3O^+(aq)$

equilibria in a sodium bicarbonate solution, $Na^+(aq) + HCO_3^-(aq)$, carbon dioxide may be assayed through the pH change. These gas sensors are not specific: they respond to all gases that have basic or acidic properties.

Ion-selective electrodes (Chapter 6) are ideally suited for sensor use. Apart from the need for periodic calibration, they require little attention and provide a convenient concentration-dependent voltage signal that can be used to actuate controls. Their lack of high precision in measuring concentrations is seldom a problem. The fluoride-ion sensor is described on page 121 and so-called isfets are discussed on page 284. An interesting link to microbiology is provided by the valinomycin **potassium-ion sensor**. **Valinomycin** is a powerful bactericidal agent isolated from cultures of the microorganism *Streptomyces fulvissimus*. This antibiotic[903] is incorporated into an organic liquid or plastic membrane that, in the manner of Figure 6-8, provides a potassium-ion ISE. It is a cyclic trimer[904],

---

[903] also described on page 295. Nonactin (page 121) is a synthetic alternative to valinomycin.

[904] that is, three identical units linked together, as in an equilateral triangle.

$(C_{18}H_{30}N_2O_6)_3$, composed of amino acid and hydroxy acid units, with a total chain length of 36 atoms. The secret of valinomycin's antibiotic properties is its ability to desolvate and engulf a potassium ion,

9:3 $\qquad K^+(aq) + (C_{18}H_{30}N_2O_6)_3(aq) \rightleftarrows K(C_{18}H_{30}N_2O_6)_3^+(aq) \qquad K' \approx 10^6\ M^{-1}$

and thereby interfere with the bacterium's metabolism. There is an excellent size match between the K^+ ion and a six-oxygen-atom-fringed void that the valinomycin molecule can develop. The valinomycin literally wraps itself around the naked cation, aided by the attraction between the positively charged K^+ and the six oxygen atoms, which have negative polarities. Other cations do not fit so well[905]. A compound such as valinomycin that is able to bind an ion, and later release it, is called an **ionophore**. There are many ionophores, some with high specificity for particular ions[906]. When complexed by an ionophore, inorganic ions are better able to enter an organic medium and cross lipophilic[907] boundaries, which is their physiological role. A related structure is an ion-selective **transmembrane channel**, the function of which will be discussed later in this chapter.

An example of a very successful potentiometric sensor is the zirconia-based oxygen sensor described on page 68. Such sensors will be found in the exhaust manifolds of most modern vehicles, where they monitor the oxygen content of the exhaust gases as a means of ensuring the most fuel-efficient operation of the internal combustion engine. The solid electrolyte used in these cells is ZrO_2 heavily doped with yttrium (or niobium) to ensure that the zirconia retains the cubic crystal modification that alone has a high electrical conductivity (about 0.03 S m^{-1} at 1000 K) resulting from oxide-ion migration. Working on a similar principle, another high-temperature potentiometric gas sensor measures hydrogen sulfide in flue gases through a cell that employs a sulfide-ion-conducting solid electrolyte, consisting of mixed sulfides of a lanthanide (Sm or Y) metal and an alkaline earth (Ca or Ba) metal.

At standard pressure and temperature, oxygen dissolves in water to a concentration of 1.27 mM[908] and therefore air-saturated water has c_{O_2} = 0.27 mM. However, water exposed to air can be undersaturated or oversaturated. The actual concentration vitally affects aquatic organisms and can be measured with a Clark[909] cell, of which Figure 9-1 overleaf shows a cross-section. The **Clark oxygen sensor** is by no means limited to measuring oxygen concentrations in water; indeed Clark himself invented the sensor as a means of

---

[905] In particular, the Na^+ analogue of reaction 9:3 has an equilibrium constant of only 10 M^{-1}. However, the NH_4^+ complex is stable enough to interfere significantly in ISE measurements of potassium ion content.

[906] For example the ionophore **calcimycin**, also known by the uninformative name A23187, is an excellent carrier of Ca^{2+}. It, too, is an antibiotic obtained originally from the fungus *Streptomycin chartreusensis*.

[907] literally "fat-loving", having a preference for an oily environment over an aqueous locale. Lipophilic ≈ hydrophobic; hydrophilic ≈ lipophobic.

[908] or 1.97 mM at 0.0°C. Calculate the solubility (25.0°C) as a percentage by mass and in ppm (by volume). See Web#908.

[909] Charles Leland Clark, 1918–2005, U.S. chemist, biochemist, physiologist and surgeon.

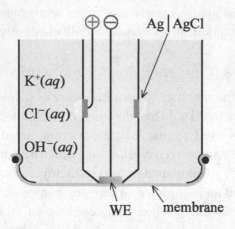

Figure 9-1 A typical Clark cell, an amperometric oxygen sensor employing a working cathode of platinum and an Ag⎢AgCl reference anode. A thin film of solution occupies the space between the WE and the membrane. The oxygen in external contact with the membrane may be in a gaseous or an aqueous phase.

measuring oxygen levels in blood when he was developing an early heart-lung machine. The electrolyte is generally a mixture of potassium chloride and hydroxide, $K^+(aq) + Cl^-(aq) + OH^-(aq)$. Oxygen permeates through the membrane[910], then diffuses through the thin layer of solution that wets the electrode and is reduced

9:4 $\qquad O_2(blood) \rightleftarrows O_2(aq) \qquad O_2(aq) + 2H_2O(\ell) + 4e^- \rightarrow 4OH^-(aq)$

under conditions of total transport polarization. The current is almost completely determined by the rate of permeation through the membrane. Clark oxygen sensors are easily calibrated using air.

The Clark cell is classified as an **amperometric sensor** because it generates a *current* output that is *directly* proportional to the concentration of the target species. In this, it differs from potentiometric sensors, such as ISEs, which deliver *voltage* signals that respond *logarithmically* to the target's activity. There are many amperometric sensors, operating similarly to the Clark cell. Among gases that can be sensed in this way are Cl_2, HCl, HCN, NH_3, PH_3, CO, CO_2, O_2, O_3, NO, NO_2, N_2O, and several organic gases and vapors. Generally, the reaction at the working electrode is a straightforward oxidation or reduction of the dissolved gas, for example

9:5 $\quad NO(g) \rightleftarrows NO(aq) \qquad NO(aq) + 6H_2O(\ell) \rightarrow 3e^- + NO_3^-(aq) + 4H_3O^+(aq)$

and

9:6 $\qquad O_3(g) \rightleftarrows O_3(aq) \qquad O_3(aq) + 2H_3O^+(aq) + 2e^- \rightarrow O_2(aq) + 3H_2O(\ell)$

but sometimes more sophisticated schemes are employed. One **carbon dioxide sensor** relies on the CO_2 to lower the *p*H of an aqueous solution, thereby shifting the equilibrium of the homogeneous reaction

---

[910] Membranes can be selected to favor specific permeants; polyethene, polypropene, and polytetrafluoroethene are frequent membrane materials. The last, known as Teflon®, is most common; its permeability to oxygen is 1.0×10^{-14} mol m⁻¹ s⁻¹ Pa⁻¹. As in Web#910, calculate the current from a Clark oxygen sensor with a gold disk cathode of 3.3 mm diameter, covered by a Teflon membrane of 13 µm thickness, when exposed to air.

9:7
$$4H_3O^+(aq) + Cu(NH_2C_3H_6NH_2)_2^{2+}(aq) \rightleftharpoons$$
$$Cu^{2+}(aq) + 2NH_3C_3H_6NH_3^{2+}(aq) + 4H_2O(\ell)$$

and liberating uncomplexed copper ions. It is the amperometric reduction of these free ions to $CuCl_2^-(aq)$ that provides access to the carbon dioxide concentration.

Our senses of smell and taste, and more potently those of other animals[911], are based on a contact interaction between a target molecule and a receptor. There is a complementarity, both sterically and electrically, between the receptor and the target, so that, when the fit occurs, there is an electrical stimulation of nerves adjacent to the receptor. Thereby we "smell" or "taste" the target molecule. A feeble attempt, at least in its current stage of development, to emulate nature's efficient sensing technology, is the field of **biosensors**. There are two elements to biosensor development: first, the identification of a receptor suitable for selectively recognizing the target; and, second, the transduction of that recognition into a signal proportional to the target concentration. The second element may be mechanical, optical, electrical, or electrochemical. This is a field of intense research activity, accompanied by much ballyhoo, but commercial realization at present is limited to very few targets, including glucose, lactate and glucosamine.

The story of the **electrochemical glucose sensor** is interesting. In 1962 during his development of the heart-lung machine, and following his invention of the amperometric blood oxygen sensor, Clark[909] introduced a small quantity of the enzyme glucose oxidase inside the sensor. Glucose oxidase, Glox, is always associated with a moiety of flavin adenosine dinucleotide, and the combination exists either in the oxidized form as GloxFAD or in the reduced GloxFADH$_2$ form. Clark found that, along with the oxygen, glucose $C_6H_{12}O_6$ from the blood was able to permeate through the sensor membrane into the aqueous solution. Here it was readily oxidized by the glucose oxidase to gluconolactone $C_6H_{10}O_6$

9:8
$$C_6H_{12}O_6(aq) + GloxFAD(aq) \rightarrow C_6H_{10}O_6(aq) + GloxFADH_2(aq)$$

The reduced enzyme was, however, immediately reoxidized by the oxygen present:

9:9
$$GloxFADH_2(aq) + O_2(aq) \rightarrow GloxFAD(aq) + H_2O_2(aq)$$

with the production of hydrogen peroxide. The net result of these two homogeneous[912] reactions is to decrease the oxygen current in direct proportion to the glucose content of the blood. In effect Clark had developed a rudimentary glucose sensor. Because oxygen and hydrogen peroxide reduce at different potentials and by different numbers of electrons, the simultaneous determination of oxygen and glucose is possible. Commercialization of Clark's discovery followed and glucose sensors were marketed by U.S. and German firms as early as the 1970s. These are very successful instruments that are still in daily use in

---

[911] Even the human nose can detect malodorous substances at concentrations of one part per billion.

[912] The glucose oxidase molecule is huge in comparison to an inorganic molecule such as O_2, so one may validly regard a reaction such as 9:9 as homogeneous *or* heterogeneous.

clinical laboratories throughout the world.

Diabetics suffer from defective carbohydrate metabolism, but knowledge of their blood glucose levels makes them better able to manage their condition. Is it feasible to adapt the commercial electrochemical glucose sensor and make it not only convenient for individual diabetic use, but also competitive with rival colorimetric methods? This challenge was first successfully overcome in 1987 when an electrochemical glucose sensor specifically designed[913] for home use came on the market. There are at least three competing designs available today. They use test strips, onto which a drop of blood is placed. The strip may incorporate three electrodes: a reference electrode and two working electrodes. The glucose from the blood permeates into a microcell that contains the first working electrode, glucose oxidase and a mediator[914]. In one version of the sensor, the mediator is a water-soluble derivative of ferrocene[611], here represented by Fcd. The corresponding ferrocenium ion is produced at the electrode

9:10 $$\mathrm{Fcd}(aq) \rightleftharpoons \mathrm{e}^- + \mathrm{Fcd}^+(aq)$$

and reacts with the reduced form of glucose oxidase produced by reaction 9:8

9:11
$$\mathbf{GloxFADH_2}(aq) + 2\mathrm{Fcd}^+(aq) + 2\mathrm{H_2O}(aq)$$
$$\to \mathbf{GloxFAD}(aq) + 2\mathrm{Fcd}(aq) + 2\mathrm{H_3O}^+(aq)$$

regenerating the ferrocene derivative and the enzyme. In this catalytic scheme, the overall process of which is illustrated in Figure 9-2, each glucose molecule eventually liberates two electrons, which are assayed amperometrically. The cathode is a straightforward $\mathrm{Ag}\,|\,\mathrm{AgCl}$ electrode. Indirectly, the current flowing from the first working electrode reflects the glucose content of the blood. The microcell serving the second working electrode, into which solutes from the blood also permeate, contains mediator but no enzyme. The role of this second WE is to compensate for any reducing agent[915] other than glucose that might

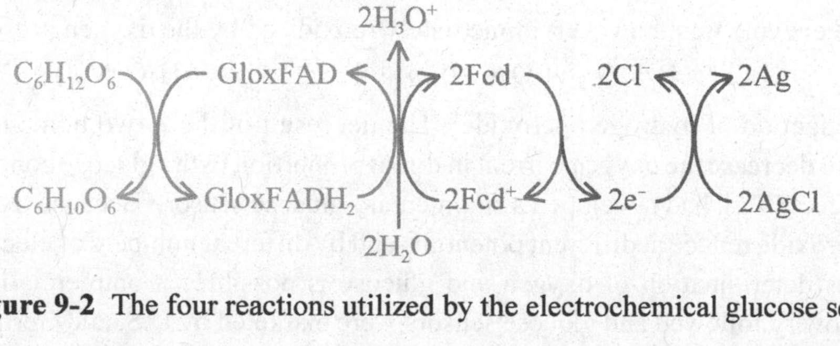

Figure 9-2 The four reactions utilized by the electrochemical glucose sensor.

---

[913] It was based on research carried out at the University of Oxford by H.A.O. Hill and his group.

[914] A **mediator** is a member of a redox couple that can carry out an oxidation (or a reduction) of a species A, being itself then regenerated by an oxidation (or a reduction) by B in circumstances where the direct oxidation (reduction) of A by B is kinetically unfavorable. As here, B may be an electrode.

[915] such as ascorbic acid, Vitamin C.

be present. The output, when the test strip has been loaded into a measuring device, is proportional to the difference between the anodic currents at the two working electrodes. These sensors work well, and have been steadily improved over the years.

Stripping Analysis: assaying pollutants in water at nanomolar levels

Environmentalists generally use parts per million as their concentration scale, whereas chemists prefer subunits of the molar scale[916]. Both sets of units are included in the accompanying table that lists guidelines for the maximum acceptable levels[917] of certain elements found as their ions in drinking water. Some anions of concern are included in the table. To ensure potability, drinking water supplies must be analyzed frequently and the need to perform analyses repetitively at the submicromolar level provides a severe economic challenge. An electrochemical strategy that can address such low concentration levels is called **stripping analysis**[918]. There are several varieties of this technique, but they all share the following principles: a substance derived from the target species is accumulated

	ppm	μM
Cu	1	16
Cr	0.05	1
U	0.02	0.08
Se	0.01	0.13
Cd	0.005	0.04
Pb	0.003	0.014
Hg	0.001	0.005
NO_3^-	45	720
F^-	1.5	80
CN^-	0.2	8

from aqueous solution onto or into a metal electrode during a **preconcentration stage**, which is followed by a **measurement stage** during which the accumulated deposit is "stripped" back into solution.

Variations in the stripping analysis technique, some of which are charted overleaf, arise from different choices of electrode, different accumulation strategies during the preconcentration stage, and different stripping signals during the measurement stage. The nomenclature of the various options is not closely standardized. The two options marked with asterisks, which are probably the most frequently employed, are usually implied when the respective names **anodic stripping voltammetry** and **adsorptive cathodic stripping**

---

[916] The scales interconvert through the formula (molarity) = (parts per million)/(1003 M) where M is the molar mass of the solute in g mol⁻¹. Derive this formula or see Web#916.

[917] However, an element's toxicity does not correlate solely with its concentration. Chromium in its highest oxidation state, Cr(VI), for example, is much more toxic than Cr(III). Conversely, the toxicity of As(III) exceeds that of As(V).

[918] For further information see M. Lovrić in: F. Scholz (Ed.), *Electroanalytical Methods*, 2nd edn, Springer, 2010, pages 201–222; or the book *Stripping Analysis,* 2nd edn, by J. Wang, VCH Publishers, 1995.

voltammetry are used without further qualification.

$$
\text{stripping analysis}
\begin{cases}
\text{solid}
\begin{cases}
\text{cathodic deposition - anodic ramp} \\
\text{anodic deposition - cathodic ramp}
\end{cases} \\[2em]
\text{Hg(drop or film)}
\begin{cases}
\text{anodic deposition}
\begin{cases}
\text{simple cathodic ramp} \\
\text{modulated cathodic ramp}
\end{cases} \\[1em]
\begin{matrix}\text{cathodic deposition \&} \\ \text{amalgamation}\end{matrix}
\text{ anodic ramp}
\begin{cases}
\text{simple} \\
\text{modulated *}
\end{cases} \\[1em]
\text{adsorption}
\begin{cases}
\text{anodic ramp} \\
\text{cathodic ramp}
\begin{cases}
\text{simple} \\
\text{modulated *}
\end{cases} \\
\text{constant current}
\end{cases}
\end{cases}
\end{cases}
$$

Target metal ions such as those of silver, gold or mercury may be cathodically electrodeposited as metals from stirred solutions onto a solid electrode (of platinum or glassy carbon, for example).

9:12
$$ M^{n+}(aq) + ne^- \underset{\text{stripping}}{\overset{\text{deposition}}{\rightleftharpoons}} M(s \text{ or } \ell) $$

After a waiting period to allow the convective motion to subside, a positive-going ramp voltage signal is applied, as in Figure 9-3. When the potential of the electrode reaches a value close to the standard potential of reaction 9:12, the deposit redissolves and a burst of anodic current appears, as in the lower trace in the figure. The same concept can be applied the other way around: such ions as $Br^-(aq)$, can be preconcentrated anodically on a silver electrode

9:13
$$ Ag(s) + Br^-(aq) \underset{\text{stripping}}{\overset{\text{deposition}}{\rightleftharpoons}} e^- + AgBr(s) $$

and then cathodically stripped off; mercury can replace silver.

Stripping analysis is rarely practiced with bare solid electrodes, but it has been discussed here as a convenient introduction to stripping analysis at mercury or mercury-coated electrodes. Interestingly, though, the first commercial application of electrochemical analysis was to determine the thickness of tin coatings on copper wire by stripping the tin deposit.

As the only metal liquid at room temperature, mercury is unique as an electrode in stripping analysis. For cathodic deposition, the process differs from that described above

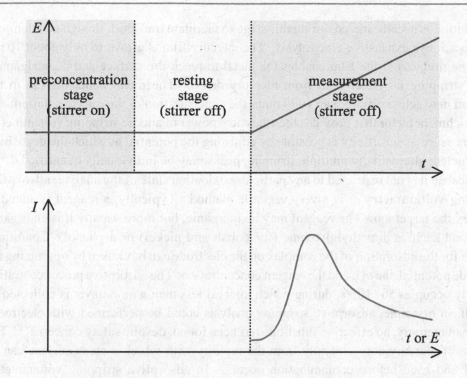

Figure 9-3 Electrode potential and current traces during stripping voltammetry.

only in that the metal produced dissolves in the electrode rather than being plated onto it:

9:14
$$M^{n+}(aq) + ne^- \underset{\text{stripping}}{\overset{\text{deposition}}{\rightleftharpoons}} M(amal)$$

In consequence there is a diffusion process within the amalgam involved in both the deposition and stripping stages. Cations of eight metals can be accurately determined at mercury by the method known as **anodic stripping analysis**, though only three of these[919] – those of copper, cadmium and lead – are of serious environmental concern. At one time, hanging mercury drop electrodes (page 118) were used and this is the only electrode for which an established body of theory exists. Nowadays a solid electrode substrate, such as glassy carbon, is more often employed and mercury is codeposited by adding a mercury(II) salt to the analyte solution, so that the mercury is being formed

9:15
$$Hg^{2+}(aq) + 2e^- \xrightarrow{\text{codeposition}} Hg(\ell)$$

at the same time as, but at a rate perhaps 100 times faster than, the amalgamating metal.

---

[919] The other five, Bi, Tl, In, Zn, and Sn, seldom occur in dangerous concentrations in drinking water, though they may be found in industrial effluents. Other metals present in water cannot be analyzed by this method, either because they do not amalgamate with mercury or because their aqueous ions do not reduce in the narrow potential window between water reduction and mercury oxidation.

The solution is usually stirred during this stage to facilitate transport, though no attempt is made to achieve exhaustive electrolysis. The mercury film is grown to only about 10 nm thick; the thinness of the film enables the metal to reach the surface and deamalgamate during stripping far faster than from mercury drops. The metal concentration in the amalgam may achieve more than 1000 times the concentration of the ions in solution; it is this enrichment factor that provides the analytical power to anodic stripping voltammetry. Some measure of specificity is possible by adjusting the potential at which the deposition is conducted; alternatively multiple stripping peaks may be individually quantified.

Because it is not restricted to any particular oxidation state of the analyte, **adsorptive stripping voltammetry**[920] is a very versatile method. Typically, a reagent is added to complex the target ion. The reagent may be inorganic, but more usually it is an organic compound such as dimethylglyoxime (for cobalt and nickel) or a phenol. Conditions suitable for the adsorption of the complex on the electrode can be chosen by optimizing the electrode potential, the pH, and the reagent concentration. The adsorptive preconcentration typically occupies 50–100 s, during which interval less than a monolayer is collected[921]. Though, in principle, adsorptive stripping analysis could be performed with electrodes other than mercury, no effective substitute has been found, despite safety concerns[922]. The benefits of mercury derive mainly from the rapidity with which a clean surface can be formed and used before contamination occurs. In adsorptive stripping voltammetry, hanging drop, or thin film electrodes may be used. The solution may be still, or stirring may be employed, sometimes even during the measurement stage. Though *anodic* stripping has been used, generally the ramp used is negative-going and *cathodic* peaks are observed and quantified. The reduction process may involve either the target species or the complexing reagent.

Irrespective of the particular variety of stripping voltammetry practiced, a stripping peak results. The area of this peak, or its height, is evidently proportional to the concentration of target ions in the original solution, but the constant of proportionality depends on many factors, most of which are imperfectly known. This is a common paradigm in analytical chemistry and it is addressed by one or other of two well-known strategies. One is called **standardization**: solutions of known concentration are prepared and analyzed in *exactly* the same way as is the sample of unknown content. The unknown concentration is then established by the procedure illustrated in Figure 9-4, or more simply

---

[920] The common distinguishing names – anodic stripping voltammetry versus adsorptive stripping voltammetry – can be confusing. Whereas the "anodic" adjective refers to the measurement stage, the "adsorptive" refers to the preconcentration stage.

[921] Assuming the analyte molecule to be a cube of edge length 1.0 nm and to have a diffusivity D of 5×10^{-10} $m^2 s^{-1}$, calculate what fraction of a monolayer accumulates in 60 s from a solution of 1.0 μM concentration, if the rate of adsorption is given by the Cottrell expression $c^b\sqrt{D/\pi t}$ (page 157). See Web#921.

[922] for the analyst's health and the disposal of laboratory waste. Bismuth has been advocated as a safer replacement.

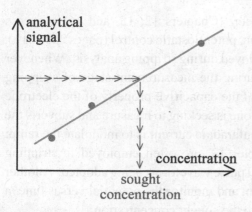

Figure 9-4 In the standardization procedure, several samples of known concentration are determined in exactly the same way as is the unknown sample. The "best straight line" is drawn by eye through the known points on a graph of signal versus concentration (or a "least squares" analysis is performed numerically – see Web#653). The sought concentration is then found by the interpolation procedure indicated by the blue arrows.

by direct proportionality[923]. Alternatively the method of **standard additions** is adopted. In this, the original sample is reanalyzed after the addition of a small known amount of the target species. Figure 9-5 shows the procedure[924] in which this "spiking" is carried out with several successive amounts. Often the volume change resulting from the spiking is small enough to be ignored. Inasmuch as interpolation is more reliable than extrapolation, it might be imagined that the standardization strategy would be superior to standard additions. But this is not necessarily so. Because determinations made in the latter strategy use the same medium, errors arising from the presence of interferants are lessened.

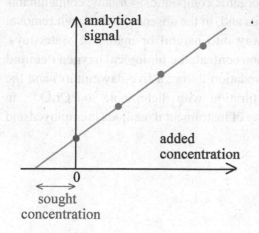

Figure 9-5 In the standard addition procedure, the sample of unknown concentration is first determined. Then a small known amount of the analyte is added and the determination is repeated. Several more additions are made. After each addition, the "added concentration" is calculated, and a best-straight-line graph is made, as shown. The sought concentration is then found from the "negative intercept" as the diagram indicates.

---

[923] Stripping analysis of an aqueous solution containing an indium In(III) salt, gave a peak of height 9.3 µA at a potential of −0.35 V. Under similar conditions a standard solution of 0.0550 mM In(NO$_3$)$_3$ gave a peak of height 11.3 µA at the same potential, whereas an indium-free solution gave no peak, but at −0.35 V the current was 0.8 µA. Estimate the indium concentration in the first solution. See Web#923.

[924] **Background** is the signal provided by an analytical method when a sample contains no analyte whatsoever. How would you adjust the procedure shown in Figure 9-5 to take account of background? A stripping analysis of polluted river water for copper shows a peak of height 230 nA, with a background of 60 nA. Successive additions, each increasing the Cu^{2+}(aq) concentration by 0.20 µM without a significant change in sample volume, produced peak heights of 310 nA, 380 nA, 480 nA, and 540 nA. What is the copper content of the river water in parts per million? See Web#924.

Conditions that are standard in voltammetry (Chapters 12, 15, and 16) – excess supporting electrolyte, strict exclusion of oxygen, potentiostatic control (pages 208–211), and carefully controlled transport – may be employed during stripping analysis. Whenever the potential of an electrode is changed, as during the measurement stage of stripping analysis, a nonfaradaic current flows because of the capacitive property of the electrode interface. This adds to the faradaic current that one is seeking to measure and subverts the analysis. One way of discriminating against nonfaradaic current is to modulate the ramp; such strategies, which are explained in Chapter 13, are often employed in stripping analysis, square-wave and especially differential pulse waveforms being adopted. Another stripping procedure is to apply a constant current and monitor the potential versus time; a plateau is observed of a duration proportional to the sought concentration.

Electrochemical Purification of Water: getting the nasties out

Analysis is only half of the solution to the problem caused by pollution; **remediation** is needed whenever pollutants exceed acceptable levels. Of course, the destination of the purified water is a consideration; the high purity requirements for drinking water need not apply to "grey" water for cooling or irrigation purposes. Water in urban and agricultural regions often has a high content of deleterious organic compounds. Organic contaminants are also present in the wastes of many industries and, in the absence of thorough removal before leaving the facility, they find their way into natural or manmade waterways. Traditionally, there were two measures of carbon content: the **biological oxygen demand** (measured by the oxygen required for biodegradation during a five-day culture) and the **chemical oxygen demand** (determined by titration with dichromate ion, $Cr_2O_7^{2-}$, in strongly acidic conditions). Nowadays, a variety of instrumental methods is employed and a finer classification of water quality is achieved

$$
\text{total carbon}
\begin{cases}
\text{inorganic}
\begin{cases}
\text{volatile } (CO_2) \\
\text{dissolved } (HCO_3^-) \\
\text{particulate (carbonates)}
\end{cases} \\
\text{organic}
\begin{cases}
\text{volatile (small molecules)} \\
\text{disssolved} \\
\text{particulate}
\end{cases}
\end{cases}
$$

Though it causes **hardness**[925], inorganic carbon is of little concern and is not removed during water treatment, other than by water softeners at the point of use. The quantity and type of organic carbon determine the appropriate remedial measures. Electrochemical methods are useful whenever the pollutant is known to be susceptible to destruction at

---

[925] "Hard" water contains salts, largely bicarbonates, of the metals Ca, Mg, and Fe. Though harmless (in fact, nutritive) such salts lead to unpleasant deposits in kettles, water heaters and laundered clothing.

electrodes or by an electrogenerated mediator.

The classical dichromate titration is being replaced by electrochemical methods because of concern for the hazards posed by chromium and other reagents. One of these methods uses coulometry. It employs an illuminated titanium dioxide electrode to generate electron/hole pairs. The holes react with water to generate hydroxyl radicals which react with the organic content of the water. The electrons are measured to provide a quantitative measure of the pollution.

Of the very many types of organic pollutants, we choose phenols as an example. The formulas of two typical phenolic compounds are illustrated to the right. Phenols are present in waste water from many industries: coal, petroleum, paper, plastics, dyes, pharmaceuticals, photography, etc. They are toxic and malodorous, but can be decomposed by vigorous anodic oxidation. Mild oxidation must be avoided because the quinones thereby formed are more toxic than the phenols. Rather than oxidizing the phenols directly at an anode, which can lead to a polymeric coating, it is better to oxidize via electrochemically generated mediators, such as hydroxyl radical $HO^{\bullet}$, hypochlorite ion OCl^-, or ozone O_3. These potent oxidizers are formed only at very positive potentials and therefore anodes with a very large overvoltage for O_2 generation are required to bypass that more thermo-dynamically favored anodic reaction. Lead dioxide

OH

resorcinol or
1,3- dihydroxybenzene

OH

beta-naphthol or
2-hydroxynaphthalene

PbO_2 once filled this role, but fear of introducing lead into the water has led to replacement by SnO_2 and, most recently, boron-doped diamond. One stratagem is to create an oxidizing environment at *both* electrodes by blowing air through an **air electrode**, in the form of a porous carbon cathode, thereby generating hydrogen peroxide H_2O_2 cathodically.

9:16   cathode: $O_2(air) + 2H_2O(\ell) + 2e^- \rightarrow H_2O_2(aq) + 2OH^-(aq)$

as well as ozone anodically. Electrodes such as the air electrode, in which a gaseous reactant meets the liquid ionic conductor within the pores of an electronic conductor, are known as **gas diffusion electrodes**.

Quite apart from its role in changing the chemical identity of pollutants, electro-chemistry plays other roles in water treatment. The adsorption of pollutants on bulky precipitated hydroxides of iron and aluminum is a standard nonelectrochemical process in water purification. The production of these hydroxides from metal anodes, for example

9:17 $\begin{cases} \text{anode:} \quad 2Al(s) \rightarrow 6e^- + 2Al^{3+}(aq) \\ 2Al^{3+}(aq) + 6OH^-(aq) \rightarrow 2Al(OH)_3(s) \end{cases}$

is now being used as an effective means of introducing these hydroxides into industrial waste waters. Because such hydroxides are hydrated and colloidal, they readily adsorb pollutants as diverse as oil droplets, heavy metal ions, and dyestuffs. This technique, termed **electrocoagulation**, is often carried out in conjunction with a second process known as **electroflotation**. Figure 9-6 illustrates the joint process. The hydroxide ions needed for reaction 9:17 are produced at the cathodes[926]

9:18 $\qquad$ cathode: $\qquad 6H_2O(\ell) + 6e^- \rightarrow 3H_2(g) + 6OH^-(aq)$

and evolve hydrogen as bubbles which attach to the colloidal particles and lift them up (a "flotation") into a sludge that forms on the surface of the cell, from where it can be skimmed off. Industrial waste water can often be converted by this electroflotation-electrocoagulation process, followed by filtration, into grey water fit for irrigation and cooling.

For reasons of health, potable water must meet biological, as well as chemical standards; meeting the former is described as **disinfection**, the eradication of live

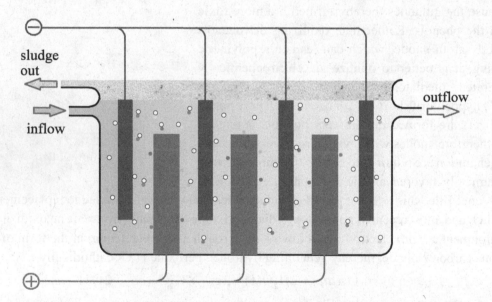

sludge out

inflow

outflow

Figure 9-6 In a combined process, the colloidal aluminum hydroxide particles produced from dissolving aluminum anodes adsorb pollutants and are buoyed by the hydrogen bubbles produced at the inert cathodes. During its tortuous path through the cell, the water loses many of its contaminants.

[926] Demonstrate that, *in principle*, a water treatment plant employing reactions 9:17 and 9:18 could operate without requiring an external supply of electricity. Why is this not so *in practice*? See Web#926.

microorganisms[927]. Chemical treatment (with Cl_2, ClO_2, or O_3), or exposure to ultraviolet radiation are the chief methods of disinfection, but electrolysis, often with a.c., across closely-spaced electrodes has also proved effective. It appears that three mechanisms are jointly involved: the electrogeneration of "killer" species such as $HO^\bullet$, direct inactivation ("electrocution") of the organisms, and the adsorption of the microorganisms on the electrodes.

Chlorine-containing oxidants are falling into disfavor because incomplete oxidation of certain organic pollutants is known to lead to the formation of traces of chloroform, $CHCl_3$, and similar toxic compounds. Ozone, an attractive alternative oxidant, is a powerful disinfectant, but its instability requires that it be produced on site. The established method of ozone production involves a high-voltage "coronal" electrical discharge through dry air, but electrochemical methods are also in use. One of these is illustrated in Figure 9-7. An **air cathode** is used, in which air diffuses into the pores of a porous electrode composed of carbon and poly(tetrafluoroethene), where the reaction

$$9:19 \qquad \text{cathode:} \qquad O_2(aq) + 4H_3O^+(aq) + 4e^- \rightarrow 6H_2O(\ell)$$

occurs. At the anode, the reaction

$$9:20 \qquad \text{anode:} \qquad 6H_2O(\ell) \rightarrow 4e^- + O_2(g) + 4H_3O^+(aq)$$

is thermodynamically favored over the sought reaction

$$9:21 \qquad \text{anode:} \qquad 9H_2O(\ell) \rightarrow 6e^- + O_3(g) + 6H_3O^+(aq)$$

and therefore conditions are chosen that optimize the latter reaction at the expense of the former. These include using a glassy carbon electrode, the tetrafluoroborate supporting anion, and such a high current density that the anode must be cooled. Nevertheless, the fraction of ozone in the anode gases seldom exceeds one-third. Even this concentration needs an air diluent to avoid the danger of an ozone explosion. Ozone is believed to form

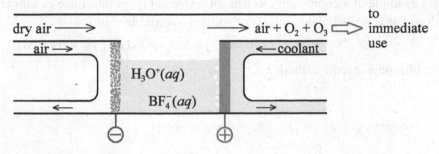

Figure 9-7 One design of an electrochemical ozone-generation cell.

[927] The standard method of biological assessment of water is a **coliform count**. If the number of bacterial colonies after incubation is less than one per 100 mL of water, the water is considered fit to drink, whereas unpolluted lake water typically gives a count of 30, and raw sewage more than 10^5. Sadly, passing the coliform test is no guarantee that the water is free of such potent pathogens as *guardia*, *cryptosporidium* and viruses.

through the following mechanism:

9:22
$$\begin{cases} 2H_2O(aq) \rightarrow e^- + H_3O^+(aq) + HO^{\bullet}(ads) \\ O_2(aq) + HO^{\bullet}(ads) \rightarrow HO_3^{\bullet}(ads) \\ HO_3^{\bullet}(ads) \rightarrow e^- + HO_3^+(aq) \\ HO_3^+(aq) + H_2O(\ell) \rightarrow H_3O^+(aq) + O_3(g) \end{cases}$$

that involves hydroxyl radicals adsorbed on the glassy carbon surface.

The electrogeneration, and immediate use, of ozone at water-treatment plants is but one example of an increasing trend towards the use of small-scale electrosynthetic cells located at the site of product use. Another example from the water-treatment industry is chlorine dioxide which, being liable to explode on maltreatment, is unsuited to transportation. It is prepared on site, either chemically by the reaction[928] of chlorine with a chlorite:

9:23 $$Cl_2(g) + 2ClO_2^-(aq) \rightarrow 2Cl^-(aq) + 2ClO_2(g)$$

or by electrosynthesis in a cell in which the reactions are

9:24   cathode: $H_2O(\ell) + e^- \rightarrow OH^-(aq) + \frac{1}{2}H_2(g)$

and

9:25   anode: $ClO_2^-(aq) \rightarrow e^- + ClO_2(g)$

Though chlorine is transportable, it too is often electrosynthesized at its point of use, sometimes from seawater. Yet another example is hydrogen peroxide, which may be electrogenerated from air, on site, by reaction 9:16 at a carbon electrode.

The semiconductor industry also makes use of chemicals electrosynthesized as needed at chip fabrication plants. Commercial hydrogen is not of adequate purity for semi-conductor applications; ultrapure $H_2(g)$ is synthesized by the electrolysis of purified potassium hydroxide solution when and where needed. Arsine, an extremely toxic gas, is a second example of a compound electrosynthesized at its point of use in semiconductor technology. It is made by electrolysis of sodium hydroxide solution

9:26   cathode: $As(s) + 3H_2O(\ell) + 3e^- \rightarrow AsH_3(g) + 3OH^-(aq)$

using an ultrapure arsenic cathode.

Electrochemistry of Biological Cells: nerve impulses

Animal cells (as well as the organelles inside them) are enclosed by bilayers that serve as the basic "bricks" of biological membranes. These bilayers, about 5 nm thick, consist

---

[928] What are the oxidation states (page 29) adopted by the element chlorine in the species involved in reaction 9:23? Web#928 has the answer.

of two sheets of phospholipid molecules, positioned tail-to-tail, with hydrophillic head groups in contact with the extracellular and intracellular aqueous fluids. The lipid bilayers are internally hydrophobic and do not allow the passage of water or ions. As mentioned on page 172, the antibiotic ionophore valinomycin is able to defeat the ion impermeability of phospholipid bilayers by wrapping the ion within a molecule that possesses a lipophyllic exterior. With the ion's lipophobicity disguised in this way, penetration of the bilayer by the ionophore can occur, with subsequent release of the ion on the far side. This is illustrated on the left-hand side of Figure 9-8. Certain other antibiotics, such as gramicidin-A[929], adopt a different strategy for getting ions through phospholipid bilayers. Pairs of these protein molecules, which can adopt a spiral shape, insert themselves across the bilayer, spanning it from one side to the other. Their spiral arrangement is such that the gramicidin-A molecules form a channel linking the outside of the cell to its interior. There is a geometric and electrical match, akin to the properties of ionophores, between the channel's inner surface and specific ions, allowing those ions – primarily H_3O^+ ions in the

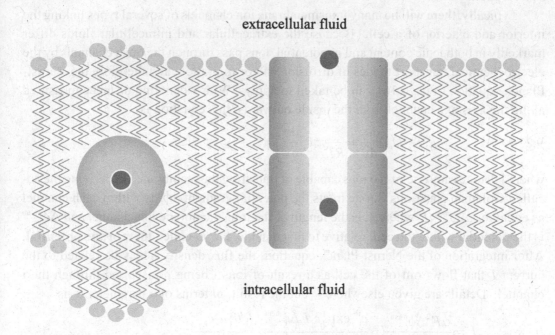

extracellular fluid

intracellular fluid

Figure 9-8 The "twin-tailed tadpoles" represent phospholipid molecules forming a bilayer, here shown in cross section, separating the exterior and interior of a biological cell. The diagram illustrates two ways in which specific ions may transit the bilayer. Ions can shuttle back and forth within an ionophore carrier or stream through a transmembrane channel. The molecules that confer ion permeability adopt hollow sphere or hollow cylinder shapes and have hydrophillic and lipophillic regions, as do the phospholipid molecules themselves.

---

[929] a 15-amino-acid polypeptide, $HCONH(amino\ acids)_{15}CONH(CH_2)_2OH$.

case of gramicidin-A – to pass through easily. The right-hand structure in Figure 9-8 illustrates this property.

The gramicidin-A transmembrane channel provides a good model of several structures that occur naturally in cell walls and permit various exchanges between the exterior and interior. Embedded in the bilayer are very many such gateways that service the diverse needs of the cell, but our interest here is confined to those **transmembrane channels** that allow the passage of simple ions. Such channels are composed of polypeptide molecules that are implanted in, and span, the lipid bilayer, similarly to the channel in Figure 9-8. Only an ion of a specific size and charge may travel through. The matching between the channel's inner surface and its conforming ion is so good that foreign ions rarely defy the specificity of the ionic channels[930]. The ions travel in single file through the channels, which are "passive" in the sense that no energy is expended by the cell in their operation. The protein molecules forming the channel have some flexibility, however, and can close on sensing a voltage change, either by squeezing shut or by means of an end "flap" that acts as a stopper.

Typically, there will be many transmembrane ion channels of several types linking the interior and exterior of a cell. Because the extracellular and intracellular fluids differ markedly in both ionic content and in potential, ions pass through the open channels by the electrochemical transport modes of diffusion and/or migration, and therefore the Nernst-Planck equation of page 161 can be taken to apply. With x denoting distance measured along the axis of a channel from the inside outwards, this equation is

9:27
$$j_i = -D_i \left[\frac{dc_i}{dx} + \frac{F}{RT} z_i c_i \frac{d\phi}{dx} \right] \quad \text{Nernst-Planck equation}$$

where i denotes any one of the ions capable of transiting the membrane. In deriving the so-called Goldman equations[931], one treats the potential gradient $d\phi/dx$ within each channel as equal to $-\Delta\phi^{mem}/L_i$ where L_i is the length of the transmembrane ion channel and $\Delta\phi^{mem}$ is the potential within the cell, relative to that outside, the so-called **membrane potential**. After integration of the Nernst-Planck equation, the flux density j_i may be related to the current I_i that flows out of the cell as a result of ions i being transported through the i channel. Details are given elsewhere[932] but the result, in terms of a parameter α, is

9:28
$$I_i = \frac{\alpha F^2 \Delta\phi^{mem}}{RT} G_i \frac{c_i^{in} \exp\{z_i F \Delta\phi^{mem} / RT\} - c_i^{out}}{\exp\{z_i F \Delta\phi^{mem} / RT\} - 1} \quad \begin{array}{l} \text{Goldman's} \\ \text{current equation} \end{array}$$

---

[930] For example, sodium ions transit potassium-ion channels with only 0.1% the frequency of K^+ ions.

[931] David E. Goldman, 1910–1998, U.S. biophysicist. With the **G** representing Goldman, the equations are also known as **GHK equations**. **H** denotes Sir Alan Lloyd Hodgkin, 1914–1998, English physiologist, Nobel laureate 1963. The **K** is for Sir Bernard Katz, 1911–2003, whose Jewish heritage led him to flee Germany to Australia and England; Nobel laureate 1970, for his neurophysiology discoveries.

[932] See Web#932 for details of the derivation, and the assumptions, of Goldman's current equation.

where c_i^{in} and c_i^{out} are the concentrations of ion i in the intracellular and extracellular fluids respectively. G_i is the "Goldman permeability", a quantity characterizing the ease with which ions i pass through their dedicated transmembrane channels. Both $\Delta\phi_i$ and G_i are liable to change with time as the transmembrane ion channels open and close, but other terms are treated as constants. Equation 9:28 is **Goldman's current equation**.

Though there may well be several ion currents flowing into and out of a cell, there cannot normally be any *net* current. Thus the sum of the I_i currents over *all* the permeant ions will be zero, so that

$$9{:}29 \qquad\qquad 0 = \sum_i G_i \frac{c_i^{in} \exp\{z_i F \Delta\phi^{mem} / RT\} - c_i^{out}}{\exp\{z_i F \Delta\phi^{mem} / RT\} - 1}$$

This equation, which enables the membrane potential difference to be expressed in terms of the ion concentrations inside and outside the cell, is called **Goldman's voltage equation**. It expresses the membrane potential implicitly in terms of the intracellular and extracellular ionic concentrations. Muscle cells are permeable to calcium Ca^{2+} ions but, for most other cells, the only values of z_i that are encountered are +1 and −1 and, in that eventuality, formula 9:29 solves[933] to

$$9{:}30 \qquad \Delta\phi^{mem} = \frac{RT}{F} \ln\left\{ \frac{\sum_{cations} G_i c_i^{out} + \sum_{anions} G_i c_i^{in}}{\sum_{cations} G_i c_i^{in} + \sum_{anions} G_i c_i^{out}} \right\} \qquad \begin{array}{l}\text{Goldman's} \\ \text{voltage equation}\end{array}$$

This equation becomes simpler still for the important case in which the permeant ions are predominantly K^+ and Na^+, for then the potential inside the cell with respect to its surroundings is

$$9{:}31 \qquad \Delta\phi^{mem} = \frac{RT}{F} \ln\left\{ \frac{G_{K^+} c_{K^+}^{out} + G_{Na^+} c_{Na^+}^{out}}{G_{K^+} c_{K^+}^{in} + G_{Na^+} c_{Na^+}^{in}} \right\} = (26.7\,\text{mV}) \ln\left\{ \frac{G_{K^+} c_{K^+}^{out} + G_{Na^+} c_{Na^+}^{out}}{G_{K^+} c_{K^+}^{in} + G_{Na^+} c_{Na^+}^{in}} \right\}$$

Here the stated millivoltage refers to the human metabolic temperature of 37°C.

Typical concentrations of potassium and sodium ions inside and outside neurons (nerve cells) are close to those tabulated to the right. The large disparities between the internal and external concentrations of each ion are maintained by ATP-fueled[934] sodium ion out-pumps and potassium ion

	concentration / mM	
	$Na^+(aq)$	$K^+(aq)$
inside	1.5	140
outside	120	5

---

[933] Derive equation 9:30, Goldman's voltage equation, or see Web#933.

[934] Adenosine triphosphate, ATP, is the body's energy currency. A ΔG of 1.7 kJ mol^{-1} is provided by each ATP molecule. In comparison, the transfer of K^+ ions against the concentration differential reported in the table corresponds to a Gibbs energy difference of 8.6 kJ mol^{-1}. Derive this last quantity and then compare with Web#934.

in-pumps, which work continuously[935]. The resting membrane potential difference is close to -70 mV, which shows that the Goldman permeability for K^+ exceeds that for Na^+ about 35-fold[936]. It is the large differences in potassium and sodium ion concentration between the inside and the outside of nerve cells that provide the key to their operation.

Nerve impulses are transmitted within a single nerve cell electrically and between two nerve cells chemically. Each nerve cell has several axons radiating from its nucleus-housing "soma", as suggested in Figure 9-9. Electrical signals are conveyed along the axons by an **action potential**. On being triggered by an incoming stimulus, the action potential starts with a potential rise near the soma and propagates along the axons. Small stimuli are ineffective, but if a rise in the membrane potential to a level more positive than a threshold of about -30 mV is experienced, a succession of events occurs at this site:

(a) The stimulus causes an opening of the many closed sodium-ion channels that are found in the soma near the nucleus.

(b) In response to the huge concentration gradient across the newly opened channels, Na^+ ions flood into the axon.

(c) As a result, the membrane potential, previously about -70 mV, climbs quickly to a maximal value of about $+40$ mV.

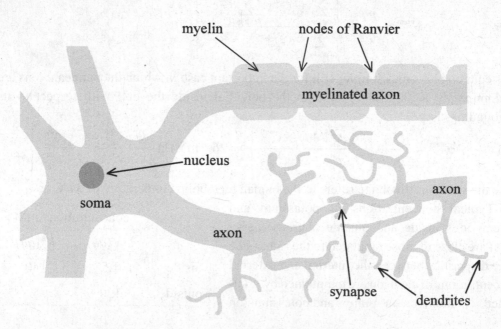

Figure 9-9 Schematic diagram of neuron morphology.

---

[935] Some pumps are dual acting; one of these expels three sodium ions while admitting two potassium ions. "Active transport" of this kind is estimated to consume about 25% of the metabolic energy of the human body.
[936] Demonstrate that this is so. Also show that, for the tabulated concentrations, the maximum and minimum possible values of $\Delta\phi^{mem}$ are $+117$ mV and -89 mV. See Web#936.

(d) The sudden rise in potential has two effects. One of these is to trigger closed sodium-ion channels further along the axon to open, thereby propagating the action potential along the axon.

(e) The second effect of the 110 mV rise in membrane potential is to trigger closed potassium-ion channels to open.

(f) K^+ ions exit through the opened channels in response to the 30-fold concentration differential favoring their travel outwards into the extracellular fluid.

(g) The outward flow of potassium ions restores the membrane potential to negative values.

(h) The negative $\Delta\phi^{mem}$ signals the local transmembrane ion channels – sodium and potassium – to return to their resting closed states.

(i) The site is now ready to respond to another stimulus, the entire (a)–(h) sequence having taken place within an interval of about 30 ms.

The brief spike in potential – the so-called action potential – moves rapidly[937] as a wave-like signal along the axon. Recognize that it is a rather small number of Na^+ ions that move into the axon and a similar small number of K^+ ions that subsequently move out, certainly not enough to significantly affect concentrations. The effect resembles the charging of a very small capacitor in that a small flow of charge causes an appreciable potential change. Even if the active pumps are disabled by drugs, the nerve cell is able to provide an action potential many thousands of times before becoming inactivated by the homogenization of interior and exterior ion concentrations. Nerve impulses are digital, rather than analog, signals: an intense stimulus delivers a higher *frequency* of action potential spikes than a weaker stimulus, not spikes of higher voltage or longer duration.

Large animals that must respond fast to capture prey or escape predators benefit from rapid nervous impulses. In consequence, axons through which the action potential moves faster have evolved. Such axons are "myelinated". **Myelin** is a lipid that insulates some axons in vertebrates, preventing the normal ionic ingress and egress except at evenly spaced gaps (nodes of Ranvier[938]) in the myelin sheathing. The action potential then hops from node to node, a faster[939] and less energy-consumptive motion than the usual wave. Figure 9-9 illustrates myelinated and unmyelinated axons.

Action potentials are not limited to the neurons of animals. They govern muscle action (where calcium Ca^{2+} ion channels play a role) and heartbeats (in which the action potential is more like a square-wave than a spike). Even fungi and plants (in the *Mimosa* genus, for example) employ action potentials that arise from chloride Cl^- ion channels.

Communication *between* nerve cells usually occurs by transfer of **neurotransmitters** across gaps – called **synapses** or synaptic clefts and about 50 nm wide – where the

---

[937] A typical spike in unmyelinated axons has a duration of less than one millisecond and a speed of 25 m s^{-1}.

[938] Louis-Antoine Ranvier, 1835 – 1922, French physician and pathologist.

[939] Speeds as high as 120 m s^{-1} have been measured in the spinal motor axon of cats.

dendrites of two neurons come very close together. Neurotransmitters, of which acetylcholine and dopamine[940] are examples, are small cations stored in presynaptic vesicles at the axon terminal. When an action potential reaches the terminal, calcium ion channels open, which releases the contents of the vesicle into the synaptic cleft. The neurotransmitters diffuse across the gap and penetrate the postsynaptic membrane, triggering an action potential in the receptor axon.

The timescale associated with diffusion is about L^2/D, where L and D are the distance traveled and the diffusivity. For diffusion across the synaptic cleft, these parameters have values of about 50 nm and 2.5×10^{-10} m^2 s^{-1}, from which one estimates the remarkably brief transit time of 10 μs. Nevertheless, where transit times are crucial, as in the retina of the human eye, synapses are replaced by **connexins**; these are transmembrane ion channels permitting direct ion flow from one neuron to another.

Summary

In this chapter we have merely touched on the ways in which electrochemistry can help in assaying pollutants and in the remediation of water[941]. Electrochemistry makes contact with environmental concerns in many diverse ways, beyond those discussed in this chapter. Electric vehicular propulsion is an obvious way in which rechargeable electrochemical power sources can help by displacing the pollution-emitting fossil-fuel-burning Carnot-efficiency-limited internal combustion engine. In lamenting the slow pace at which this conversion to electric fuel is taking place on our roads, don't forget that conversion to electric propulsion is almost complete for railways. Of course, as long as electricity is made from fossil fuels, the environment benefits only modestly from the electricity route. Electric propulsion, using batteries recharged from renewable sources, is thoroughly "green". It is to the shame of electrochemistry that better secondary batteries are still awaited.

---

[940] $H_3CCOO(CH_2)_2N(CH_3)_3^+$ and $(HO)_2C_6H_3(CH_2)_2NH_3^+$ respectively.
[941] For much more on these topics, and others facets of pollution, see K. Rajeshwar and J.G. Ibanez, *Environmental Electrochemistry: fundamentals and applications in pollution abatement*, Academic Press, 1997.

10

Electrode Polarization

When current flows across an electrode, its behavior ceases to be governed solely by thermodynamics. Several factors then influence the interdependence of current and electrode potential and their collective effect is known as **electrode polarization**. Three contributors to electrode polarization, each arising from a distinct cause, are examined, individually and collectively, in this chapter. These effects may, or may not, change with the passage of time; such temporal considerations will not greatly concern us here.

Although they are deeply entrenched in the history of electrochemistry, the concepts of "electrode polarization" and "overvoltage" are bypassed by some authors, who exclude them from their writings. Nevertheless, these concepts endure, and prove fruitful, in many branches of electrochemistry. They clarify the varied factors that influence the behavior of electrodes, independently of any particular experiment.

Three Causes of Electrode Polarization: sign conventions and graphs

An electrode is said to be **totally depolarized** if it remains at its null potential when current of either sign and any magnitude is passed through it. No electrode meets this ideal perfectly, but secondary battery electrodes (Chapter 5) and well designed reference electrodes (Chapter 6) come close. The vertical red line in Figure 10-1 represents total depolarization. The blue line, in contrast, represents the behavior of a **totally polarize**d electrode, one through which no current passes at any potential. Again, this represents an idealization, but many electrodes pass no significant steady current at any potential within the "window" (page 130) allowed by the solvent. It is the green line in Figure 10-1 that represents typical behavior: the electrode potential E adopts a value more positive than the null potential when the electrode behaves as an anode, but becomes more negative than E_n when passing cathodic current. **Overvoltage**[1001] is the name given to the difference

---

[1001] or commonly **overpotential**. Our preference for the name "overvoltage" is based on our usage of "voltage" to mean "electrical potential difference". η is certainly a potential *difference* rather than a potential.

Electrochemical Science and Technology: Fundamentals and Applications, First Edition. Keith B. Oldham, Jan C. Myland, Alan M. Bond.
© 2012 John Wiley & Sons, Ltd. Published 2012 by John Wiley & Sons, Ltd.

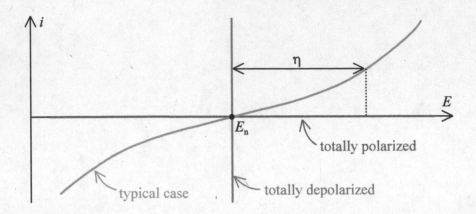

Figure 10-1 Overvoltage η is the change in electrode
potential caused by the passage of current.

between the actual potential E and the null potential E_n and it is given the symbol η:

10:1 $\eta = E - E_n$ overvoltage

Overvoltage measures the extent of electrode polarization, sharing the sign of the current
that causes it. Here cathodic overvoltages are negative but you may encounter the absolute
magnitude $|E-E_n|$ used as the definition of η. Of course, the overvoltage depends on the
current density i. Though it may be intuitive to imagine current flow being caused by a
voltage, it may be more fruitful in discussing electrode polarization, to think of the current
density as being responsible for the overvoltage.

In this chapter, we discuss three sources of electrode polarization, each associated with
its own overvoltage

$$
\left.
\begin{array}{r}
\text{ohmic} \\
\text{kinetic} \\
\text{transport}
\end{array}
\right\}
\text{polarization leading to}
\left\{
\begin{array}{l}
\text{ohmic overvoltage } \eta_{ohm} \\
\text{kinetic overvoltage } \eta_{kin} \\
\text{transport overvoltage } \eta_{trans}
\end{array}
\right.
$$

As its name suggests, **ohmic polarization** is manifested as a resistance; it arises from the
slowness of ionic migration. **Kinetic polarization**[1002] has its origin in the inherent
slowness of many electrode reactions. It is slowness in the supply of reactants to, and/or
the removal of products from, the electrode that causes **transport polarization**[1003]. A
fourth effect, **crystallization polarization**, arising from the slowness of crystal growth, is
not addressed here. Thus each type of electrode polarization reflects a slowness of one
kind or another. Extra voltage – the overvoltage – can often compensate for the slowness.

[1002] or **activation polarization** or **charge-transfer polarization**.

[1003] or **concentration polarization**.

Because many graphs and equations accompany this chapter, this is a good point at which to review a topic that bedevils electrochemical science: **signs**. The International Union of Pure and Applied Chemistry (IUPAC) prescribes such matters as definitions, symbols, and sign conventions. However, some of these run counter to established usage and may not be adhered to by all electrochemists, including the present authors. Here our purpose is to describe the sign and other conventions that are adopted in this book, and to alert the reader to other usages that may be encountered elsewhere.

The flow of electricity in an electronic conductor is in the direction opposite to that in which electrons travel, which sometimes leads to confusion; *the sign of the current reflects the motion of electricity, not electrons*. We represent current by I; others use i. We represent current density by i; others use j. We write Ohm's law as $I = -\Delta E/R$ or as $i = -\kappa d\phi/dx$ because, in a conductor, *current flows in the direction in which electrical potential decreases*; elsewhere the negative sign may be omitted.

Inevitably, ambiguity exists in discussing the sign of cell currents, unless one electrode is clearly identified as the working electrode. Nowadays, most electrochemists allocate a *positive sign to currents arising from an anodic reaction* at the working electrode, as does this book; however the opposite convention – cathodic currents being positive – may be encountered elsewhere. Figure 10-2 usefully illustrates some of these conventions.

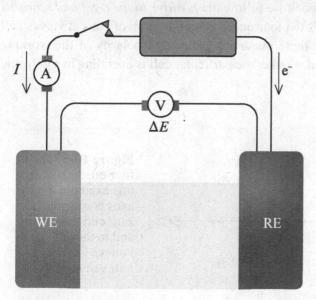

Figure 10-2 Signs. This diagram conforms to the definitions of cell voltage (its sign matches that read by the voltmeter), cell current (I is positive when WE is anodic), and electrode potential (in aqueous solution only, $E = \Delta E + E_{\text{RE(versus SHE)}}$; otherwise specify RE).

In this book, currents, overvoltages, and mobilities are allowed to adopt either sign; elsewhere they may be defined as unsigned quantities, with appropriate compensatory changes being made to equations. On the other hand, we do not allow stoichiometric coefficients to be negative, as some others do.

We use E for electrode potentials and ΔE for cell voltages; others use E for both. Allocating signs to electrode potentials is straightforward: the potential takes the sign of the voltmeter reading in the circuit diagrammed in Figure 10-2. The cited electrode potential is the voltmeter reading versus whatever reference electrode is specified. The ferrocene/ ferrocenium couple is frequently used as the reference in nonaqueous solutions. Alternatively, if the ionic conductor is an aqueous solution at 25°C, the electrode potential may be reported as $\Delta E + E_{RE}$, where ΔE is the voltmeter reading and E_{RE} is the potential of the reference versus a SHE reference[1004]. This latter is the convention we follow in listing standard electrode potentials in the table on page 392 and elsewhere in this book; however you may find standard electrode potentials cited in older publications, and especially in physical chemistry texts, with signs opposite to these.

To compound the ambiguity of the signs to be associated with current and potential, there are also different customs as to how those quantities are plotted in drawing a current-voltage graph. Following the mathematical convention used in cartesian graphs, it would seem natural to plot electrode potential (or cell voltage, or overvoltage) increasing positively to the right and current (or current density) increasing positively upwards, as was done in Figure 10-1. However, other ways of plotting were once in vogue, and linger still. Such counterintuitive plotting was practiced because, at one time, cathodic reactions were far more commonly studied than their anodic counterparts and it was then more convenient to plot experimental results in the opposite fashion, with negative potential to the right and cathodic currents upwards. In this book we follow the *positive to the right and upwards* rule, as in Figure 10-3, which labels the four quadrants of a graph of current versus cell overvoltage. As the figure shows, these quadrants reflect the polarity of the working electrode, the sign of the current, and whether the particular cell is operating in a galvanic or an electrolytic mode.

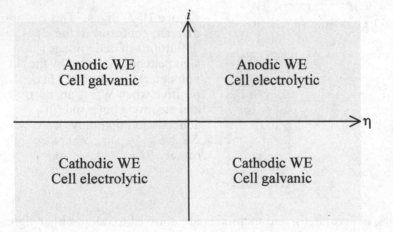

Anodic WE
Cell galvanic

Anodic WE
Cell electrolytic

Cathodic WE
Cell electrolytic

Cathodic WE
Cell galvanic

Figure 10-3 The four quadrants. In this example the two axes correspond to zero current density and to the cell voltage having its null value.

[1004] This is liable to introduce inconsistencies if the temperature is other than 25°C.

Ohmic Polarization: countered by adding supporting electrolyte

For current to flow across a cell, ions must move through the ionic conductor and to maintain this motion requires an electric field. The field is supplied by a difference of the electrical potentials, within the ionic conductor, between the surface adjacent to the working electrode and that adjacent to the reference electrode. Thus the **ohmic overvoltage** is[1005]

$$10{:}2 \qquad \eta_{ohm} = \phi_{WE} - \phi_{RE} = \int_{RE}^{WE} d\phi = \frac{-1}{\kappa}\int_{RE}^{WE} i\,d\ell = \frac{-I}{\kappa}\int_{RE}^{WE} \frac{d\ell}{A(\ell)} \qquad \text{ohmic overvoltage}$$

Here ϕ is the local potential in the solution, or other ionic conductor. In 10:2, use has been made of Ohm's law in the form $i = -\kappa d\phi/d\ell$ where κ is the conductivity of the ionic conductor. The ℓ symbol denotes the length coordinate measured from the working electrode in the direction of current flow and $A(\ell)$ is the area of the **equipotential surface**[1006] at that distance, as illustrated in Figure 10-4. To perform the integration in

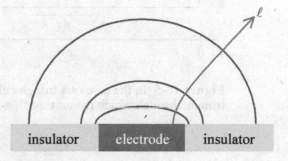

Figure 10-4 The coordinate ℓ is in the direction of current flow. Equipotential surfaces, of which three are shown in cross section, are orthogonal (at right angles) to the coordinate and have areas $A(\ell)$.

equation 10:2 requires the specification of the cell geometry, which in practice may often be rather irregular. For cells of certain simple geometries the integration is readily accomplished.

The "trough" geometry in Figure 10-5, overleaf, is especially simple because, in that case, A is a constant and integration gives

$$10{:}3 \qquad \eta_{ohm} = \frac{-I}{\kappa}\int_{L}^{0} \frac{dx}{A} = \frac{IL}{\kappa A} = IR_{cell} \qquad \text{trough geometry}$$

Here, Ohm's law in the form of equation 1:23 was used in the final step[1007] and R_{cell} denotes the resistance of the cell. In fact, the equations

---

[1005] This ignores any resistance in the electronic-conductor portion of the circuit, which may sometimes be important.

[1006] Read more about equipotential surfaces at Web#1006.

[1007] Calculate the ohmic overvoltage of a cell of the geometry of Figure 10-5 when the cell, containing 11 mM aqueous copper(II) nitrate $Cu(NO_3)_2$ solution, is a cube of edge length 19 mm and the current is 123 μA. See Web#1007.

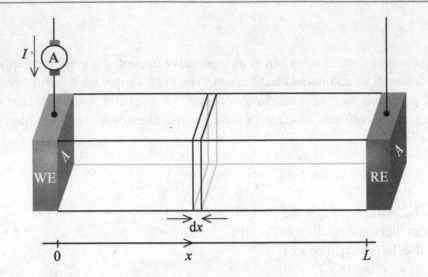

Figure 10-5 In the case of a trough cell, the cross-sections of the
trough, through which the current flows, are of constant area A.

10:4 $$\eta_{ohm} = IR_{cell} = AR_{cell}i \qquad \text{ohmic overvoltage}$$

apply to a two-electrode cell of any shape, R_{cell} being the resistance of the ionic conductor
between the two electrodes. One sees why **IR drop** is often used as a synonym for "ohmic
overvoltage".

Apart from the simple "trough"
geometry, two other geometries that are
important in electrochemistry, and
particularly in voltammetry (Chapters
12, 15, and 16), are those shown in
Figure 10-6. The left-hand diagram is a
cross section of a **hemispherical
electrode** and the adjacent solution. In
this case, the equipotential surfaces (one
is shown in red) are hemispherical shells
that surround the working electrode
rather like the layers in half an onion,

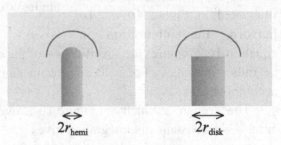

Figure 10-6 Small hemispherical and disk
working electrodes shown in cross section. A
typical equipotential surface is shown in red.

and each has an area of $2\pi r^2$ where r is the radius measured from the center. Were the
reference electrode to be a large hemispherical shell of radius r_{RE}, then the integral in
equation 10:2 evaluates as follows:

10:5 $$\eta_{ohm} = \frac{-I}{\kappa} \int_{r_{RE}}^{r_{hemi}} \frac{dr}{2\pi r^2} = \frac{I}{2\pi\kappa}\left[\frac{1}{r_{hemi}} - \frac{1}{r_{RE}}\right] \qquad \text{hemispherical WE}$$

where r_{hemi} is the radius of the hemispherical working electrode. Of course, a large hemispherical reference electrode is seldom if ever employed, but equation 10:5 shows that if r_{hemi} is small compared with the dimensions of the cell, then the second term in the brackets is unimportant and the equation reduces to

10:6 $\qquad \eta_{ohm} = IR_{cell}$ where $\qquad R_{cell} \approx \dfrac{1}{2\pi\kappa r_{hemi}}$ hemispherical WE

irrespective of the shape of RE or its position, provided that it is large and remote. The corresponding integration in the case of a small **disk electrode**[1008], shown on the right in Figure 10-6, is less straightforward and is addressed elsewhere[1009]; the result, however, is

10:7 $\qquad \eta_{ohm} = IR_{cell}$ where $\qquad R_{cell} \approx \dfrac{1}{4\kappa r_{disk}}$ disk WE

and is seen to differ from the hemispherical case only in a numerical coefficient. However, there is a further distinction between the two cases that is not apparent in the equations; whereas the current density is uniform on the surface of the hemisphere, this is not the case for the disk, which carries an elevated current density near its edge.

Because ohmic overvoltage is almost invariably an unwelcome property, one seeks to diminish it. The inverse proportionality of η_{ohm} to conductivity is the prime reason for adding supporting electrolyte to cell solutions, but there are other repercussions. Briefly, the effects of the addition are:

(a) to diminish the resistance and hence the ohmic polarization.

(b) to make migration such a minor transport mode (Chapter 8) that it may be ignored in comparison with diffusion. This massively simplifies the modeling of cell behavior.

(c) to make activity coefficients (Chapter 2) – and thereby such quantities as formal potentials and formal rate constants – less variable. This is because ionic activity coefficients reflect the ionic strength of the solution. With excess supporting electrolyte, the ionic strength is virtually uniform and constant; without it, the ionic strength may vary from point to point in the solution and from one instant to the next.

(d) to provide an environment conducive to the study of a particular reaction; for example supporting electrolytes are often buffer solutions designed to enforce a known H_3O^+ or OH^- concentration[1010].

(e) to make natural convection (Chapter 8) less troublesome. This ameliorating effect arises because the density of a concentrated ionic solution is less affected by small changes in the concentration of electroactive ions than is a pure solvent.

(f) to make the double layer (Chapter 13) thinner and less populated by the electroactive ions, thereby making its capacitance more reproducible and Frumkin effects (page 280) less obtrusive.

[1008] The term **inlaid disk electrode** is often used to distinguish a disk whose surface is flush with a surrounding insulator from one that has an exposed rim.

[1009] See Web#1009 where, in addition to the disk electrode, resistances associated with some other working electrode geometries are discussed.

[1010] In such cases, the synonyms "inert electrolyte" and "indifferent electrolyte" are inappropriate.

(g) to introduce ions that may complex the electroactive species, making the latter of less certain composition.

(h) to render the solution less comparable to the media used in nonelectrochemical studies (such as spectroscopies of various kinds), making it difficult to intercompare results.

(i) to call into doubt the results of electrochemical measurements of such dilute-solution properties as equilibrium constants or solubility products.

(j) to place an upper limit on the concentration of the electroactive species, because of the need to have supporting electrolyte in great excess.

In items (g)–(j), the effect of supporting electrolyte is adverse. However, the advantages conferred by items (a)–(f) are great enough that, unless there is some cogent reason[1011] not to, supporting electrolyte is almost always added in excess.

Kinetic Polarization: currents limited by electrode reaction rates

Some electrode reactions, like ordinary chemical reactions, are inherently slow. It is this slowness that gives rise to kinetic polarization. The sluggishness of all reactions, chemical or electrochemical, can be lessened by increasing the temperature, but electrode reactions can be accelerated also by changing the electrode potential.

Equation 7:27 demonstrates that the current density due to the simple reaction $R(soln) \rightleftarrows e^- + O(soln)$ is given by the Butler-Volmer equation

$$10:8 \qquad i = Fk^{o\prime}(c_R^b)^\alpha (c_O^b)^{1-\alpha} \left[\frac{c_R^s}{c_R^b} \exp\left\{ \frac{(1-\alpha)F}{RT}(E - E_n) \right\} - \frac{c_O^s}{c_O^b} \exp\left\{ \frac{-\alpha F}{RT}(E - E_n) \right\} \right]$$

where $k^{o\prime}$ is the formal rate constant and α is the reductive transfer coefficient. The two concentration quotients in this equation reflect disparities between surface concentrations and bulk concentrations, which is a symptom of *transport* polarization. When these disparities are absent[1012], equation 10:8 may be contracted to

$$10:9 \quad i = i_n \left[\exp\left\{ \frac{(1-\alpha)F}{RT}\eta_{kin} \right\} - \exp\left\{ \frac{-\alpha F}{RT}\eta_{kin} \right\} \right] \qquad \text{where} \quad i_n = Fk^{o\prime}(c_R^b)^\alpha (c_O^b)^{1-\alpha}$$

where η_{kin} is the kinetic overvoltage. The preexponential factor in 10:9 is the **exchange current density**; the name is appropriate because i_n represents the current flowing in each direction (being "exchanged") at the null potential E_n. Figure 10-7 has graphs[1013] showing the $\alpha = 0.35$ and $\alpha = 0.50$ examples of this relationship.

[1011] for example, if one is seeking equilibrium constants or analyzing river water in situ.

[1012] Perhaps because the solution is well stirred, or because the reaction rate is so slow that natural diffusion makes the ratio differ insignificantly from unity.

[1013] Notice that the curves lie entirely within the blue areas in Figure 10-3; we are discussing electrolytic electrochemistry. The null potential represents equilibrium; we are using external energy to drive reactions.

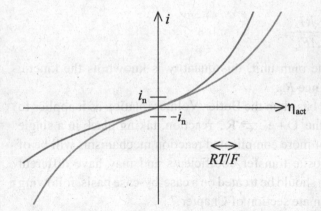

Figure 10-7 Kinetic polarization depicted as a graph showing the current density associated with η_{kin} for two values of the transfer coefficient. The shape of the green curve is that of a hyperbolic sine[1014]: $i = 2i_n \sinh\{F\eta_{kin}/2RT\}$.

Equation 10:9, which expresses the current density as a function of the overvoltage, may be inverted to give η_{kin} as a function of i only if α takes the value ½. In that case[1014]

$$10\text{:}10 \qquad \eta_{kin} = \frac{2RT}{F}\text{arsinh}\left\{\frac{i}{2i_n}\right\} \qquad \text{when} \qquad \alpha = \tfrac{1}{2} \qquad \text{kinetic overvoltage}$$

where the exchange current density i_n is then equal to $Fk^{o'}\sqrt{c_R^b c_O^b}$. However, approximate[1015] inversions are available where the overvoltage is small and in the two Tafel regions (page 137)[1016]:

$$10\text{:}11 \qquad \eta_{kin} \approx \frac{RT}{F}\frac{i}{i_n} \qquad\qquad\qquad \text{when } \eta_{kin}\text{ small}$$

$$10\text{:}12 \qquad \eta_{kin} \approx \frac{-RT}{\alpha F}\ln\left\{\frac{-i}{i_n}\right\} \qquad \text{when } \eta_{kin}\text{ large and negative} \left.\vphantom{\begin{array}{c}1\\1\\1\\1\\1\end{array}}\right\} i_n = Fk^{o'}(c_R^b)^\alpha(c_O^b)^{1-\alpha}$$

$$10\text{:}13 \qquad \eta_{kin} \approx \frac{RT}{(1-\alpha)F}\ln\left\{\frac{i}{i_n}\right\} \quad \text{when } \eta_{kin}\text{ large and positive}$$

It can be shown[1017] with the help of equation 10:9 that, in the vicinity of the null potential, the electrode potential varies with current at a rate

---

[1014] The hyperbolic sine function, and its inverse, are defined in the Glossary.

[1015] Demonstrate that, irrespective of α, approximations 10:12 and 10:13 lead to less than 5% error, as replacements for equation 10:9, provided that the overvoltage has a magnitude of at least 77 mV. See Web #1015.

[1016] Construct a polarization curve for an electrode reaction meeting the following criteria, when both ohmic and concentration polarization are absent: $c_R^b = 1.50$ mM, $c_O^b = 0.50$ mM, $E^{o'} = -0.0500$ V, $k^{o'} = 4.0\times10^{-6}$ m s^{-1}, $\alpha = 0.60$, $T = 280$ K. See Web#1016.

[1017] Calculate the charge-transfer resistance during the experiment described in Footnote 727. Check your result at Web#1017.

10:14
$$\frac{dE}{dI} = \frac{1}{A}\frac{d\eta_{kin}}{di} = \frac{RT}{AFi_n}$$
charge transfer resistance

and, reflecting the fact that it has the ohm unit, this quantity is known as the kinetic resistance or **charge-transfer resistance** R_{ct}.

All of equations 10:8 – 10:14 are based on the Butler-Volmer relation as it applies to the simple $R \rightleftarrows e^- + O$, or equally the $O + e^- \rightleftarrows R$, reaction, taking place in a single step. The corresponding equations for more complicated reaction mechanisms will be of similar form, but will involve composite transfer coefficients and may have different concentration terms. Each mechanism should be treated on a case-by-case basis, following the procedures outlined in the penultimate section of Chapter 7.

Transport Polarization: limiting currents

Transport polarization results from any slowness in the supply of reactant to the electrode and/or in the removal of product from the electrode. Three different modes of transport were discussed in Chapter 8, but we shall not be greatly concerned here with which ones are operative. Appreciate, though, that reactant and product may be propelled by different modes of transport, as when only one of the two is ionic. Be aware, too, that sometimes transport may not be required at all for certain species, as for the reactant in equation 4:17 or for one of the two redox partners, as in the reactions displayed on page 73. Even when both R and O are transported by the same mechanism, there is an important distinction between the transport of a reactant and of a product in that there is a limit to the rate of transport of the reactant, but not of the product.

We shall continue to address the generic one-step $R(soln) \rightleftarrows e^- + O(soln)$ reaction and consider it to be proceeding anodically.

Because the anodic reaction consumes species R at the electrode surface, this species must be replenished by transport from the bulk solution. However, transport is a sluggish process and a disparity therefore develops between the bulk and surface concentrations in the sense that

10:15
$$c_R^s < c_R^b$$

where the superscript ^{b} or ^{s} relates to the bulk or surface site. On the other hand, species O is produced at the electrode surface and, to provide a driving force for it to disperse into the surrounding solution, there needs to be a concentration gradient in the sense that

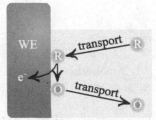

10:16
$$c_O^s > c_O^b$$

which is an inequality in the opposite sense to that in 10:15.

We here treat only the **nernstian condition**, slowness of the electron-transfer process having been addressed already in the previous section. Nernst's law

$$10:17 \qquad E = E^{o\prime} + \frac{RT}{F} \ln\left\{\frac{c_O^s}{c_R^s}\right\} \qquad \text{Nernst's law}$$

relates the electrode potential to two *surface* concentrations. It is the *bulk* concentrations, however, that determine the potential that the electrode adopts in the resting state; thus, the null potential is

$$10:18 \qquad E_n = E^{o\prime} + \frac{RT}{F} \ln\left\{\frac{c_O^b}{c_R^b}\right\} \qquad \text{null potential}$$

Subtraction of 10:18 from 10:17 leads immediately to a formula[1018] for the transport overvoltage

$$10:19 \qquad \eta_{\text{trans}} = E - E_n = \frac{RT}{F} \ln\left\{\frac{c_O^s}{c_O^b}\frac{c_R^b}{c_R^s}\right\} \qquad \text{transport overvoltage}$$

Inequalities 10:15 and 10:16 show that each of the concentration quotients[1019], c_O^s/c_O^b and c_R^b/c_R^s, that appear in 10:19 are greater than unity for the anodic process being addressed, leading, as expected, to a positive overvoltage.

Thus far in this section, we have discussed only the way in which transport polarization affects concentrations at the electrode surface and thereby, through the medium of Nernst's law, determines the overvoltage, as illustrated towards the right-hand side of Figure 10-8. Yet to be explained is how the current density interacts, through the medium of Faraday's law, to determine the values of the concentration quotients. The full picture is clarified by Figure 10-8, which shows how the various laws governing the behavior of the system conspire to create a dependence of transport overvoltage on current density.

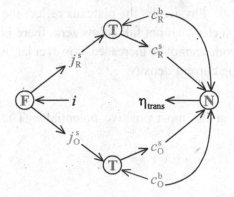

Figure 10-8 Illustrating how the transport laws, **T**, by providing linkage between the surface concentration and the surface flux density of each electroactive species, thereby interrelate the current density i to the transport overvoltage η_{trans}. The letters **N** and **F** respectively represent the roles played by Nernst's and Faraday's laws.

[1018] Show, as in Web#1018, that equation 10:19 may be derived directly from 10:8 by letting the formal rate constant approach infinity.

[1019] For an electrooxidation at 25°C, what is the transport overvoltage if the disparity between the surface and bulk concentrations of both R and O is 1%?, 50%?, 99%? See Web#1019.

It was shown in Chapter 8 that the transport laws provide a linkage between the concentration of a reactant or product at the surface of the electrode and the flux density there. Often the relationship is as simple as

10:20 $$j_i^s = m_i \left(c_i^s - c_i^b \right) \qquad i = R, O$$

where m_i is a transport coefficient that may, or may not, incorporate time as a factor. Moreover, for the reaction $R(soln) \rightleftarrows e^- + O(soln)$, the simple stoichiometry tells us that the arrival of each R at the electrode coincides with the release of an electron into the electrode and the creation of an O. It follows that

10:21 $$\frac{i}{F} = -j_R^s = j_O^s \qquad \text{Faraday's law}$$

is the simple consequence of Faraday's law[1020]. Though we have focused on the oxidation $R(soln) \rightleftarrows e^- + O(soln)$, the equations hereabouts, and indeed throughout the chapter, are equally valid for the $O(soln) + e^- \rightleftarrows R(soln)$ reduction.

Equations 10:19, 10:20, and 10:21 may be combined into[1021]

10:22 $$\frac{F\eta_{trans}}{RT} = \ln\left\{1 + \frac{i}{Fm_O c_O^b}\right\} - \ln\left\{1 - \frac{i}{Fm_R c_R^b}\right\} \qquad \text{transport overvoltage}$$

If the transport law is not as simple as 10:20, or if the electrode reaction has a different stoichiometry, then equation 10:22 will need modification, but the principles implied by Figure 10-8 remain valid whenever the overvoltage reflects transport polarization alone.

Because the arguments of logarithms can never be negative, equation 10:22 shows that the current density cannot exceed $Fm_R c_R^b$ or be less than $-Fm_O c_O^b$. That is

10:23 $$-Fm_O c_O^b \leq i \leq Fm_R c_R^b \qquad \text{current limits}$$

Thus the current density is bracketed between two values. These bounding values are known as **limiting current densities** and correspond to plateaus on a graph of current density versus electrode potential, as in Figure 10-9. Physically, the plateaus reflect the fact that, because a concentration at the electrode surface cannot fall below zero, there is a limit beyond which the rate of supply to the electrode cannot be increased, however large the magnitude of the potential. The anodic limiting current density,

10:24 $$i_{lim}^{an} = Fm_R c_R^b \qquad \text{anodic limiting current}$$

is the largest possible current density, acquired at the most positive potentials. The cathodic limiting current

10:25 $$i_{lim}^{cath} = -Fm_O c_O^b \qquad \text{cathodic limiting current}$$

[1020] Write the Faraday-law equations corresponding to 10:21 for (a) reaction 9:15 and (b) reaction 9:14. See Web#1020.

[1021] As in Web#1021, carry out the algebra to derive equation 10:22.

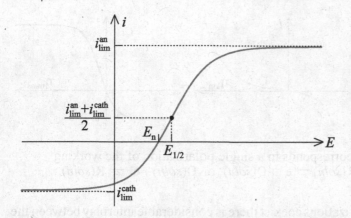

Figure 10-9 Current-voltage curve arising from transport polarization. If other polarizations are present, the wave becomes less steep, but the limiting-current plateaus are unaltered.

is the most negative current accessible[1022], acquired at the most negative potentials.

Equation 10:22 may be written in terms of the limiting current densities as

$$10:26 \qquad E = E_{1/2} + \frac{RT}{F}\ln\left\{\frac{i - i_{lim}^{cath}}{i_{lim}^{an} - i}\right\}$$

where

$$10:27 \qquad E_{1/2} = E_n - \frac{RT}{F}\ln\left\{\frac{i_{lim}^{cath}}{i_{lim}^{an}}\right\} = E_n - \frac{RT}{F}\ln\left\{\frac{m_O c_O^b}{m_R c_R^b}\right\} \qquad \text{half-wave potential}$$

is the **half-wave potential**, the central point of the "wave".

Multiple Polarizations: the big picture

Each of the three most recent sections has addressed the effect of one of the three polarizations – ohmic, kinetic, or transport – in the absence of the other two. Figure 10-10 overleaf compares the shapes of the i versus η curves caused by a single polarization of each kind. Evidently the effect of each polarization is quite distinctive. But what if there are several simultaneous polarizations?

We may think of the three kinds of electrode polarization as distinct and the three overvoltages as more-or-less additive. The additivity is strict if one of two polarizations is ohmic:

$$10:28 \qquad \eta = \eta_{ohm} + \eta_{kin} \qquad \text{if} \qquad \eta_{trans} = 0$$

$$10:29 \qquad \eta = \eta_{ohm} + \eta_{trans} \qquad \text{if} \qquad \eta_{kin} = 0$$

[1022] If the product O is absent from the bulk solution, the concept of overvoltage is meaningless because there is no null potential. Nevertheless, the current-voltage curve retains the shape of Figure 10-9, with a cathodic limiting current of zero.

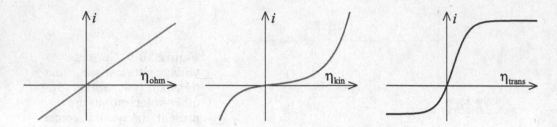

Figure 10-10 Each curve corresponds to a single polarization of the working electrode for the reaction $R(soln) \rightleftarrows e^- + O(soln)$ or $O(soln) + e^- \rightleftarrows R(soln)$.

but if kinetic and transport polarizations coexist there is considerable interplay between the two because concentrations are involved, via the Butler-Volmer equation, in both. Figure 10-11 shows the complicated routes involved by which the current density determines the overvoltage when all three polarizations conspire to polarize the electrode.

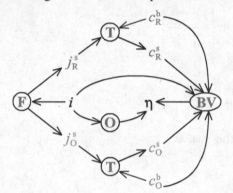

Figure 10-11 When an electrode is polarized by the joint action of ohmic, kinetic, and transport polarizations, there are several routes by which the current density i influences the overall over-voltage. Ohm's law **O**, Faraday's law **F**, transport laws **T**, and the Butler-Volmer equation **BV**, are all involved.

Despite the complexities evident in Figure 10-11, it is nevertheless possible to write a single equation[1023], namely

10:30
$$\frac{F\left[\exp\{F(\eta - AR_{cell}i)/RT\} - 1]\right]}{i} =$$
$$\frac{\exp\{\alpha F(\eta - AR_{cell}i)/RT\}}{k'^o (c_R^b)^\alpha (c_O^b)^{1-\alpha}} + \frac{1}{m_O c_O^b} + \frac{\exp\{F(\eta - AR_{cell}i)/RT\}}{m_R c_R^b}$$

interrelating the current density i and the overall tripartite overvoltage η. The expression is not pretty and it cannot be made explicit[1024]. Figure 10-12 shows the effect of adding

---

[1023] Show that equation 10:30 reduces (a) to 10:4 when kinetic and transport polarizations are negligible [let k'^o and each m approach ∞]; (b) to 10:9 when ohmic and transport polarizations are negligible [set R_{cell} to zero and each m to infinity]; and (c) to 10:22 when ohmic and kinetic polarizations are negligible [let $R_{cell} = 0$ and $k'^o = \infty$). See Web#1023.

[1024] By an "explicit" interrelation is meant one in the form $i = f(\eta)$ or $\eta = f(i)$. Equation 10:30 can be written in the latter form when $\alpha = \frac{1}{2}$.

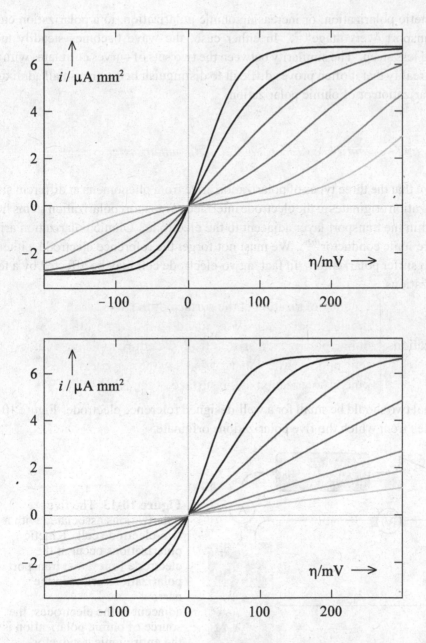

Figure 10-12 In the black polarization curves the sole cause of overvoltage is transport polarization, with limiting currents at 6.75 and -2.89 A m^{-2}. The other curves represent the addition of kinetic polarization (upper graph) or ohmic polarization (lower graph). Equation 10:30 was used with the following data: c_R^b = 0.7 mM; c_O^b = 0.3 mM; α = 0.6; $T = T°$; $m_R = m_O = 10^{-4}$ m s^{-1}. In the upper graph R_{cell} = 0; $k°' = \infty$, **100**, **31**, **10**, 3, 1 μm s^{-1}. In the lower graph $k°' = \infty$; $A = 10^{-9}$ m^2; R_{cell} = **0**, **10**, **25**, **50**, 100, 150, 250 MΩ.

increasing kinetic polarization, or increasing ohmic polarization, to a polarization curve with pure transport overvoltage[1025]. In either case, the wave becomes steadily more prolonged and less steep. The similarity between the two sets of curves correlates with the experimental reality that it often proves difficult to distinguish between a small admixture of kinetic polarization or of ohmic polarization.

Polarizations in Two- and Three-Electrode Cells: the potentiostat

Recognize that the three types of polarization arise from phenomena at different sites. Kinetic polarization originates at the electrode interface. Transport polarization stems from processes within the transport layer adjacent to the electrode. Ohmic polarization arises from the entire ionic conductor[1026]. We must not forget the reference electrode, either; it too is liable to suffer polarization. In fact, a two-electrode cell may be affected by a total of five polarizations:

$$\text{polarizations} \begin{cases} \text{kinetic polarization at the surface of the WE} \\ \text{transport polarization associated with transport to the WE} \\ \text{ohmic polarization arising from the entire ionic conductor} \\ \text{transport polarization associated with transport to the RE} \\ \text{kinetic polarization at the surface of the RE} \end{cases}$$

though the final two should be small for a well-designed reference electrode. Figure 10-13 locates the sites from which the five polarizations originate.

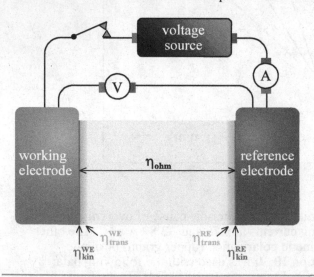

Figure 10-13 The five polarizations associated with a two-electrode cell. Kinetic polarizations occur at the electrode **junctions**; transport polarizations arises in the narrow **transport layers** adjacent to the electrodes; the source of ohmic polarization is the entire ionic conductor.

[1025] $10^{-4}\,\mathrm{m\,s^{-1}}$ was chosen for the m's because this value is appropriate to steady-state voltammetry (Chapter 12) at an microdisk electrode of area $10^{-9}\,\mathrm{m^2}$, for solutes with the typical diffusivity $1.4 \times 10^{-9}\,\mathrm{m^2\,s^{-1}}$.

[1026] With electrode materials of low conductivity, the electronic conductor may also contribute to ohmic overvoltage.

Notice that, in this two-electrode cell, the RE fills two roles: (a) it provides a location with respect to which the potential of the WE may be measured; and (b) it provides an exit/entrance for the current flowing through the cell when the WE serves as an anode/cathode. In Figure 10-14, these roles have been separated. The smaller portion has adopted role (a), whereas role (b) is assumed by the larger portion, which takes the name[1027] **counter electrode** (CE). The last two polarizations of the five listed above now effectively disappear! This is because overvoltages are caused by current flow and, though these still exist at the CE, their effects are not measured by the voltmeter, which allows no current to flow across the RE. The penalty paid for this improvement in performance is that a constant voltage source no longer ensures that the potential experienced by the WE is constant. Technology, however, comes to our aid in the form of a **potentiostat**[1028]. This is a complicated device described elsewhere[1029] that adjusts the current through the CE pathway to whatever is needed to make the $E_{WE} - E_{RE}$ voltage equal to the value that the experimenter commands.

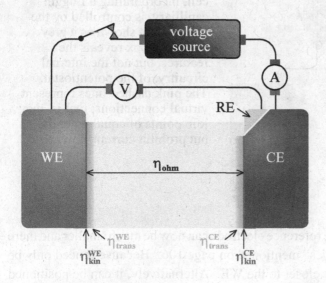

Figure 10-14 Division of the reference electrode into two portions, the larger now being named the counter electrode, is the principle behind three-electrode cells.

Nowadays, commercial potentiostats incorporate many features beyond those described above, and frequently operate digitally with computer control. Facilities may be included whereby the user can impose simple or complex potential waveforms, and record the time dependences of both the electrode potential and the cell current. Thereby, both the voltmeter and ammeter of traditional instrumentation are, in effect, incorporated into the potentiostat. Current-voltage and other graphs are plotted automatically and are often also provided as digitized time series, suitable for numerical manipulations. Many potentiostats

---

[1027] **auxiliary electrode** is also used.

[1028] Invented in 1942 by Archie Hickling, 1908–1975, English electrochemist.

[1029] See Web#1029 for details.

are able to function also as **galvanostats**, controlling the current, rather than the potential.

The use of potentiostat-supported three-electrode cells has become the norm in investigative electrochemistry and, in most of the remainder of this book, their use will be assumed. From the viewpoint of the experimenter, the salient features of a potentiostat are: (a) input connections from the voltage source[1030]; (b) output connections to the three electrodes; (c) the ability to measure the voltage between the working and reference electrodes; (d) the ability to measure the current flowing into the working electrode; and (e) a switch by which the WE can be isolated from the potentiostat. These features are the only ones shown on our symbolic representation of a potentiostat, as in Figure 10-15.

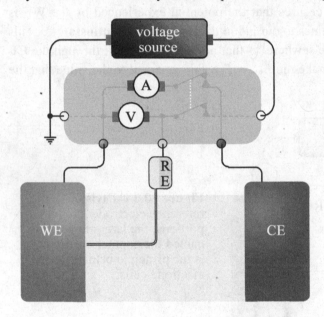

Figure 10-15 A three-electrode cell, incorporating a Luggin capilliary, is controlled by the potentiostat shown as a grey box. The box reveals the features, but not the internal circuitry, of the potentiostat. The pink dashed lines represent virtual connections; that is, they join points of equal potential but prohibit current flow.

Relieved of passing current, the reference electrode can now be much smaller and there is less call for the "abundant supplies" mentioned on page 106. Because it need only be small, the RE may be moved much closer to the WE. Alternatively, it can be positioned externally and connected to the cell by a narrow **Luggin capillary**[1031] as in Figure 10-15. Both the external and internal diameters of the capillary can be very small, the latter because no current is carried by the solution inside. The mouth of the capillary is brought close to the WE surface, thereby reducing the ohmic overvoltage to a small, and often insignificant, value. The residual resistance is the **uncompensated resistance** R_u, so called because the potentiostat is said to have "compensated" for most of the cell resistance[1026]. Understand that R_u is *not* the resistance between the WE and the RE. The potential sensed

---

[1030] sometimes called a **function generator**. Though the voltage source and the potentiostat are often integrated into a single commercial instrument, we regard them as distinct devices.

[1031] also known as a **Luggin probe**, or colloquially "a luggin". Hans Luggin, 1863 – 1899, Austrian physical chemist.

by the Luggin capillary is that of the equipotential surface that traverses its mouth[1032]. For example, the equipotential surfaces in the case of the trough cell illustrated in Figure 10-5 are simply planes parallel to the working electrode. If the mouth of the Luggin probe is at a distance x_{RE} from such an electrode surface, the uncompensated resistance will be

$$10:31 \qquad\qquad R_u = \frac{x_{RE}}{\kappa A} \qquad \text{trough cell}$$

where A is the area of the working electrode. Troughs are seldom employed but, unlike that of the cell as a whole, the geometry of the space between the WE and the Luggin tip is usually simple, permitting an estimate of R_u to be calculated from the conductivity. Bear in mind though that if the Luggin capillary is too large, or positioned too close, it will interfere with the electric field in its vicinity, because it occupies space that would otherwise be occupied by the ionic conductor.

The transition from two to three electrodes reduces the demands placed on the reference electrode and, with a carefully positioned RE, massively reduces the ohmic polarization. Demands on the counter electrode are even less, since polarization there is immaterial[1033]; a short length of platinum wire often suffices[1034]. There are circumstances in which two electrodes must be used, as in the systems addressed in Chapters 4 and 5, but otherwise the use of three electrodes has been adopted by electrochemists, even when experimental goals can be met equally well by two electrodes, despite the added instrumentation that the third electrode necessitates.

The transition from two to three electrodes is without effect on the working electrode, except that it becomes virtually grounded[1029]. It continues to experience kinetic and/or transport polarization but, with a well-designed three-electrode cell, experiments benefit from a greatly reduced ohmic polarization. This confers greater authority on measured electrode potentials and is beneficial in reducing the R_uC time constant (page 276) in several voltammetries. Not only is the ohmic polarization reduced by the three-electrode configuration, but the residual uncompensated resistance may become open to calculation from the conductivity of the ionic conductor and the geometry of the space between the working electrode and the reference electrode (page 278).

One important consequence of the introduction of the potentiostat-controlled three-electrode cell is the ease with which many solvents other than water can now serve as ionic conductors. This development[224] has had a major impact, allowing the expansion of electrochemistry into many hitherto inaccessible arenas.

[1032] The presence of the capillary will, to some extent, distort the equipotential surfaces. Moreover, if its mouth is placed too close to the WE surface, the Luggin capillary will shield a portion of the surface. In practice, one strives to position the capillary as close as possible to the electrode without seriously interfering with the WE's operation.

[1033] Recognize, however, that some electrochemical process will occur there. Products of that process may interfere with the WE reaction in some prolonged experiments.

[1034] Too small a CE, however, places voltage demands on the potentiostat that it may be unable to fulfill.

Summary

Three distinct polarizations shift the potential of an electrode when current flows. Ohmic polarization leads to an overvoltage, the "*IR* drop", proportional to the resistance; its effect can be ameliorated by addition of supporting electrolyte, which has other benefits too. Kinetic polarization originates in the slowness of the kinetics of the electron-transfer reaction: it leads to the relation

10:32
$$\frac{i}{Fk^{o\prime}(c_R^b)^\alpha(c_O^b)^{1-\alpha}} = \exp\left\{\frac{(1-\alpha)F}{RT}\eta_{act}\right\} - \exp\left\{\frac{-\alpha F}{RT}\eta_{act}\right\}$$

for the simple $R(soln) \rightleftarrows e^- + O(soln)$ reaction, when Butler-Volmer kinetics are espoused. Transport polarization for this same reaction is given by

10:33
$$\frac{F\eta_{conc}}{RT} = \ln\left\{\frac{c_O^s c_R^b}{c_O^b c_R^s}\right\} = \ln\left\{\frac{i_{lim}^{an}}{i_{lim}^{cath}} \times \frac{i - i_{lim}^{cath}}{i_{lim}^{an} - i}\right\}$$

and arises from concentration disparities which originate in the slowness of transport of electroactive solutes from, and particularly to, the electrode. When several polarizations cooperate, as they frequently do, polarization curves adopt more complicated shapes. The use of a thee-electrode cell, controlled by a potentiostat, eliminates polarizations suffered by the second electrode of a two-electrode cell and almost negates the effect of ohmic polarization at the working electrode.

11

Corrosion

Though the term is sometimes used in other contexts, the primary meaning of "corrosion" is the unwanted oxidation of metals by their environment. By gradually destroying infrastructure, corrosion is a major economic impediment to our metal-based civilization. The usual oxidizing agents are atmospheric oxygen and/or water. The mechanism by which these agents corrode metals generally involves electrochemistry.

Vulnerable Metals: corrosive environments

With the sole exception of gold, the aerobic oxidation,

11:1 $\qquad M(s) + \frac{b}{2a}O_2(air) \rightarrow \frac{1}{a}M_aO_b(s)$

of *all* metals is thermodynamically favorable, as the adjacent table[1101] illustrates for eight metallic elements. In reality, however, only the alkali and alkaline earth metals (Na, K, Ca, etc.) react at all rapidly with dry air. Protection from rapid corrosion for other metals is provided by the slowness of the heterogeneous oxidation reaction or, for metals such as Al, Cr, and Ti, by the formation of an adherent oxide layer that is impermeable to oxygen.

Airborne pollutants, such as compounds of sulfur, can increase corrosion, in part because of the increased thermodynamic driving force:

M	a,b	$\dfrac{\Delta G_{1:1}}{\text{kJ mol}^{-1}}$
Au	2,3	83
Ag	2,1	−5
Pt	1,1	−46
Cu	1,1	−129
Zn	1,1	−319
Fe	2,3	−368
Mg	1,1	−568
Al	2,3	−790

11:2 $\quad Ag(s) + \frac{1}{4}O_2(air) + \frac{1}{2}H_2S(1\,ppm) \rightarrow \frac{1}{2}Ag_2S(s) + \frac{1}{2}H_2O(g) \quad \Delta G = -121\ \text{kJ mol}^{-1}$

11:3 $\qquad Cu(s) + O_2(air) + SO_2(1\,ppm) \rightarrow CuSO_4(s) \qquad \Delta G = -409\ \text{kJ mol}^{-1}$

Comparison with the table above of the Gibbs energy changes for reactions 11:2 and 11:3

[1101] Addressing each most stable oxide, the data recognize the nonunity activity of oxygen in air because of its 0.21 bar partial pressure. Check one of these data, using $G°$ values from page 391, or see Web#1101.

Electrochemical Science and Technology: Fundamentals and Applications, First Edition. Keith B. Oldham, Jan C. Myland, Alan M. Bond.
© 2012 John Wiley & Sons, Ltd. Published 2012 by John Wiley & Sons, Ltd.

shows the marked effect of these gases, even at the parts per million level.

Corrosion by water may occur in the absence of oxygen

11:4 $$M(s) + nH_3O^+(aq) \rightarrow M^{n+}(aq) + \tfrac{n}{2}H_2(g) + nH_2O(\ell)$$

but often oxygen is an accomplice[1102]:

11:5 $$M(s) + \tfrac{n}{4}O_2(air) + nH_3O^+(aq) \rightarrow M^{n+}(aq) + \tfrac{3n}{2}H_2O(\ell)$$

The decrease in Gibbs energy accompanying reaction 11:5 is generally of a similar magnitude to that for reaction 11:1, as inspection of the accompanying table[1103] will confirm. Despite the comparable thermodynamic driving forces, the corrosion by oxygen in an aqueous environment is generally faster than in dry air, particularly when the presence of acids lowers the pH. Recognize that it is not necessary for the metal to be submerged in water for accelerated corrosion to occur; a film of water, or even a humid atmosphere, will suffice. In a marine environment, or where road salt[1104] is used to remove winter ice, chloride ions facilitate corrosion by complexing metal ions,

M	n	$\dfrac{\Delta G_{11:5}}{\text{kJ mol}^{-1}}$
Au	3	$81 + 17.1pH$
Ag	1	$-40 + 5.7pH$
Pt	2	$19 + 11.4pH$
Cu	2	$-169 + 11.4pH$
Zn	2	$-382 + 11.4pH$
Fe	3	$-357 + 17.1pH$
Mg	2	$-930 + 11.4pH$
Al	3	$-838 + 17.1pH$

11:6 $$Fe(s) + \tfrac{3}{4}O_2(air) + 3H_3O^+(aq) + 2Cl^-(aq)$$
$$\rightarrow FeCl_2^+(aq) + \tfrac{9}{2}H_2O(\ell)$$

for example, or by attacking protective oxide layers:

11:7 $$Al_2O_3(s) + 8Cl^-(aq) + 6H_3O^+(aq) \rightarrow 2AlCl_4^-(aq) + 9H_2O(\ell)$$

The effect of the concentration of complexing anions can be dramatic[1105].

The adjective **noble** is used to describe a metal that is inherently difficult to oxidize, and thereby resists corrosion. The tables in this section are arranged roughly with the most noble metals towards the top. Aluminum[1106] resists corrosion well, as evidenced by the longevity of aluminum cookware, but its nobility is superficial, being entirely due to its oxide layer. In air, the layer forms immediately on exposure of a new surface, but oxidation soon ceases because of the impermeability of that layer.

[1102] We have chosen to write equations 11:4 and 11:5 in a way that is appropriate when the aqueous environment has a pH of less than 7. Note, however, that the corrosion reaction will increase the pH. Write the reactions that take over from 11:4 and 11:5 when the pH exceeds 7. See Web#1102.

[1103] Using $E°$ values from page 392, check one of the entries from this table. See Web#1103.

[1104] Salt also detrimentally improves the conductivity of the aqueous layer in Figure 11-1.

[1105] Use the data from the table on page 391 to decide whether platinum is liable to corrode at 20°C in air-saturated hydrochloric acid [ions $H_3O^+(aq)$ and $Cl^-(aq)$] of concentration (a) 5 µM or (b) 5 mM, as a result of the formation of the chloroplatinum(II) ion, $PtCl_4^{-}$. See Web#1105.

[1106] For a dramatic demonstration of the inherent corrodibility of aluminum, see Web#1106.

Corrosion Cells: two electrodes on the same interface

The singular equations of the last section do not accurately describe how corrosion occurs. In most corrosion events, the oxidation of a metal in the presence of water takes place through two electrochemical reactions, both occurring at the *same* metal | aqueous-solution interface. There are two electrodes – one anode and one cathode – occupying adjacent sites at the junction of an electronic and an ionic conductor, as illustrated in Figure 11-1. The system represents a rudimentary two-electrode cell, a **corrosion cell**, or **local cell**.

$$2H_3O^+(aq) + 2e^- \rightleftharpoons 2H_2O(\ell) + H_2(g)$$

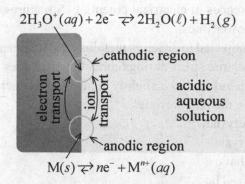

Figure 11-1 Anodic and cathodic regions share the surface of a corroding metal in contact with an aqueous solution.

acidic aqueous solution

$$M(s) \rightleftharpoons ne^- + M^{n+}(aq)$$

The diagram is misleading in suggesting that there is a single anode and a single cathode. In reality, there are many. Moreover a particular site does not permanently remain anodic or cathodic. At an anodic site, the Nernst equation

11:8 $$E = E^\circ + \frac{RT}{nF} \ln\{a_{M^{n+}}\} \qquad \text{anodic site}$$

demonstrates that a consequence of the build-up of metal ions is to make the site more positive and therefore less efficient as an anode. Conversely the Nernst equation applicable to a cathodic site

11:9 $$E = E^\circ + \frac{RT}{2F} \ln\left\{ \frac{a_{H_3O^+}^2}{p_{H_2}} \right\} \qquad \text{cathodic site}$$

shows that a small increase of pH, caused by the cathodic reaction, makes that site slightly more negative and hence a poorer cathode. What occurs in response to these small adjustments in the local potentials of the metal is a reversal of roles: what was an anode becomes a cathode and vice versa. The picture we have, then, is of an ever-changing mosaic of temporary anodes and temporary cathodes.

What has just been described is an occurrence of **generalized corrosion**, in which there are no permanent electrodes and corrosion occurs more-or-less uniformly over the entire metal surface. There are, however, very many instances of **localized corrosion**, in which an anodic region becomes a permanent anode and the site of continuous corrosion, while much or all of the remaining metal serves as the cathode. Why does one surface region of a metal object "decide" to become an anode while a nearby region, ostensibly similar, becomes a cathode? There may be several reasons:

(a) The object may, in fact, be made of two metals, as when a copper pipe joins an iron pipe. The more noble metal serves as the cathode and the less noble corrodes. Corrosion is common in such settings, keeping plumbers in business.

(b) Not infrequently, metals contain impurities in the form of "inclusions" of more noble metals. The inclusion becomes the cathode.

(c) Similarly, an alloy may not be completely homogeneous. The nobler regions become cathodes.

(d) Electric fields may induce slight differences in electrical potential. Structures are said to corrode faster close to high-voltage power lines.

(e) Portions of the surface may have been subjected to different metallurgical treatments. The head of a steel nail, for example, receives more stress during fabrication than does the nail's shaft. The iron on the head's surface therefore has a slightly greater Gibbs energy, making it destined to serve as the anode when the nail corrodes.

(f) Metal surfaces are a mosaic of small crystal faces. Not all these faces have the same crystallographic arrangement of atoms. Faces that have a high surface energy will become anodes. The adjacent diagram shows adjoining faces with two different atomic packing patterns on the surface of a metal. One of these will likely have a higher Gibbs energy than the other. Higher still will be the energy at the disordered "grain boundary" where the two crystals meet. Grain boundaries are commonly the sites of corrosive attack.

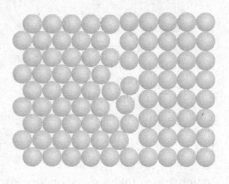

(g) Different concentrations of oxygen may exist at two sites on the metal surface, setting up a concentration cell (page 66). This **differential aeration** is a common cause of corrosion. For example, a steel pile driven into a river bed to support a jetty will experience water with a high oxygen activity at the water surface, but much lower oxygenation in the anoxic mud on the river's bottom.

(h) Corrosion itself may foster differential aeration. Thus, O_2 has difficulty diffusing into a crevice, which therefore becomes anodic. The ensuing corrosion enlarges the cavity, concurrently maintaining the difference in oxygen activities. Similarly, the rusty efflorescences that form on corroding iron impede access by oxygen to the underlying metal.

(i) Organisms living on or near a metal surface may increase or decrease the oxygen content, setting up a concentration differential and hence a corrosion cell.

(j) Likewise, biota often perturb the pH in their vicinity, so differential acidity may also set up a concentration cell, with the site of higher pH being anodic. Dissolved carbon dioxide can be the agent that creates an acidity gradient.

(k)   In the case of the corrosion of iron and its alloys, a greater accumulation of corrosion products in a particular surface region may foster faster corrosion at that site through a catalytic mechanism. The direct anodic oxidation of iron

11:10 $$Fe(s) \rightarrow 2e^- + Fe^{2+}(aq)$$

is a slow process that yields ions in the +2 oxidation state. But, as these ions accumulate, some are air-oxidized to ions in a higher oxidation state by the homogeneous reaction[1107]

11:11 $$4Fe^{2+}(aq) + O_2(aq) + 4H_3O^+(aq) \rightarrow 4Fe^{3+}(aq) + 6H_2O(\ell)$$

Because the produced Fe^{3+} ions are potent oxidizers of iron, the heterogeneous process

11:12 $$2Fe^{3+}(aq) + Fe(s) \rightarrow 3Fe^{2+}(aq)$$

then ensues. Thus Fe^{2+} ions play the role of an electrogenerated catalyst in the overall corrosion reaction, which is[1108]

11:13 $$4Fe(s) + 3O_2(aq) + 12H_3O^+(aq) \rightarrow 4Fe^{3+}(aq) + 18H_2O(\ell)$$

Understandably, but surprisingly, agitation slows corrosion by removing the catalyst.

Electrochemical Studies: corrosion potential and corrosion current

The proximity of the anode and cathode in a corrosion cell prevents any direct measurement of current, but it is straightforward to measure the **corrosion potential**, which is the null potential adopted by a corroding piece of metal, as measured by the apparatus illustrated in Figure 11-2 overleaf. The corrosion potential is an example of a **mixed potential**, one whose value is determined by two (or more) electrode processes. In the case of a metal being corroded by acid in an aqueous environment, the anodic corrosion reaction is

11:14 $$M(s) \rightleftarrows ne^- + M^{n+}(aq) \qquad \text{anodic reaction}$$

whereas the reduction process, in the absence of oxygen, is

11:15 $$2H_3O^+(aq) + 2e^- \rightleftarrows 2H_2O(\ell) + H_2(g) \qquad \text{cathodic reaction}$$

The net current, which is measured in polarization studies of corrosion, is the sum of the positive current from the metal dissolution process 11:14 and the negative current from hydronium ion reduction. These two currents sum to zero at the corrosion potential.

Corrosion reactions are slow by their very nature. They are often slow enough that neither ohmic polarization nor transport polarization (Chapter 10) is significant, whereas kinetic polarization is severe. Figure 11-3 (page 219) shows the kinetic polarization curves for reactions 11:14 and 11:15 and locates the corrosion potential. The situation is similar

---

[1107]   Reactions 11:11 and 11:12 are themselves composite processes composed of several mechanistic steps.

[1108]   Iron(III) ions go on to form the all-too-familiar rust, which is hydrated Fe_2O_3.

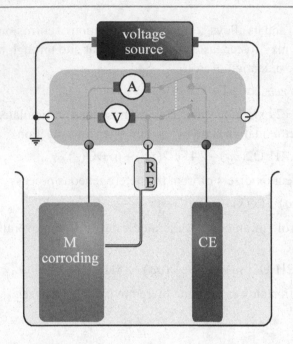

Figure 11-2 A three-electrode cell, served by a potentiostat, being used to study a corroding sample of metal. With the switch open, the corrosion potential E_{cor} is measured. Application of potentials close to E_{cor} permits a **polarization curve** to be recorded as in the green curve of Figure 11-3.

to that discussed on page 137, with the corrosion potential E_{cor} replacing the null potential, but there are two important differences. One is that currents now replace current densities, the latter being inappropriate because the areas of the metal surface devoted to the anodic and cathodic processes are unknown. The second distinction is that, whereas the upper and lower curves in Figure 7-4 represent the oxidative and reductive partial currents of the *same* reaction, the lime and violet curves in Figure 11-3 represent the net currents for each of two *distinct* reactions.

Figure 11-4 conveys the same information, but in a very different mode of presentation. Firstly, instead of current itself, it is the decadic logarithm of the absolute values of currents that are plotted against potential[1109]. Secondly, it is only the *partial* currents that are plotted and not the *net* currents, as in Figure 11-3. Because these partial currents are taken to vary exponentially with potential, in accord with the Butler-Volmer model, the lines in Figure 11-4 are all straight. Taking reaction 11:14 as an example, the partial oxidative and partial reductive currents are

$$11:16 \quad I_{ox} = A_a i_n \exp\left\{\frac{\alpha_{ox}F}{RT}[E - E_n]\right\} \quad \text{and} \quad I_{rd} = -A_a i_n \exp\left\{\frac{\alpha_{rd}F}{RT}[E - E_n]\right\}$$

where A_a is the area of the corroding sample serving an anodic role. As in Chapter 7, i_n and E_n respectively denote the exchange current density and the null potential for the metal dissolution reaction. The α's are the composite transfer coefficients (page 141). It follows

---

[1109] In the corrosion literature the potential axis is frequently plotted vertically. This explains why the "b" parameters, discussed later, are described as "slopes".

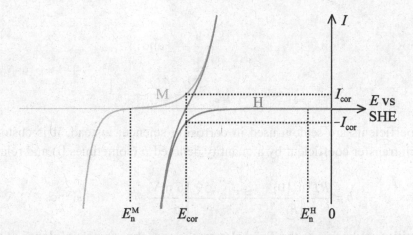

Figure 11-3 The upper curve, labeled M is the polarization curve of the metal describing the way in which the net current for metal dissolution depends on the electrode potential. The lower curve, labelled H similarly describes the polarization curve for reaction 11:15. The green curve is the overall current, as directly measurable, equal to the sum of the M and H curves; it crosses zero at the corrosion potential E_{cor}. The corrosion current I_{cor} reflects the rate of corrosion of the unpolarized metal.

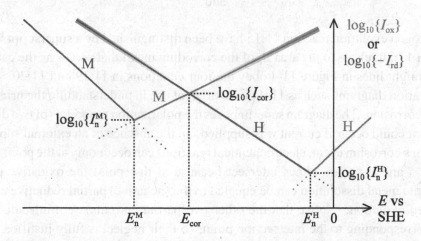

Figure 11-4 Logarithmic presentation of the dependence on electrode potential of the *partial* currents of the two reactions participating in the corrosion event. The two lines labeled M are the logarithmic magnitudes of the partial currents that sum to the M polarization curve in Figure 11-3. Similarly, the two other lines correspond to the H polarization curve in Figure 11-3. Where the lines are bordered in green, the total current closely approximates the reductive or oxidative partial currents. The slopes of the four lines are $-1/b_{rd}^{M}$, $1/b_{ox}^{M}$, $-1/b_{rd}^{H}$, and $1/b_{ox}^{H}$.

that

$$\log_{10}\{I_{ox}\} = \log_{10}\{A_a i_n\} + \frac{\alpha_{ox}F[E - E_n]}{RT\ln\{10\}} \quad \text{and}$$

11:17   Tafel equations

$$\log_{10}\{-I_{rd}\} = \log_{10}\{A_a i_n\} + \frac{\alpha_{rd}F[E - E_n]}{RT\ln\{10\}}$$

Transfer coefficients are seldom used in corrosion science; instead, it is customary to replace each transfer coefficient by a quantity denoted b (sometimes β) and related to α through

11:18 $$b = \frac{RT\ln\{10\}}{|\alpha|F} \xrightarrow{25°\,C} \frac{59.16\,\text{mV}}{|\alpha|}$$ Tafel slopes

The b quantities are known as the **Tafel slopes**; they are the *reciprocal* slopes of the lines in such graphs as Figure 11-4. The sign of the slope is generally ignored in citing b values. In this terminology, the equations in 11:17 become

11:19 $\log_{10}\{I_{ox}^M\} = \log_{10}\{A_a i_n^M\} + \dfrac{E - E_n^M}{b_{ox}^M}$ and $\log_{10}\{-I_{rd}^M\} = \log_{10}\{A_a i_{rd}^M\} - \dfrac{E - E_n^M}{b_{rd}^M}$

A superscript ^{M} has been inserted on each pertinent symbol to emphasize that this is a term relating to reaction 11:14, the M dissolution reaction. The corresponding equations

11:20 $\log_{10}\{I_{ox}^H\} = \log_{10}\{A_c i_n^H\} + \dfrac{E - E_n^H}{b_{ox}^H}$ and $\log_{10}\{-I_{rd}^H\} = \log_{10}\{A_c i_n^H\} - \dfrac{E - E_n^H}{b_{rd}^H}$

for the hydrogen evolution reaction 11:15 have been distinguished by a superscript ^{H}. The area term in 11:20 relates to the area A_c of the corroding metal that serves as the cathode. The four straight lines in Figure 11-4 obey the four equations in 11:19 and 11:20.

Polarization diagrams such as 11-4 prove of great use in understanding the principles underlying corrosion. The diagram superimposes the polarization properties of two distinct reactions that could occur if current was supplied. In the absence of an external supply of current, as in a corrosion event, electrochemical reactions can occur only at the point where the $\log_{10}\{I_{ox}^M\}$ and $\log_{10}\{-I_{rd}^H\}$ lines intersect because at that point the oxidative partial current for the metal dissolution can be equated to the (negative) partial reductive current for hydrogen evolution. Notice that the other two partial currents are miniscule at the potential corresponding to the intersection point, so their neglect is fully justified. This intersection point, therefore, serves to identify both the **corrosion potential**, E_{cor}, that the corroding metal adopts and the **corrosion current**, I_{cor}, that flows[1110]. The corrosion current resembles an exchange current (page 134) in that there is no net current flow, two partial currents being equal in magnitude and opposite in sign. Of course, knowledge of the corrosion current is technologically very valuable in being a quantitative measure of the amount of corrosion taking place.

[1110] Web#1110 derives expressions for the corrosion potential and the corrosion current.

A partial current can be measured experimentally only when its magnitude greatly exceeds that of its counterpart and therefore approximates the net current's magnitude. Thus, to find I_{cor} by the method suggested by Figure 11-4 involves extrapolation from the large-current regions shown bordered in green on the diagram. This may not always be possible. Moreover, there is no guarantee that the corrosion mechanism under high currents remains unchanged from that operative at the corrosion potential.

There are several other methods of measuring I_{cor}. In the poorly named **linear polarization technique**[1111], one applies a series of small potentials on either side of the corrosion potential to delineate the polarization curve (the green curve in Figure 11-3) that can be shown[1112] to obey the equation

11:21 $\qquad I = I_{cor} \left[\exp\left\{ \dfrac{E - E_{cor}}{0.434\,b_{ox}^{M}} \right\} - \exp\left\{ \dfrac{-(E - E_{cor})}{0.434\,b_{rd}^{H}} \right\} \right]$ $\qquad$ polarization curve

Because E_{cor} is easily found by examination of the curve, this equation has three unknowns, namely I_{cor}, b_{ox}^{M}, and b_{rd}^{H}, which can be evaluated by a nonlinear regression exercise. More crudely, one can determine the slope of the polarization curve at the corrosion potential, which has a reciprocal equal to the so-named **polarization resistance**:

11:22 $\qquad \dfrac{1}{\text{slope at } E_{cor}} = \dfrac{1}{(\mathrm{d}I/\mathrm{d}E)_{E_{cor}}} = \dfrac{0.434\,b_{ox}^{M} b_{rd}^{H}}{\left(b_{ox}^{M} + b_{rd}^{H}\right) I_{cor}} = R_{pol}$ $\qquad$ polarization resistance

The corrosion current can thus be found from the polarization resistance if the Tafel slopes are known or can be estimated. Other methods of determining the corrosion current make use of impedance spectroscopy (Chapter 15) or the loss of weight of a specimen exposed to the corrosive environment[1113].

Concentrated Corrosion: pits and crevices

Because it is locally intense, pitting corrosion can breach the integrity of a metal sheet, even though the total amount of corroded metal may be rather small. Often the pits may be of millimeter size, or smaller, and hidden from view by seemingly innocuous corrosion products, so that no warning of corrosion precedes the penetration of the sheet, with subsequent, and possibly disastrous, repercussions. Pitting corrosion of copper pipes by potable water is sometimes a problem in regions of "hard" water[925], but here we shall concentrate exclusively on the pitting corrosion of iron and steels. The chloride ion is implicated in almost all examples of the pitting of steel, so this variety of corrosion is

---

[1111] The name arose from the mistaken idea that the polarization curve is linear at the corrosion potential. The polarization curve is linear there only if $b_{ox}^{M} = b_{rd}^{H}$.

[1112] See Web#1112.

[1113] A thin square sheet of zinc, of 5.0 cm edge length, suspended in an aqueous acid solution, lost 12.4 mg of mass during a 24 h exposure. What was the corrosion current density? Check at Web#1113.

prevalent in marine environments, where road salt is used, and in chemical plants where chlorine compounds are employed.

When steels are examined in an experimental arrangement akin to Figure 11-2, with an aerated chloride-containing neutral aqueous environment, increasingly positive potentials promote some mild generalized corrosion, but only after a critical pitting potential is surpassed does pitting commence. For many steels, the **pitting potential** is about 0.24 V versus SHE, but increasing the chromium content of stainless steel pushes the pitting potential positively and delays the onset of pitting.

Neither the initiation of pitting nor its subsequent propagation has a chemistry that is perfectly understood. The passive layer that coats stainless and other steels may be regarded as FeOOH and it normally dissolves at a rate that is so slow as to be innocuous. When chloride is present, however, an accumulation of bound chloride appears on the metal surface at specific sites[1114] where pitting will subsequently occur. It is believed that dissolution is more rapid at such sites, producing the high-chloride "blanket", possibly by the reaction

11:23 $$FeOOH(s) + Cl^-(ads) \rightarrow FeOCl(s) + OH^-(aq)$$

Under the blanket, the passive layer is eventually breached and the pitting reaction takes over.

Essential to the maintenance of the pitting reaction is the cap of corrosion products, mostly gelatinous $Fe(OH)_3$, that forms over the developing pit, dissolving only slowly. This prevents the liquid within the pit from mixing with, and being diluted by, the solution outside. Migration of ions through the cap is able to occur and this is the mechanism by which chloride ions are able to enter the enlarging pit. Below the passive layer, the iron undergoes dissolution and hydrolysis, perhaps by the reaction sequence

11:24 $$Fe(s) + Cl^-(aq) \rightarrow 2e^- + FeCl^+(aq)$$

followed by

11:25 $$FeCl^-(aq) + 2H_2O(\ell) \rightarrow FeOH^-(aq) + H_3O^+(aq) + Cl^-(aq)$$

Notice that, whereas the initiation of the pitting process increases the pH through reaction 11:23, the content of the pit becomes increasingly acidic as a result of reaction 11:25. The dissolved Fe(II) species oxidizes as it leaves the acidic pit and some is subsequently precipitated, augmenting the cap with additional iron(III) hydroxide

11:26 $$FeOH^-(aq) + 2H_2O(\ell) + \tfrac{1}{2}O_2(aq) \rightarrow Fe(OH)_3(s) + OH^-(aq)$$

The electrons liberated by reaction 11:24 are dissipated by reduction

11:27 $$\tfrac{1}{2}O_2(aq) + H_2O(\ell) + 2e^- \rightarrow 2OH^-(aq)$$

at distant sites on the surface of the metal.

Figure 11-5 summarizes the processes that occur as pitting proceeds. Essentially, the

---

[1114] There is some evidence that pitting is initiated preferentially at sites where sulfide inclusions exist.

growing pit contains hydrochloric acid, sheltered by the cap from the outside solution and maintained by the entry of chloride ions through the cap and by the hydrolysis of the corrosion products within the pit. Ultimately the engorged pit penetrates to the opposite surface.

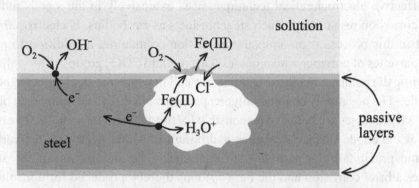

Figure 11-5 Putative mechanism of pitting corrosion.

The chemistry of **crevice corrosion**, illustrated in Figure 11-6, is similar to that of pitting corrosion. Crevice corrosion benefits from some geometric feature, such as an insulating washer or gasket, that shields a portion of the metal surface and provides a ready-made initiation site for corrosion. The corrosion removes oxygen from the sheltered site, so the differential aeration can then foster the breakdown of a passivating layer. Once initiated, crevice corrosion proceeds in much the same way as pitting corrosion, chloride ions again being part of the story.

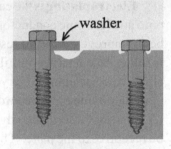

Figure 11-6 Sites of crevice corrosion.

Other mechanisms may be responsible for initiating crevice corrosion. For example, rainwater will remain longer at certain sites, such as ledges and beside bolts, and promote conditions that favor a discrimination between nascent cathodic and anodic regions, the latter being potential sites for crevice corrosion. Likewise scratching, denting and similar mechanical damage provides sites that eagerly lead to crevice and other forms of corrosion.

Fighting Corrosion: protection and passivation

Obvious ways to counter the corrosion of a vulnerable metal are to coat it with grease or paint, or to plate it with a less corrodible metal. These methods are in common use, and are effective, up to a point. **Powder coating** with plastics is a more permanent alternative.

Tin-plated sheet steel resists corrosion well, but should the plating be scratched, a site for anodic corrosion is exposed and, since the entire remaining surface is available to serve as the cathode of a corrosion cell, vigorous corrosion occurs at the damage site and the structure may fail more rapidly than were there no plating.

An effective electrochemical technique, used extensively in the metals industry for applying corrosion resistance to such steel products as car bodies, is **electropainting**[1115]. The paint in this process is an aqueous suspension containing, in addition to pigments, colloidal particles of polymer which possess carboxylate COO^- groups. The walls of large tanks holding the paint are made the cathode of a cell in which the steel object to be painted is the anode. The negatively charged polymer particles migrate to the immersed object and adhere. Two processes are believed responsible for the adherent deposition. The oxidation of water at the anode surface generates hydronium ions that neutralize the carboxylate groups, and precipitate the particle onto the steel. The anodization of the steel also encourages a brief corrosion and the $Fe^{2+}(aq)$ ions thereby produced form insoluble salts with the carboxylates. The electropainted coating is extremely uniform because deposition occurs preferentially on poorly coated areas, while well coated areas are severely insulated and attract little current[1116].

Electroplating is the cathodic deposition of a metal, usually from an aqueous solution, onto an electronic conductor, usually a different metal. The purpose of the metal layer may be to improve corrosion resistance, but electroplating is used for many other reasons, for instance to construct multilayer solid-state devices in the electronics industry, to reduce abrasive wear in machinery, to reduce thermal or electrical contact resistance, to deposit material for magnetic recording, or to impart an attractive luster to jewelry. The item to be plated must be scrupulously clean and may require pretreatment to ensure the surface will be receptive to the plate. The anode used in electroplating may be of the metal to be plated, or it may be an inert electrode from which oxygen evolves. In the latter case, but not the former, the plating bath will gradually become depleted. In addition to a salt of the metal in question (or salts of the metals, if an alloy is to be plated), the plating bath may include some or all of the following: acids or buffers (to control the pH), thickeners (to increase the viscosity), complexing agents (to reduce the proportion of free metal ions), brighteners (to affect the luster of the plate), wetting agents (surfactants to make the metal more hydrophilic), stress reducers (to make it less likely that the plate will flake off), and grain refiners (to improve the crystal structure). The temperature, salt concentration and current density are also factors that must be closely controlled. The precise role of these various factors is not always clearly understood, as electroplating is still in the process of emerging from an art to a science. As with all **electrocrystallization** (page 281), there are two stages in the formation of electroplates. Nanometer-sized nuclei are first formed. Then these

---

[1115] or "electrophoretic painting". The technique is akin to electrophoresis (page 152).

[1116] The ability of a bath to form uniform coatings is called its **throwing power** and is important also in both electropainting and electroplating.

nuclei aggrandize and become crystals. The rate of nucleus formation and the rate of their growth affect the quality of the plated metal and it is clear that the role played by many of the plating parameters is to influence these rates. When awkward shapes (such as a vase or tube) are to be plated, there is a tendency for the more accessible regions (corners, for instance) to receive a greater thickness than the more remote regions (such as inner surfaces), but this tendency can be countered by careful attention to the plating conditions[1116]. If the problem is insurmountable, **electroless plating** may be used instead of electroplating. In this nonelectrical alternative, the reduction to a metal is carried out chemically, in a bath that contains, in addition to the metal salt, a mild reducing agent such as formaldehyde or glucose.

Compounds termed **corrosion inhibitors** are valuable weapons in the battle against corrosion. These compounds, often amines or other nitrogen-containing organics, adsorb on metal surfaces, hampering either or both of the anodic or cathodic corrosion reactions. Importantly, they absorb especially well on the same "active sites", such as grain boundaries and inclusions, that are preferred locations for corrosive attack. Other inhibitors act by reacting chemically with the metal to produce adherent layers of salts, such as phosphates and chromates, that then confer protection. "Phosphating" – immersion in a hot (about 75°C) acidic (about pH 2.5) phosphate bath – is a common anti-corrosion pretreatment for steel. Oxidizing agents, such as concentrated nitric acid, can be used to thicken a preexisting oxide layer on iron; or the thickening may be brought about electrochemically, as in "anodized" aluminum.

Another approach is to add alloying ingredients to a vulnerable metal to improve corrosion resistance. Examples include the creation of stainless steel from iron, and the addition of manganese and magnesium to aluminum to improve its corrosion resistance. Of course, the metallurgist has many other criteria to meet (strength, flexibility, cost, etc.) besides resistance to corrosion, in designing an alloy for a particular purpose.

One electrochemical strategy to limit corrosion is named **cathodic protection**. The idea is to connect the corroding item to a less noble metal and let the latter corrode instead. Imagine that the corroding metal is iron, with zinc as the substitute victim. Drawn in the style of Figure 11-4, Figure 11-7 overleaf logarithmically plots the magnitudes of the partial currents when two metals – iron and zinc – corrode in the presence of an acid, serving as oxidizer. Because zinc is less noble than iron, its null potential is more negative than that of iron and, in consequence, its corrosion current is larger. When iron and zinc are both present, and electrically coupled, the mixed potential that results is that at which

11:28 $$I_{ox}^{Zn} + I_{ox}^{Fe} + I_{rd}^{H} = 0 \qquad \text{tripartite mixed potential}$$

This mixed potential lies between those of the individual metals but, because the corrosion current is larger, much closer to that of the zinc. In fact the potential is little changed from the value E_{cor}^{Zn} that the zinc would have adopted in the absence of the iron. Thus, as a result of the cathodic shift in its potential caused by the presence of the zinc, the iron corrodes at

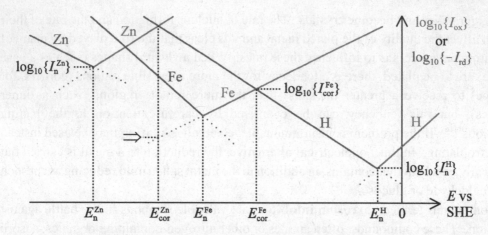

Figure 11-7 A graph versus potential of the logarithms of the absolute values of the partial currents for three reactions: $2H_3O^+(aq)+2e^- \rightleftarrows H_2(aq)+H_2O(\ell)$, $Fe(s) \rightleftarrows 2e^- +Fe^{2+}(aq)$, and $Zn(s) \rightleftarrows 2e^- +Zn^{2+}(aq)$. In each case, the oxidative partial current is shown in red and the reductive partial current in blue. The symbols $\log_{10}\{I_{cor}\}$ and E_{cor} show, for each metal, where the corrosion current and the corrosion potential would have been located in the absence of the other metal. When both metals are present and electrically connected, the corrosion potential is close to that for zinc alone, resulting in a greatly diminished corrosion current for the iron.

a rate close to that indicated by the $\Rightarrow$ arrow in Figure 11-7. The zinc, on the other hand, corrodes at a much greater rate than the iron did on its own, but the corrosion of iron is much less, which is the objective. The zinc has protected the iron at its own expense: it is known as a **sacrificial anode**. The zinc of galvanized iron serves this purpose for the humble trash can; unlike the case of tin plating, the zinc coating retains its protective martyrdom even if it is abraded. The hulls of steel ships and the metalwork of bridges are similarly protected by sacrificial anodes of magnesium, aluminum or zinc. In suitable circumstances, instead of a sacrificial anode, cathodic protection may be provided by electrically imposing a negative potential on the vulnerable metal.

As we have noted previously, the presence of tenacious and impermeable oxide layers[1117] prevents the corrosion of certain metals – notably aluminum, titanium, chromium, and nickel – that would otherwise corrode rapidly. These metals are said to be naturally **passive**. Passivity may also be conferred, at least temporarily, on other metals. As mentioned earlier, steels can acquire a passivating oxide layer by treatment with concentrated nitric acid, and iron may be passivated also by anodic polarization. Figure 11-8 shows how, during progressive positive polarization of iron, passivation occurs dramatically at a particular potential, the **Flade potential**[1118] or **passivation potential**.

[1117] Surprisingly, these layers are often electronically conducting.

[1118] Friedrich Flade, 1880–1916, German chemist killed in World War I.

Passivity is maintained as long as the potential remains, and this is the idea behind **anodic protection**. This strategy preserves the metal's passivity by applying a positive potential, either chemically or electrochemically. Sufficiently positive potentials destroy the passive layer and corrosion resumes.

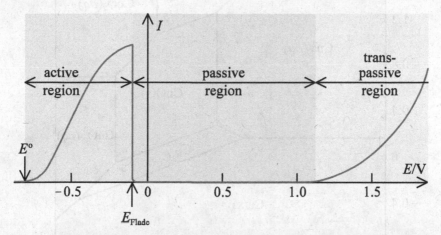

Figure 11-8 On progressively making the potential of an iron electrode more positive, corrosion suddenly ceases and the metal becomes passivated.

It can be argued that a passivating oxide layer must prevent the ingress of atmospheric oxygen to the underlying metal and this will be fostered if the oxide can pack tightly onto the surface lattice of the metal. Such a tight lattice might be expected to form if the molar volume of the oxide matches, or is somewhat larger than, that of the basal metal. Such considerations led Pilling and Bedworth to suggest in 1923 that a quantitative measure of the protectiveness of a particular oxide M_aO_b might be provided by the ratio[1119]

11:29 $$\frac{M_{oxide}\rho_{metal}}{aM_{metal}\rho_{oxide}}$$ Pilling-Bedworth ratio

where molar masses and densities are represented by subscripted M and ρ symbols. The two metallurgists argued that a value of the ratio of less that 1.0 would signify a layer that was open and unprotective, that a ratio between 1.0 and, say, 2.0 would correspond to a tight layer offering protection, and that if the ratio exceeded 2.0, the layer would buckle and then detach. The prediction is not without success as the values for Mg/MgO(0.8), Cd/CdO(1.2), Zn/ZnO(1.6), Ti/TiO$_2$(1.6), Al/Al$_2$O$_3$(1.7), Cu/CuO(1.8), Fe/Fe$_2$O$_3$(2.1) and V/V$_2$O$_5$(3.2) indicate, but the model on which the hypothesis is based is rather simplistic, and there are frequent exceptions to the rule.

The occurrence of passivation is often pH-dependent. The case of copper is illuminating in this respect. Copper oxidizes in both acidic and basic media. Figure 11-9

[1119] This is the ratio adopted by Pilling and Bedworth although, because areas are more pertinent than volumes, the ratio $(b/a)(M_{oxide}\rho_{metal}/M_{metal}\rho_{oxide})^{2/3}$ would seem more appropriate.

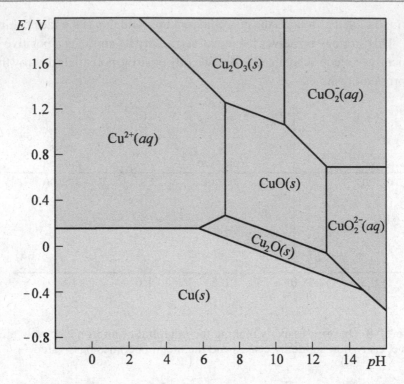

Figure 11-9 Pourbaix diagram for copper, showing regions in which copper is thermodynamically immune from corrosion, passivated, or liable to corrode.

shows the Pourbaix diagram[1120] (page 115) for copper. It predicts, for example, that as copper is progressively polarized positively at pH 5, it will start to corrode at a potential of 0.34 V, but will passivate at a potential of about 1.7 V. At other pH values, passivation may occur at one potential and be lost at another[1121]. Limitations of diagrams such as this are that they ignore kinetic factors and assume the absence of complexing agents. The Pourbaix diagram for copper tells quite a different story if chloride ions, for example, are present.

Extreme Corrosion: stress cracking, embrittlement, and fatigue

Three related phenomena are responsible for the failure of metal structures under the joint effects of tensile stress and electrochemical corrosion. The cracks that these engender

---

[1120] The diagram relates wholly to unit activities. For clarity, lines for other activities have not been included.

[1121] Find such a pH. At what pH values should copper be free from corrosion at any potential, according to the diagram? See Web#1121.

develop rapidly and have been the cause of many disastrous episodes, such as bridge collapses and aircraft crashes. In all three types of extreme corrosion, the crack develops perpendicularly to the direction in which the stress is applied, so that the residual metal experiences a steadily increasing stress, and crack propagation accelerates.

Stress corrosion cracking is failure caused by a steady tensile stress in conjunction with a corrosive environment. It was first reported in brass structures in the presence of ammonia, but has since been found to occur in a wide variety of alloys in diverse chemical conditions. Interestingly, pure metals are rarely affected. **Hydrogen embrittlement** is the name often given to a condition that leads to cracking induced by the presence of hydrogen. Hydrogen gas H_2 dissociates and the atoms dissolve in many metals, entering and weakening the lattice structure. Because hydrogen is often a byproduct

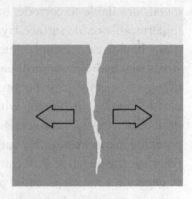

of electroplating or cathodic protection, these protective measures may ironically render the metal prone to hydrogen-provoked cracking. **Corrosion fatigue** cracking results from a fluctuating tensile stress. It is the number of fluctuations, rather than their frequency, that appears to be the paramount factor, though very high frequencies can be ineffectual.

Though there are similarities between the three crack-inducing phenomena, there are also marked differences. The chemical environment is all-important with stress corrosion cracking, whereas the other modes are less pernickety, and fatigue cracking can occur even in a vacuum. The morphology of the crack is also different: branched cracks are common only for stress corrosion cracking; the crack tip tends to be sharp (an acute angle) for a hydrogen embrittled metal, but blunt when fatigue is the cause. Already mentioned is the observation that pure metals are not usually susceptible to stress corrosion cracking, but no such immunity extends to the other modes. Further information is available elsewhere[1122].

The mechanism of crack propagation continues to be a matter of debate, and no one explanation may fit all examples. Undoubtedly, the metal atoms at the crack tip have a much enhanced activity, which may manifest itself in several ways. These atoms may dissolve anodically. They may even undergo chemical reactions that unstressed atoms do not experience. They may migrate outwards along the periphery of the crack, seeking a less stressful location. The metal lattice at the crack tip may deform, creating a region of plasticity and allowing local flow. Atoms at grain boundaries are more prone to migrate, even in the absence of stress, and a recrystallization of the grains may occur, with the developing crack following the new grain boundaries. The embrittling role of hydrogen may be more widespread, and it may be a factor also in stress corrosion cracking.

[1122] See R.N. Parkins, *Predictive approach to stress corrosion cracking* in: R.P. Gangloff and M.B. Ives (Eds.), *Environment Induced Cracking of Metals*, National Association of Corrosion Engineers, Houston, TX, 1990, pages 1–119.

Summary

Corrosion cells are unwanted galvanic cells in which the oxidation of a metal is brought about by the reduction of oxygen and/or water at an adjacent site. Almost all metals are liable to corrode, but some metals have natural passivity on account of their impermeable oxide surface layers. Differential aeration, contact with more noble metals, grain boundaries, complexing anions, inclusions, sulfurous environments, mechanical stress, and hydrogen embrittlement are among the factors that exacerbate corrosion. Ways in which metals may be protected from corrosion include alloying, electroplating, electropainting, treatment with corrosion inhibitors, cathodic protection, and anodic protection. Cathodic protection occurs by shifting the metal's potential from its corrosion potential to a more negative value, thereby decreasing the rate of the $M(s) \rightarrow ne^- + M^{n+}(aq)$ reaction; the negative shift can be achieved electrically or by attaching a less noble metal to generate a mixed potential and serve as a sacrificial anode. Anodic protection, which can be provided electrically or by strong oxidizing agents, provides or preserves a passivating oxide layer on susceptible metals.

Because the economic and environmental consequences of corrosion are considerable, major research effort is being devoted both to gaining a deeper understanding of the many facets of corrosion and to mitigating its effects.

12

Steady-State Voltammetry

The distinction between an equilibrium state and a steady state was drawn on page 160. An extended classification of electrochemical experiments according to their temporal behavior is:

Equilibrium state: No concentrations change with time in an equilibrium experiment and there are no fluxes of any kind. Cells at equilibrium were discussed in Chapter 3.

Steady state: No pertinent concentrations change with time when an experiment is in a steady state. Fluxes do occur, but their densities do not change with time within the space of interest. This chapter is devoted to electrochemical cells in a steady state.

Periodic state: When a cell is in a periodic state, concentrations change with time, as do fluxes. However, all important concentrations and fluxes return to their original values after integer multiples of a specific time interval, the period. Periodic electrochemical experiments[1201] are discussed in Chapter 15.

State	Concentrations	Flux densities
equilibrium	constant	none
steady	constant	constant
periodic	recurrent	recurrent
transient	varying	varying

Transient state: Concentrations and fluxes change nonperiodically with time in transient studies. Transient electrochemical experiments are discussed in Chapter 16, but already in Chapter 8 one transient experiment – the potential-leap experiment at a planar electrode – was introduced.

Steady states develop, rather than being established immediately. Thus a steady state is always preceded by a pre-steady interval, during which transience exists. In truth, a steady state is never strictly attained, but only approached as a limit. Nevertheless, a well designed experiment will have a very brief pre-steady interval prior to there being no detectable departure from the steady state.

The absence of a time variable is a valuable asset in steady-state experiments. Because

---

1201 Some authors classify periodic experiments as "steady state".

Electrochemical Science and Technology: Fundamentals and Applications, First Edition. Keith B. Oldham, Jan C. Myland, Alan M. Bond.
© 2012 John Wiley & Sons, Ltd. Published 2012 by John Wiley & Sons, Ltd.

the electrode potential is unchanging, there is no nonfaradaic current (page 126) to worry about. Moreover, because the electrical variables are constant, they can be measured with more certitude than can transient variables. The lack of a time variable makes modeling electrochemical phenomena simpler because there may be no partial differential equations to solve.

Before discussing steady-state voltammetry as such, some general features of voltammetry will be addressed.

Features of Voltammetry: purpose and classification

The goals of the electrochemistry discussed in Chapters 4, 5, 9, and 11 are eminently clear and practical. The purpose of the broad branch of electrochemistry that has acquired the name "voltammetry" is less specific. Primarily, the goal of voltammetry is to reach an understanding of the ways in which materials behave and interact electrochemically, usually without any direct practical application[1202]. Experiments are carried out to probe these behaviors and interactions, but an important part of voltammetry is the attempt to interpret the outcome of experiments quantitatively. Thus, much of voltammetry is concerned with **modeling** electrochemical phenomena, either to try to explain experimental results, or to suggest further experiments. The modeling is carried out by mathematical analysis or by computer simulation, or sometimes by a hybrid of the two.

There are many different forms of voltammetry, but common to most are the features that we cite below. Bear in mind that there will be exceptions.

(a) Voltammetry is carried out in three-electrode cells, served by a potentiostat (or galvanostat).

(b) The size, shape, and material of the working electrode are carefully chosen.

(c) The ionic conductor is a liquid: either an ionic liquid or a molecular solvent (often not water) with a high concentration of an appropriate supporting electrolyte.

(d) A single electroreactant species is predissolved in the ionic conductor, at a known uniform concentration, typically in the millimolar range.

(e) Attention is directed to excluding other electroactive species: high purity reagents are used; atmospheric oxygen is removed and excluded; the range of applied voltages is carefully chosen to avoid oxidation or reduction of the solvent, the electrode material, or the supporting electrolyte.

(f) Except when a rotating disk (or other convective) electrode is being used, the electrolyte solution is allowed to become quiescent prior to the experiment, and care is taken to avoid sources of natural convection.

[1202] An exception is provided by stripping voltammetry, discussed in Chapter 9, the goals of which are decidedly practical.

(g) Some means of referencing the working electrode's potential is provided[1203]: either a traditional reference electrode (pages 105 – 109), or an **internal reference** (page 353).

(h) Except in thin-layer voltammetry, the counter electrode is at least one millimeter from the working electrode and the space between the two is unimpeded, so that the transport field is essentially semiinfinite[1204].

(i) The cell is large enough that there is no significant depletion of the electroactive species overall. Thus, after stirring and a subsequent rest period, the result of a repeated experiment should duplicate the original[1205].

(j) Because the product of the electrode reaction under study is usually absent, the initial potential of the working electrode is not well controlled. Often the cell is in an open-circuit state prior to the experiment.

(k) The initial absence of the product also implies that the otherwise useful concepts of "null potential" and "overvoltage" become meaningless and are seldom used in discussions of voltammetry.

(l) An accurately generated potential program – such as a step, many steps, ramp(s), or more complex waveforms – is applied to the working electrode with respect to the reference electrode, starting at a well-defined instant. The resulting current is accurately recorded. Less often, the converse procedure is followed: a current program is applied and the cell voltage recorded.

(m) The experiment concludes before there is a danger of natural convection. 100 seconds is sometimes considered an upper limit to the duration of a well designed voltammetric experiment.

(n) Close attention is paid to two common voltammetric interferences addressed in Chapter 13: uncompensated resistance and capacitive current. These are reduced as far as possible and the residual effects are taken into account, or shown to be negligible.

(o) The results are displayed as a **voltammogram** – a graph of current versus potential – though other graphical representations of the data are sometimes preferable.

(p) The experimental voltammogram is compared with a model, which may be analytical, semianalytical, or simulated. Concurrence between the experiment and the model permits some conclusion, qualitative or quantitative, to be drawn.

(q) Frequently, a comparison of several voltammograms, measured with slightly changed conditions, is needed to reach some conclusion.

[1203] Occasionally, in difficult circumstances, a **quasireference**, such as a small platinum wire is employed. The electrochemistry at this electrode is unclear, but it is hoped that its potential remains constant.

[1204] **Semiinfinite transport** refers to conditions in which the ionic conductor may be treated as if it were of infinite extent. The "semi" prefix signifies that the ionic phase extends to infinity in one direction (outwards from the electrode) but not in the other. A distance of one millimeter is adequate to be treated as "infinite" because this length exceeds $\sqrt{Dt}$ in a voltammetric experiment.

[1205] However, there is now a minute amount of the product present, whereas there was none initially. This might sometimes be significant, especially if the product adsorbs.

An incomplete classification of voltammetry is shown below. Some voltammetric techniques have acquired distinctive names; these are noted in the chart, though they are not always fully descriptive.

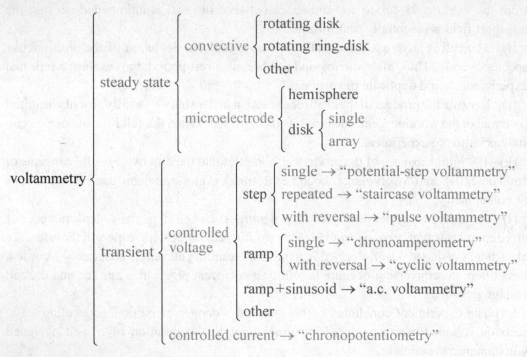

voltammetry

- steady state
  - convective
    - rotating disk
    - rotating ring-disk
    - other
  - microelectrode
    - hemisphere
    - disk
      - single
      - array
- transient
  - controlled voltage
    - step
      - single → "potential-step voltammetry"
      - repeated → "staircase voltammetry"
      - with reversal → "pulse voltammetry"
    - ramp
      - single → "chronoamperometry"
      - with reversal → "cyclic voltammetry"
    - ramp+sinusoid → "a.c. voltammetry"
    - other
  - controlled current → "chronopotentiometry"

Microelectrodes and Macroelectrodes: size matters

The establishment of a voltammetric steady state can be mediated by convection, but in the absence of this transport mode, a *steady state can be established only at an electrode surface that is small in all its linear dimensions.* What "small" means will be illustrated by making a detailed examination of the voltammetric behavior of the hemispherical working electrode pictured in Figure 12-1.

It was shown in Chapter 8 that, for the reaction $R(soln) \rightleftarrows e^- + O(soln)$ taking place under conditions of *planar*

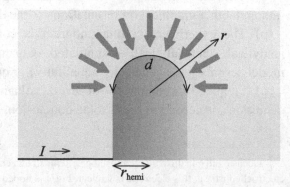

Figure 12-1 A hemispherical working electrode in cross section. The length d is the "superficial diameter" of the electrode, equal to πr_{hemi}. Arrows show the direction taken by electroreactant on its way to the electrode surface, which has as area of $A = 2\pi r_{hemi}^2$.

diffusion, the equations describing the concentration profile and the current in response to a potential leap are

12:1
$$c_R(x,t) = c_R^b \mathrm{erf}\left\{\frac{x}{\sqrt{4D_R t}}\right\}$$

and

planar electrode

potential-leap experiment

12:2
$$I(t) = FAc_R^b \sqrt{\frac{D_R}{\pi t}}$$

comprehensive solutions

Recall that, in a potential-leap experiment, the working electrode's potential is suddenly changed to a destination potential sufficiently positive that the concentration of R at the electrode surface is immediately diminished to zero, remaining at zero thereafter.

How are equations 12:1 and 12:2 modified if the electrode is a hemisphere, as in Figure 12-1? Repeating equation 8:23, Fick's second law takes the extended form

12:3
$$\frac{\partial}{\partial t} c_R(r,t) = D_R \frac{\partial^2}{\partial r^2} c_R(r,t) + \frac{2D_R}{r} \frac{\partial}{\partial r} c_R(r,t) \qquad \text{Fick's second law}$$

when the diffusion takes place with spherical symmetry. This equation must be solved subject to the following three boundary conditions:

12:4
$$c_R(r > r_{\text{hemi}}, 0) = c_R^b$$

12:5
$$c_R(r_{\text{hemi}}, t > 0) = 0$$

boundary conditions

and

12:6
$$c_R(r \to \infty, t) = c_R^b$$

These are strictly analogous to equations 8:24, 8:25, and 8:27 for the planar case. The solutions[1206] for the concentration profile and the current are

12:7
$$c_R(r,t) = c_R^b \left[1 - \frac{r_{\text{hemi}}}{r} \mathrm{erfc}\left\{\frac{r - r_{\text{hemi}}}{\sqrt{4D_R t}}\right\}\right]$$

hemispherical electrode

potential-leap experiment

and

comprehensive solutions

12:8
$$I(t) = FAc_R^b \left[\sqrt{\frac{D_R}{\pi t}} + \frac{D_R}{r_{\text{hemi}}}\right]$$

where erfc denotes the error function complement[829], erfc$\{y\}$ = 1−erf$\{y\}$. These

---

[1206] See Web#1206 for details of the solution, via Laplace transformation, to give equations 12:7 and 12:8.

equations[1207] are exact, irrespective of the size of the electrode.

At short enough times, the first bracketed term in expression 12:8 is dominant, whereas at long times the second term overwhelms the first, so that:

12:9 $\quad I(\text{short } t) \approx FAc_R^b \sqrt{\dfrac{D_R}{\pi t}} \qquad\qquad t \ll \dfrac{r_{hemi}^2}{\pi D_R}$ $\quad$ hemispherical electrode

and $\qquad\qquad\qquad\qquad\qquad\qquad\qquad\qquad\qquad\qquad\qquad\qquad$ potential-leap experiment

12:10 $\quad I(\text{long } t) \approx \dfrac{FAc_R^b D_R}{r_{hemi}} = 2\pi Fc_R^b D_R r_{hemi} \quad t \gg \dfrac{r_{hemi}^2}{\pi D_R}$ $\quad$ short or long times

We see that, early in the experiment, the current at the hemispherical electrode exactly matches that at a planar electrode of the same area (compare with equation 12:2), whereas eventually the current becomes time-independent, signifying a steady state. What exactly "early" and "eventually" imply is made clear in the table below. A **macroelectrode** is an electrode that is, or behaves as if it were, large and planar; a **microelectrode**[1208] is one at which a diffusion-mediated voltammetric steady state can be established conveniently. In constructing this table, the typical value $D_R = 8 \times 10^{-10}$ m^2 s^{-1} was assumed; moreover, the "$\ll$" and "$\gg$" symbols in 12:9 and 12:10 were interpreted as meaning that the linked terms differ by at least a factor of 100. Only the times shown in red are easily accessible in a typical voltammetric experiment. The table suggests that a particular hemispherical electrode can serve as a macroelectrode or a microelectrode in an appropriate time range, but never as both. If the radius of the hemisphere is greater than about 30 μm, the curvature of the electrode is of no account in a potential-leap experiment and the electrode functions as a macroelectrode, at least early in the experiment. If the hemispherical radius is less than

Radius of hemisphere	Time up to which the electrode behaves as a **macroelectrode**	Time beyond which the electrode behaves as a **microelectrode**
0.1 μm	40 ns	0.4 ms
1 μm	4 μs	40 ms
10 μm	0.4 ms	4 s
100 μm	40 ms	400 s
1 mm	4 s	11 h

[1207] Integrate equations 12:2 and 12:8 with respect to time, thereby deriving expressions showing how the charge $Q(t)$ increases. Show that at long times the charge increases linearly with t for the hemisphere but as $\sqrt{t}$ for the plane. Draw a graph showing curves for the charge following a potential leap for a planar electrode, and for a hemispherical electrode of equal area. Check at Web#1207.

[1208] The terms **nanoelectrode** and **ultramicroelectrode** are also in use.

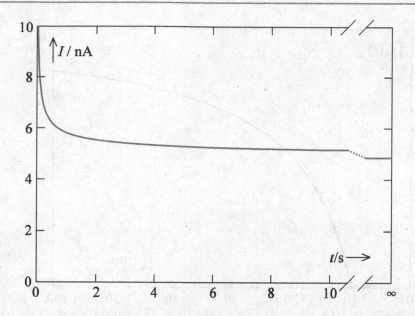

Figure 12-2 The current predicted by equation 12:8 when a potential leap is applied to a hemispherical electrode of radius 10 μm. The sudden positive change in electrode potential causes the reaction R(*soln*) → e⁻ + O(*soln*) to proceed with total transport polarization, so that R has zero concentration at the electrode. The R species is taken to have a bulk concentration of 1.00 mM and a diffusivity of 8×10^{-10} m² s⁻¹.

about 30 μm, a steady current will be attained during the time window accessible voltammetrically. Between the two times listed in the table, the electrode is in a transient state, but the current has departed from simple cottrellian behavior.

Figure 12-2 shows the response of a typical microelectrode to a potential leap. The current rapidly achieves a steady state. Bear in mind that, even though the current may have attained a steady value, there is not a true steady state uniformly throughout the entire cell. Obviously, since species R is being steadily removed from the cell, the depletion zone must continue to expand, even though the concentration profile near the electrode (and therefore the current) has become unchanging. Figure 12-3 overleaf illustrates this.

Steady-State Potential-Step Voltammetry: reversibility

Continuing to use the steady-state hemispherical microelectrode and the R(*soln*) ⇌ e⁻ + O(*soln*) electrooxidation as our exemplar, we next consider the effect of a sudden permanent positive change in the potential of a previously inactive electrode to a constant value E that is inadequate to cause total transport polarization. Such an experiment is described as a **potential-step** experiment. For the potential-*leap* experiment of the

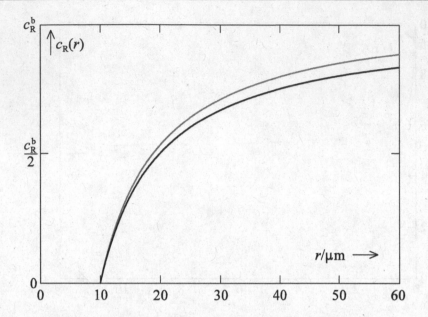

Figure 12-3 Concentration profiles during potential-leap voltammetry at a hemispherical microelectrode. The concentration gradient close to the electrode (and hence the current) has become constant, though further away the concentration continues to deplete. The graphs illustrate equation 12:7 with the following values: $r_{hemi} = 10.0\ \mu m$, $D_R = 8 \times 10^{-10}\ m^2\ s^{-1}$, $t = 10$ s and ∞.

previous section, it sufficed to examine the concentration profile of R, but now the concentration of O also must be tracked. Thus, to predict the outcome of this experiment, the equations to be solved are *two* instances of Fick's second law for spherical diffusion,

12:11
$$\frac{\partial}{\partial t}c_i(r,t) = D_i \frac{\partial^2}{\partial r^2}c_i(r,t) + \frac{2D_i}{r}\frac{\partial}{\partial r}c_i(r,t) \qquad i = R, O$$

together with a number of boundary conditions. In this section, however, our interest will be restricted to the steady state itself, to the exclusion of the pre-steady interval. That being so, the left-hand member of equation 12:11 is taken as zero and the law becomes abridged to

12:12
$$\frac{d^2}{dr^2}c_i(r) + \frac{2}{r}\frac{d}{dr}c_i(r) = 0 \qquad i = R, O$$

This is an ordinary differential equation that easily integrates[1209] to

12:13
$$c_i(r) = \frac{a_i}{r} + b_i \qquad i = R, O$$

where the a and b terms are constants. The boundary conditions $c_R(\infty) = c_R^b$ and

---

[1209] Perform two integrations of equation 12:12 to arrive at 12:13. See Web#1209.

$c_R(r_{hemi}) = c_R^s$ enable a_R and b_R to be identified, whence the concentration profile of R is

12:14
$$c_R(r) = c_R^b - (c_R^b - c_R^s)\frac{r_{hemi}}{r}$$

Similarly, from the boundary conditions $c_O(\infty) = 0$ and $c_O(r_{hemi}) = c_O^s$, it follows that

12:15
$$c_O(r) = c_O^s \frac{r_{hemi}}{r}$$

These steady-state concentration profiles[1210] are seen to have the very simple shapes[1211] displayed in Figure 12-4. From equation 12:14, the concentration gradient of the reactant R is $(c_R^b - c_R^s)r_{hemi}/r^2$ and its value at the electrode surface is therefore $(c_R^b - c_R^s)/r_{hemi}$. Similarly, from equation 12:15, the concentration gradient of O at the electrode surface is $-c_O^s/r_{hemi}$. Thence, chains of equalities

12:16
$$\frac{I}{F} = \frac{Ai^s}{F} = \begin{cases} -Aj_R^s = AD_R\left(dc_R/dr\right)^s = AD_R(c_R^b - c_R^s)/r_{hemi} = 2\pi D_R r_{hemi}(c_R^b - c_R^s) \\ Aj_O^s = -AD_O\left(dc_O/dr\right)^s = AD_O c_O^s/r_{hemi} = 2\pi D_O r_{hemi} c_O^s \end{cases}$$

lead to alternative expressions for the current[1212]. Equating these two expressions gives a

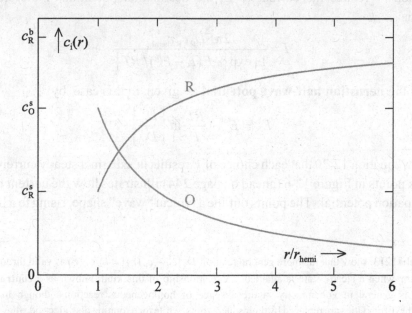

Figure 12-4 Steady concentration profiles of the reactant and product at a hemispherical microelectrode following a potential step.

[1210] Employ Fick's first law to derive expressions for $j_R(r)$ and $j_O(r)$; then use them to confirm equations 12:16 and 12:17. See Web#1210.

[1211] They are portions of rectangular hyperbolas in different orientations.

[1212] Notice that equation 12:16 incorporates the assignment of transport coefficients reported in the table on page 169.

simple relation

12:17 $$D_R[c_R^b - c_R^s] = D_O c_O^s$$

linking the concentrations[1213] of R and O. But there is a second linkage; the form that this adopts depends on whether or not kinetic polarization is present.

If kinetic polarization is absent, because the reaction is sufficiently fast, Nernst's law applies in the form of the equation

12:18 $$\frac{c_O^s}{c_R^s} = \exp\left\{\frac{F}{RT}\left(E - E^{o\prime}\right)\right\} \qquad \text{Nernst's law}$$

In voltammetry, the adjective **nernstian** is applied to reactions taking place without significant departure from the Nernst equation. It is now a matter of straightforward algebra[1214] to combine equations 12:16, 12:17, and 12:18 into

12:19 $$I = \frac{2\pi F D_O D_R c_R^b r_{\text{hemi}}}{D_O + D_R \exp\left\{-F(E - E^{o\prime})/RT\right\}}$$

nernstian steady-state voltammetry at a hemi-spherical microelectrode

This equation[1215] relates the current to E, the destination potential of the step. An alternative is

12:20 $$I = \frac{2\pi F D_R c_R^b r_{\text{hemi}}}{1 + \exp\left\{-F(E - E^h)/RT\right\}}$$

where E^h is the **nernstian half-wave potential**[1216] given, in this case, by[1217]

12:21 $$E^h = E^{o\prime} - \frac{RT}{F}\ln\left\{\frac{D_O}{D_R}\right\}$$

We see from equation 12:20 that each choice of E results in a distinct steady current I. The set of **black** points in Figure 12-6 (ahead on page 244) illustrates how the current depends on the destination potential. The points outline a typical "wave" shape, rising to a limiting-

---

[1213] As in Web#1213, show that the more general relation $D_R[c_R^b - c_R(r)] = D_O c_O(r)$ is valid throughout the ionic conductor when a steady state is attained. A relationship of this kind, known as a **contradiffusion relationship**, is general in voltammetry in the absence of homogeneous reactions, though in transient voltammetry the diffusivities are replaced by their square roots, while for a rotating disk electrode they are raised to the two-thirds power.

[1214] Carry out the algebra to derive equation 12:19 and convert it into 12:20; or see Web#1214.

[1215] Note that equation 12:19 simplifies to 12:10, when E is so positive that the "step" becomes a "leap".

[1216] Our notation distinguishes between the actual half-wave potential $E_{1/2}$ and the nernstian half-wave potential E^h. The two are identical for a reversible process, but otherwise $E_{1/2}$ is more extreme (more positive for an oxidation).

[1217] If the two diffusivities are close in value, the distinction between E^h and $E^{o\prime}$ is often insignificant. A 4% disparity between the two D values generates a one millivolt difference.

current plateau. We have more to say about the shape of such voltammetric waves in a section later in this chapter.

If kinetic polarization exists – that is, if the reaction is slow enough that its rate is pertinent – then, instead of Nernst's equation, we must use the Butler-Volmer equation 7:26. Combination of this relationship with equations 12:16 and 12:17 leads, after considerable algebra[1218], to a complicated equation similar to the relationship 10:30 on page 206

$$12:22 \quad I = \frac{2\pi F c_R^b r_{hemi} \exp\left\{ F(E - E^{o\prime})/RT \right\}}{\dfrac{\exp\left\{ \alpha F(E - E^{o\prime})/RT \right\}}{k^{o\prime} r_{hemi}} + \dfrac{1}{D_O} + \dfrac{\exp\left\{ F(E - E^{o\prime})/RT \right\}}{D_R}}$$

hemispherical microelectrode potential step

This is an equation more general than 12:19, to which it reduces as $k^{o\prime} \rightarrow \infty$. Equation 12:22 reveals terms originating from three participating polarizations: kinetic polarization, plus the transport polarizations of each species. The expression may be made more transparent by rewriting[1219] it as a **reciprocal sum formula**[1220]:

$$12:23 \quad \frac{1}{I} = \frac{1}{I_{kin}} + \frac{1}{I_{rem}} + \frac{1}{I_{lim}} \quad \text{where} \begin{cases} I_{kin} = \dfrac{F A c_R^b k^{o\prime}}{\left(D_O / D_R \right)^{1-\alpha}} \exp\left\{ \dfrac{(1-\alpha)F}{RT}(E - E^h) \right\} \\[2ex] I_{rem} = 2 F c_R^b D_R d \exp\{ F(E - E^h)/RT \} \\[2ex] I_{lim} = 2 F c_R^b D_R d \end{cases}$$

To establish uniformity with the findings of the next section, we have replaced $2\pi r_{hemi}^2$ by the area A in the expression for I_{kin}. We have also replaced πr_{hemi} by the superficial diameter d (Figure 12-1) in the I_{rem} and I_{lim} expressions.

A current obeying such a reciprocal sum formula is largely governed by whichever of the three component I's has the *smallest* magnitude. Here I_{kin} is the kinetically-controlled current; it alone would apply, so that I would equal I_{kin}, if there were no transport limitation whatsoever. I_{rem} is the removal-controlled current, the current that would flow if the diffusion of O away from the electrode provided the sole limitation to the current. I_{lim} is the limiting current; this is the maximum current possible, invariably attained at the most positive potentials; it is governed by the diffusion of R to the electrode. The electrochemistry requires three processes: supply of R to the electrode, conversion of R to O, and removal of O from the electrode; each of the subscripted I's is the current that would flow if only one of these processes were slow enough to overwhelmingly control the rate of the oxidation process.

[1218] See Web#1218 for the derivation of equation 12:22, using equation 7:26.

[1219] Convert equation 12:22 to 12:23 or consult Web#1219.

[1220] Mathematicians refer to 12:23 as a harmonic relation; I is the **harmonic sum** of I_{kin}, I_{rem}, and I_{lim}.

Note that the three contributions to the right-hand side of equation 12:23 depend differently on potential. This is brought out most clearly by plotting the logarithm of the current versus potential, as in Figure 12-5. These three diagrams display the linear graphs of the logarithms of the component currents, I_{kin}, I_{rem}, and I_{lim}, together with the resultant current I, which is a curve whose shape is determined, through equation 12:23, by the locations of three green straight lines relative to the red and blue lines. Observe that the resultant current never exceeds any of its components and is always close to the smallest of the three, as the reciprocal sum formula requires.

The three diagrams differ only in the magnitude of the important dimensionless parameter

12:24 $$\lambda = \frac{I_{kin}}{I_{rem}} \exp\left\{ \frac{\alpha F}{RT}\left(E - E^h\right) \right\} = \frac{k^{o\prime} r_{hemi}}{D_R^\alpha D_O^{1-\alpha}} \quad \begin{array}{l} \text{microhemisphere} \\ \text{voltammetry} \end{array}$$

that we term the **reversibilty index**[1221]. **Reversibility**[1222] reflects the importance of kinetic polarization compared with other polarizations in a particular experiment. If kinetic polarization is trivial, because the reaction is sufficiently fast, the experiment occurs reversibly.

In diagram (a), λ has a large value. Notice that here I_{kin} is never close to becoming the smallest of the three component currents and therefore has almost no influence on the resultant current I, which obeys $(1/I) \approx (1/I_{rem}) + (1/I_{lim})$. Electrochemists use the term **reversible** to describe this type of behavior, in which the kinetics is fast enough to be unimportant. The resulting voltammogram is identical to that described in the nernstian case; in fact the terms "reversible" and "nernstian" are often used interchangeably.

In diagram (b), the reversibility index has an intermediate value such that each of the three component currents – I_{kin}, I_{rem}, and I_{lim} – serves as the smallest in different potential ranges, so none of them may be ignored. The full equation 12:23 must be used in this circumstance, to which the term **quasireversible** is given.

Except at the most negative potentials (where the current is miniscule anyway), I_{rem} is always much larger than the other two components in diagram (c) of Figure 12-5 and therefore plays no significant role in the resultant current, which obeys $(1/I) \approx (1/I_{kin}) + (1/I_{lim})$. This scenario is described as **irreversible** voltammetry, and is encountered when the reversibility index is small.

Figure 12-6, on page 244, shows a voltammogram corresponding to each of the three diagrams, (a), (b), and (c), in Figure 12-5 providing an example of a **reversible**, a quasireversible and an irreversible steady-state voltammogram at a hemispherical

---

[1221] Derive an expression for λ in terms of $E_{1/2}-E^h$ and thereby, as in Web#1221, show that $1/\lambda = 2\sinh\{F(E_{1/2}-E^h)/2RT\}$ if $\alpha = \frac{1}{2}$.

[1222] Confusingly "reversibility" has other meanings in some branches of chemistry, including electrochemistry. Reversibility is a property of an experiment, not of a reaction. When electrochemists incautiously speak of a "reversible electrode reaction" they mean a reaction that behaves reversibly in a particular experiment.

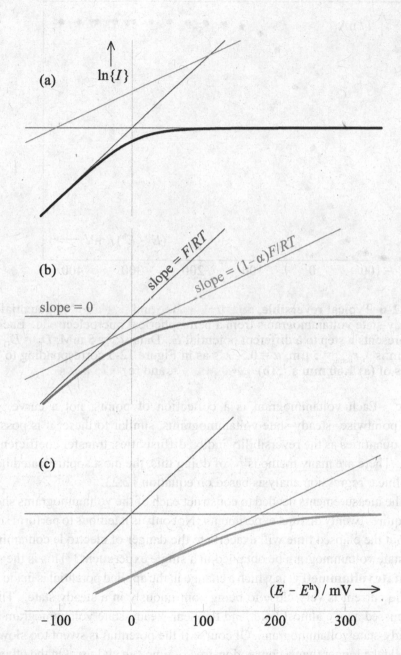

Figure 12-5 Logarithmic illustration of the reciprocal sum formula 12:20. On each diagram, the green line represents I_{kin}, the blue line represents I_{rem}, and the red line represents I_{lim}. The fourth curve in each diagram logarithmically represents the reciprocal sum of the three straight lines. The three diagrams, scaled for $\alpha = \frac{1}{2}$, differ only in the magnitude of the reversibility index: (a) $\lambda = 8$, (b) $\lambda = 1/2$, (c) $\lambda = 1/32$.

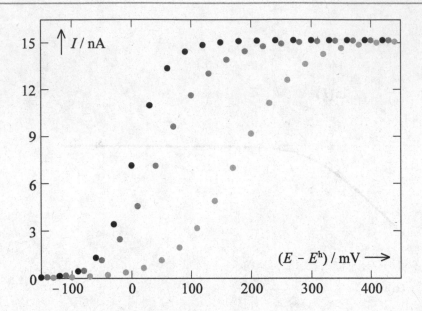

Figure 12-6 Typical **reversible**, quasireversible and irreversible potential-step steady-state voltammograms from a hemispherical microelectrode. Each point represents a step to a different potential E. Data: $c_R^b = 5$ mM, $D_R = D_O = 1 \times 10^{-9}$ m^2 s^{-1}, $r_{hemi} = 5$ μm, $\alpha = 0.5$, λ's as in Figure 12-5 corresponding to $k^{o\prime}$ values of (a) **1.60 mm s^{-1}**, (b) 0.100 mm s^{-1}, and (c) 6.25 μm s^{-1}.

microelectrode. Each voltammogram is a collection of points, not a curve. From experimental pointwise steady-state voltammograms, similar to these, it is possible to calculate such quantities as the reversibility index, diffusivities, transfer coefficients, and rate constants. There are many methods[1223] of doing this, the most sophisticated being to perform a nonlinear regression analysis based on equation 12:23.

To make the measurements needed to construct each of the voltammograms shown in Figure 12-6 requires twenty distinct experiments. Not only is it tedious to perform so many experiments, but the elapsed time will exacerbate the danger of electrode contamination. Can a steady-state voltammogram be obtained in a single experiment? This is the goal of **near-steady-state voltammetry**, in which a change in the applied potential is made slowly enough that the current is very close to being continuously in a steady state. Then the principles discussed above almost hold, and the near-steady-state voltammogram almost matches a steady-state voltammogram. Of course if the potential is swept too slowly, the experiment will take longer than a convection-free regime can endure. On the other hand, too fast a potential sweep will lead to a transient current close to those discussed in

[1223] See Web#1223 for a graphical approach to analyzing steady-state voltammograms. Analysis is simple in reversible cases; of course $k^{o\prime}$ cannot be found. Analysis is also straightforward for irreversible voltammograms; α can be found but $k^{o\prime}$ and E^h cannot be disentangled. Information content is greatest, but hardest to extract, in the quasireversible case.

Chapter 16, producing a voltammogram very different from the sought steady-state version. In practice, a compromise is struck.

As might be expected, if one starts at a potential at the foot of the steady-state oxidation wave, and gradually increases the potential, one obtains a near-steady-state voltammetric wave that is somewhat lower than the true steady-state wave, as in the **red** curve of Figure 12-7, but which eventually reaches the correct plateau. If one now reverses the direction of potential change[1224], the current returns along a path that mostly lies somewhat above the true steady-state wave, as does the **blue** curve. If the **red** and **blue** experimental curves nearly overlap, the inference is that a steady state was almost attained and that a curve interpolated between the two gives a good representation of the true steady-state voltammogram.

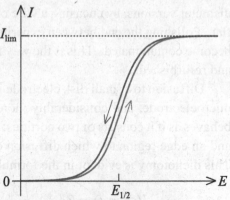

Figure 12-7 Near-steady-state voltammetry. The **red curve** is obtained as the potential slowly becomes more positive. The **blue curve** is obtained after reversal, as the potential slowly returns to its starting value.

One feature of steady-state voltammetry is that it does not matter whether one applies a constant potential and measures the ultimate current, or vice versa (apply I and measure E). Thus a rapid way of finding $E_{1/2}$ is to apply a current one-half the size of I_{lim}.

The Disk Microelectrode: convenient experimentally, awkward to model

As we have just seen, the modeling of electrochemistry at a hemispherical microelectrode is straightforward. Unfortunately such electrodes are extremely difficult

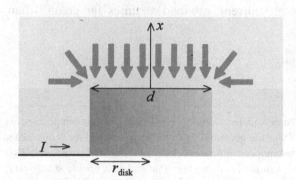

Figure 12-8 A disk working electrode in cross section. The length d is the "superficial diameter" of the electrode, equal to $2r_{disk}$. Arrows show the direction taken by electroreactant on its way to the electrode surface, which has an area of $A = \pi r_{disk}^2$.

[1224] The potential program is the same as that in the popular cyclic voltammetry (pages 346–362), with a slow sweep rate.

to fabricate. Moreover, once fabricated, they are difficult to clean. In comparison, the **disk electrode**[1008], shown in cross section in Figure 12-8, is rather easily made, even in small diameter versions, by encasing a wire or rod in glass or plastic, and then grinding the end flat. Polishing the end between experiments can restore the electrode after its surface has become contaminated. This is the way that the vast majority of microelectrodes are made and refurbished.

Diffusion to a small disk electrode is quite different from diffusion to a hemispherical microelectrode, and considerably more difficult to model[1225]. At short times, the disk behaves as if it consists of two portions: the central region, to which the diffusion is planar, and an edge region to which diffusion occurs convergently, as suggested in Figure 12-8. This dichotomy is evident in the formula[1226]

12:25 $$I(\text{short } t) \approx FAc_R^b \left[\sqrt{\frac{D_R}{\pi t}} + \frac{D_R}{r_{\text{disk}}} \right] \qquad \begin{array}{l} \text{disk electrode at short times} \\ \text{potential-leap experiment} \end{array}$$

that describes the *early* response of the disk electrode to a large potential step. Notice the similarity of this result to that for the current given in equation 12:8. However, unlike the behavior of the hemispherical electrode, this initial formulation does not endure. Instead the current comes to be described by the precise but complicated bipartite formulas

12:26 $$\left\{ \begin{array}{l} \dfrac{I(t \le 1.3)}{\pi F c_R^b D_R r_{\text{disk}}} = \dfrac{1}{\sqrt{\pi t}} + 1 + \sqrt{\dfrac{t}{4\pi}} - \dfrac{3t}{25} + \dfrac{3t^{3/2}}{226} \\[4mm] \dfrac{I(t \ge 1.3)}{\pi F c_R^b D_R r_{\text{disk}}} = \dfrac{4}{\pi} + \dfrac{8}{\sqrt{\pi^5 t}} + \dfrac{25t^{-3/2}}{2792} - \dfrac{t^{-5/2}}{3880} - \dfrac{t^{-7/2}}{4500} \end{array} \right\} \; t = \dfrac{D_R t}{r_{\text{disk}}^2} \quad \begin{array}{l} \text{microdisk} \\ \text{potential leap} \\ \text{for all times} \end{array}$$

The long-time formulation is seen to be

12:27 $$I(\text{long } t) \approx 4 F c_R^b D_R r_{\text{disk}} \left[1 + \frac{2r_{\text{disk}}}{\sqrt{\pi^3 D_R t}} \right] \qquad \begin{array}{l} \text{disk electrode at long times} \\ \text{potential-leap experiment} \end{array}$$

This equation predicts that the steady-state current, attained at times far greater than $4r_{\text{disk}}^2 / \pi^3 D_R$, is

12:28 $$I = 4 F c_R^b D_R r_{\text{disk}} \qquad \text{steady-state disk current following a potential leap}$$

[1225] Mathematical analysis of steady diffusion to a disk under *reversible* conditions, though complicated, is tractable, as described in Web#1225. The lack of uniform accessibility to the electrode surface is not then a difficulty, because the electrode surface serves as an equiconcentration surface. In the absence of reversibility, however, this is not so. Only numerical methods can handle quasireversible and irreversible steady-state voltammetry at microdisks.

[1226] Rewritten as $I(\text{short } t) / F c_R^b D_R = (A/\sqrt{\pi D_R t}) + (P/2)$, where P is the perimeter (edge length), this equation applies to an inlaid electrode of *any* size or shape, not only one in which the inlay is circular.

This important result is know as the **Saito equation**[1227].

Saito's result is one of the items listed in the table below. Features of currents at disk and hemispherical microelectrodes subjected to a potential leap are compared in this table. It might be expected that the two kinds of microelectrode would have the same current if their areas were identical. This is true of the initial current, but not when the steady state is reached. To have equal steady currents, the two electrodes must have identical **superficial diameters**. The superficial diameter d is the distance measured across the electrode, as illustrated in Figures 12-1 and 12-8. The table shows that the steady-state response to a potential leap is identical for a microdisk and a microhemispherical electrode, provided that their superficial diameters are the same, even though the disk then has an area about 23% larger than the hemisphere.

	Hemisphere	Disk	In terms of d
Superficial diameter, d	πr_{hemi}	$2r_{disk}$	
Area, A	$2\pi r_{hemi}^2$	πr_{disk}^2	
Initial current	$FAc_R^b\sqrt{\dfrac{D_R}{\pi t}}$	$FAc_R^b\sqrt{\dfrac{D_R}{\pi t}}$	
Time for current to fall to twice the steady value	$0.0507\dfrac{A}{D_R}$	$0.0479\dfrac{A}{D_R}$	
Time to reach within 1% of the steady current	$3183\dfrac{r_{hemi}^2}{D_R}$	$1290\dfrac{r_{disk}^2}{D_R}$	$323\dfrac{d^2}{D_R}$
Steady current	$2\pi Fc_R^b D_R r_{hemi}$	$4Fc_R^b D_R r_{disk}$	$2Fc_R^b D_R d$

Because of the difficulty in modeling steady-state voltammetry at a disk microelectrode other than when the behavior is reversible, it is common practice to treat the electrode as if it were a hemisphere of identical superficial diameter, in effect applying the reciprocal sum formula 12:23. Even though this is known to be inexact, the errors thereby introduced may often be less than other errors inherent in the experiment. However, the formulation

12:29
$$\frac{1}{I}=\frac{4}{\pi I_{kin}}\left[\frac{\dfrac{12}{\pi I_{kin}}+\dfrac{1}{I_{rem}}+\dfrac{1}{I_{lim}}}{\dfrac{6}{I_{kin}}+\dfrac{1}{I_{rem}}+\dfrac{1}{I_{lim}}}\right]+\frac{1}{I_{rem}}+\frac{1}{I_{lim}}$$

steady state

microdisk

potential-step

comprehensive

[1227] Y. Saito, *Review of Polarography (Japan)*, 15, 1968, 177.

which is precise[1228], corrects the errors in the reciprocal sum formula 12:23 when it is applied to a disk microelectrode. The subscripted I symbols have definitions unchanged from those in 12:23. All degrees of reversibility are embraced by equation 12:29.

Small electrodes necessarily generate small currents, which may be difficult to measure accurately. For this reason, and to improve reproducibility, **microelectrode arrays** are sometimes used. The spacing of the individual microdisks is made sufficient to avoid interference between the depletion zones surrounding each disk. They generate volt-ammograms resembling Figure 12-9, with half-wave potentials that match Figure 12-10.

Rotating Disk Voltammetry: a spinning disk electrode without and with a ring

The rotating disk electrode was mentioned on page 118 and is described in detail on pages 161 – 165. This electrode is occasionally used in the transient or periodic state, but its main feature is its ability to provide robust steady-state data without a time constraint. The rotating disk electrode provides more powerful access to voltammetric steady states than is attainable with microelectrodes. One advantage over microelectrodes is that a steady current is reached rapidly and is maintained indefinitely. This means that one may scan the applied voltage without the hysteresis problem that attends near-steady-state voltammetry with microelectrodes, provided that the scan rate is sufficiently slow. Another advantage is that the variability of rotation speed provides a tool by means of which the thickness of the transport zone may be adjusted, albeit only over a restricted range. The practical outside limits on the rotation speed are roughly[1229]

12:30 $10 \text{ Hz} < \omega < 1000 \text{ Hz}$

in aqueous solution. Above the upper limit there is danger of turbulent flow developing; below the lower limit there may be interference from natural convection.

For our standard $R(soln) \rightleftarrows e^- + O(soln)$ reaction with O initially absent, equation 8:49 shows that the current is given by

12:31 $\dfrac{I}{F} = \dfrac{Ai^s}{F} = \begin{cases} -Aj_R^s = v_L AD_R^{2/3} \omega^{1/2} (\rho/\eta)^{1/6} \left(c_R^b - c_R^s \right) & \text{rotating disk electrode} \\ Aj_O^s = v_L AD_O^{2/3} \omega^{1/2} (\rho/\eta)^{1/6} c_O^s & \text{Levich equations} \end{cases}$

which confirms the transport-coefficient assignment tabled on page 169. The analogy to equation 12:16 is evident. In fact, apart from the difference in transport coefficients, a voltammogram at the rotating disk mirrors a steady-state microhemisphere voltammogram.

[1228] The formula is empirical and comes from fitting to a numerical simulation. The formula and the simulation data never differ by more than 0.3%.

[1229] Instead of Hz, the units of ω are often equivalently cited as "radians per second". In the laboratory, rotation speeds are usually expressed in revolutions per minute. Convert rpm to Hz by multiplying by $\pi/30$; for example 1000 rpm is 104.7 Hz.

In particular, a reciprocal sum formula holds that is identical in form to equation 12:23

$$12:32 \quad \frac{1}{I} = \frac{1}{I_{\text{kin}}} + \frac{1}{I_{\text{rem}}} + \frac{1}{I_{\text{lim}}}$$

$$\begin{cases} I_{\text{kin}} = \dfrac{FAc_R^b k^{o'}}{(D_O^{2/3}/D_R^{2/3})^{1-\alpha}} \exp\left\{ \dfrac{(1-\alpha)F}{RT}(E-E^h) \right\} \\[2em] I_{\text{rem}} = v_L FAc_R^b D_R^{2/3} \omega^{1/2} (\rho/\eta)^{1/6} \exp\left\{ \dfrac{F}{RT}(E-E^h) \right\} \\[2em] I_{\text{lim}} = v_L FAc_R^b D_R^{2/3} \omega^{1/2} (\rho/\eta)^{1/6} \end{cases}$$

The nernstian half-wave potential is, however, defined slightly differently:

$$12:33 \qquad E^h = E^{o'} - \frac{2RT}{3F}\ln\left\{\frac{D_O}{D_R}\right\} \qquad \text{rotating disk voltammetry}$$

Alternatively, the reciprocal sum formula may be written in terms of transport coefficients

$$12:34 \qquad \frac{FAc_R^b}{I} = \frac{1}{k^{o'}\exp\{\alpha F(E-E^{o'})/RT\}} + \frac{1}{m_R \exp\{F(E-E^{o'})/RT\}} + \frac{1}{m_O}$$

in which form it applies with wide generality to steady-state processes.

The reversibility index in the case of a rotating disk electrode is

$$12:35 \qquad \lambda = \frac{I_{\text{kin}}}{I_{\text{rem}}}\exp\left\{\frac{\alpha F}{RT}(E-E^h)\right\} = \frac{k^{o'}(\eta/\rho)^{1/6}}{v_L(D_O^{1-\alpha}D_R^{\alpha})^{2/3}\omega^{1/2}} \qquad \text{rotating disk voltammetry}$$

As before, it is the magnitude of this parameter that determines the degree of reversibility. If λ is close to unity, the reaction behaves quasireversibly; if λ is much greater than, or much less than, unity the reaction behaves reversibly[1230], or irreversibly. Exactly what "much greater" and "much less" mean depends on the quality of the experimental data but "by a factor of about 16" would be a conservative estimate.

The presence of ω in the reversibility index means that, by adjusting the rotation speed, one can "tune" the behavior. Only to a small extent however, because ω appears in 12:35 as its square root, and is constrained by the inequalities in 12:30. Less than a tenfold variation in λ is feasible. Figure 12-9 overleaf illustrates how λ affects the shape of voltammograms. Notice how, initially, decreasing λ leads to a decreasingly steep wave, but that once irreversibility is established, the wave is unchanging in shape and simply moves towards positive potentials. This movement is quantified in Figure 12-10 which plots the change[1231] in half-wave potential with the logarithm of the reversibility index.

[1230] Decide whether an electrode reaction of formal rate constant 10^{-4} m s^{-1} would behave reversibly, quasireversibly, or irreversibly at (a) a microelectrode, (b) a rotating disk electrode. Base your decision on the typical values $\omega \approx 100\,\text{Hz}$, $D \approx 10^{-9}\,\text{m}^2\text{s}^{-1}$, $(\eta/\rho) \approx 10^{-6}\,\text{m}^2\text{s}^{-1}$ and $d \approx 10^{-5}$ m. See Web#1230.

[1231] Web#1231 demonstrates that the formula $E_{1/2} = E^h + (2RT/F)\text{arsinh}\{1/(2\lambda)\}$ holds when $\alpha = \frac{1}{2}$ and that more complicated analytical formulas also exist when $\alpha = \frac{1}{3}$ or $\frac{2}{3}$. There is no general formulation.

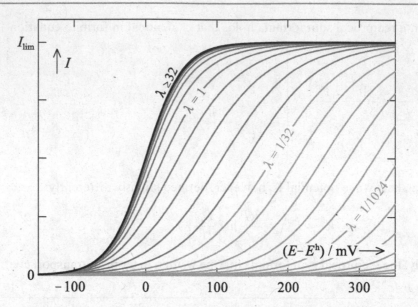

Figure 12-9 Different reversibility indices distinguish members of this family of $\alpha =$ ½ steady-state voltammograms. The λ of each curve differs by a factor of 2 from that of its neighbors. For $\lambda \geq 32$, the **reversible** curves effectively overlap. For $\lambda \leq 1/32$ the curves maintain their irreversible shape, but move to more positive potentials. The quasireversible curves are intermediate. This figure, as does 12-10, relates to steady-state voltammetry and is equally applicable to microhemispherical and rotating disk electrodes. The principles, though not the details, apply to microdisks and microelectrode arrays, too.

The rotating disk electrode is a convenient tool with which to investigate the irreversible regime. Then the I_{rem} term is large enough that its reciprocal may be ignored in the reciprocal sum formula and therefore

$$12\!:\!36 \qquad \frac{1}{I} = \frac{1}{I_{kin}} + \frac{1}{I_{lim}} = \frac{\exp\{(\alpha-1)F(E-E^{o'})/RT\}}{v_L F A c_R^b k^{o'}} + \frac{\omega^{-1/2}(\eta/\rho)^{1/6}}{v_L F A c_R^b D_R^{2/3}}$$

It follows that if one carries out the same experiment at a number of rotation speeds and then graphs $1/I$ versus $\omega^{-1/2}$, a so-called **Koutecký-Levich plot**[1232,849], the result is as shown in Figure 12-11. As is true for all voltammetric methods, investigations in the irreversible regime can yield values of α but not of $k^{o'}$ itself, the composite parameter

$$12\!:\!37 \qquad\qquad (RT/F)\ln\{k^{o'}\} - \alpha E^{o'}$$

alone being accessible. It requires analysis of the quasireversible regime to disentangle the two terms in 12:37.

[1232] Jaroslav Koutecký, 1922–2005, Czech electrochemist; member of the famous Prague school of polarography.

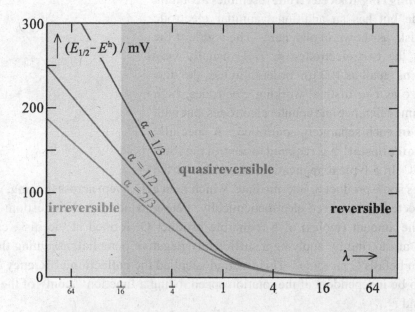

Figure 12-10 A logarithmic plot showing how, for three representative values of the transfer coefficient, the half-wave potential shifts away from E^h as the reversibility index decreases[1231]. The division of behavior into **reversible**, quasireversible, and irreversible regimes is very apparent in this graphical depiction.

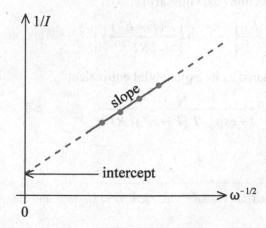

Figure 12-11 Though the rotation speed may be varied only over a limited range, a plot of $1/I$ versus $\omega^{-1/2}$ provides a straight line of slope $(\rho/\eta)^{1/6}/v_L FAc_R D_R^{2/3}$ and intercept $1/I_{kin}$. Repetition of the Koutecký-Levich experiment at a number of potentials provides access to α and to the term cited in 12:37.

The **rotating ring-disk electrode** resembles a rotating disk electrode but has an additional annular electrode outside the disk, as shown in plan here. The width of the gap between the two electrodes, r_2-r_1 is usually very narrow, even as small as 100 μm or less. In use, the disk and ring serve as two distinct working electrodes, both sharing the same reference and counter electrodes, but with the potential of each separately controlled. A specially designed **bipotentiostat**[1233] is required to control the four-electrode cell. In a typical application, a reaction at the 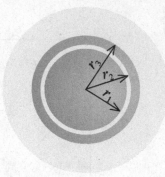 disk generates some product or intermediate, which then gets swept across the ring, where it can be detected and assayed electrochemically. We now address the question: What fraction of the amount (moles) of a reducible product O, formed at the disk, can be "recaptured" at the ring by applying a sufficiently negative potential, assuming that no other reaction befalls O en route? This fraction is called the **collection efficiency** N and it turns out to be independent of the rotation speed, being a function[1234] only of the three radii r_1, r_2, and r_3.

Shapes of Reversible Voltammograms: waves, peaks and hybrids

The **reversible wave**, diagrammed in Figure 12-12, is one of three standard shapes in voltammetry. This current-voltage relationship is described most succinctly in terms of the hyperbolic tangent function (see Glossary)

12:38 $$I(E) = \frac{I_{\text{lim}}}{2}\left[1+\tanh\left\{\frac{F(E-E^{\text{h}})}{2RT}\right\}\right] \qquad \text{reversible wave}$$

but is more often encountered as its exponential equivalent[1235]

12:39 $$I(E) = \frac{I_{\text{lim}}}{1+\exp\left\{-F(E-E^{\text{h}})/RT\right\}} \qquad \text{reversible wave}$$

or in the inverted form

[1233] For details, see A.J. Bard and L.R. Faulkner, *Electrochemical Methods*, 2nd edn, Wiley, New York, 2001, page 643.

[1234] in fact, the very complicated function $N = 3[\Phi(a)+\Phi(b)-\Phi(c)]/2\pi b^2$, where $a = (r_2^3 - r_1^3)^{1/3}/r_1$,

$b = r_1 a/(r_3^3 - r_2^3)^{1/3}$, $c = r_3 b/r_1$, and $\Phi(y) = y^2\left[\frac{\pi}{2} - \arctan\left\{\frac{2y-1}{\sqrt{3}}\right\} - \frac{1}{\sqrt{12}}\ln\left\{\frac{(1+y)^3}{1+y^3}\right\}\right]$.

For example, when $r_1 = 2.00$ mm, $r_2 = 2.01$ mm, and $r_3 = 4.00$ mm, the collection efficiency is 0.623.

[1235] Show how this equation derives from the reciprocal sum relationship, equation 12:23. See Web#1235.

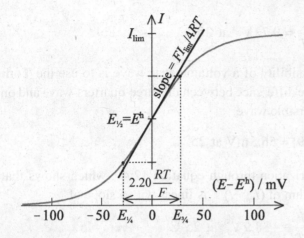

Figure 12-12 Anatomy of a one-electron reversible voltammetric wave. The scaling of the abscissa is for 25°C. Waves may be solely anodic as here, solely cathodic, or some of each, as in Figure 10-9.

12:40
$$E = E^h - \frac{RT}{F} \ln\left\{ \frac{I_{lim} - I}{I} \right\} \qquad \text{reversible wave}$$

The shape is well known in statistics, where it would be described as the cumulative function of the **logistic distribution** with a mean of E^h and a variance of $\pi RT / \sqrt{3}F$. The wave has inversion symmetry[1236] about E^h, the half-wave potential.

Waves arise from steady-state, and some other, voltammetries. Those of the reversible variety occur when Nernst's equation is obeyed, implying that the rate of the electrode reaction is inherently much faster than the rates of the transport processes. Thus a reversible wave contains no information about the kinetics of the electrode reaction. It does, however, contain information about the transport processes, primarily as the wave height, equal to I_{lim}. Moreover, the thermodynamics of the electrode process is reflected by the nernstian half-wave potential E^h, which differs from the formal potential[1237] $E^{o\prime}$ only to the extent that there may be an imbalance between the efficacies of transport to and from the electrode:

12:41
$$E^h = E^{o\prime} - \frac{RT}{F} \ln\left\{ \frac{m_O}{m_R} \right\}$$

There is interest in the steepness of a voltammetric wave, as this is the criterion by which the wave may be identified as reversible or not. The steepness may be characterized by the slope of the wave at its mid-point which, for the reversible case, is

---

[1236] Inversion symmetry of a curve about a point Q, means that if the straight line connecting any point P on the curve to Q is extrapolated a distance equal in length to PQ, to a point P′, then this third point also lies on the curve.

[1237] Electrochemists are not always careful to distinguish among the nernstian half-wave potential E^h, the formal potential $E^{o\prime}$ and the standard potential E^o.

12:42 $\qquad \dfrac{1}{I_{\text{lim}}}\left(\dfrac{dI}{dE}\right)_{E_{1/2}} = \dfrac{F}{4RT} = 9.73\,\text{V}^{-1}$ at 25° C $\qquad$ reversible wave

Another way of assessing the reversibility of a voltammetric wave is to use the **Tomeš criterion**[1238], based on measuring the difference between the three-quarters-wave and one-quarter-wave potentials. For a reversible wave:

12:43 $\qquad E_{3/4} - E_{1/4} = \dfrac{RT}{F}\ln\{9\} = 56.5\,\text{mV}$ at 25° C $\qquad$ reversible wave

A third method makes use of linearization through equation 12:40, which shows that a graph versus potential of the logarithm of $(I_{\text{lim}} - I)/I$ is linear with a slope of

12:44 $\qquad \dfrac{d}{dE}\ln\left\{\dfrac{I_{\text{lim}} - I}{I}\right\} = \dfrac{-F}{RT} = -38.9\,\text{V}^{-1}$ at 25°C $\qquad$ reversible wave

Irreversible[1239] and quasireversible waves are less steep; they have **subnernstian slopes**.

Not all voltammetric methods lead to a wave-shaped voltammogram. There are two other characteristic shapes exhibited by reversible voltammograms. The procedures that generate these shapes are not steady-state experiments, but we mention them here because of their close relationship to the classical reversible wave.

The non-steady-state experiments described on pages 270, 321, and 343 lead to peak-shaped voltammograms. If such an experiment is performed under reversible conditions, the voltammogram has a shape described by the equation

12:45 $\qquad I(E) = I_{\text{peak}}\,\text{sech}^2\left\{\dfrac{F}{2RT}(E - E^{\text{h}})\right\} \qquad$ reversible peak

or equivalently by

12:46 $\qquad I(E) = \dfrac{4I_{\text{peak}}\,\exp\{-F(E - E^{\text{h}})/RT\}}{[1 + \exp\{-F(E - E^{\text{h}})/RT\}]^2} \qquad$ reversible peak

Again, this shape has statistical importance: it is the probability function of the logistic distribution. As Figure 12-13 shows, the summit of this **reversible voltammetric peak** occurs at the nernstian E^{h} potential. The voltammogram has bilateral symmetry and inverts to

12:47 $\qquad E = E_{\text{peak}} \pm \dfrac{RT}{F}\ln\left\{\dfrac{\sqrt{I_{\text{peak}}} + \sqrt{I_{\text{peak}} - I}}{\sqrt{I}}\right\} \qquad$ reversible peak

Akin to the Tomeš criterion is the characterization of reversibility by the peak width, which is the separation between the two half-peak potentials,

---

[1238] Confirm formula 12:43; see Web#1238.

[1239] Find replacements for equations 12:42, 12:43, and 12:44 for an *irreversible* steady-state wave at a rotating disk electrode. They are given in Web#1239.

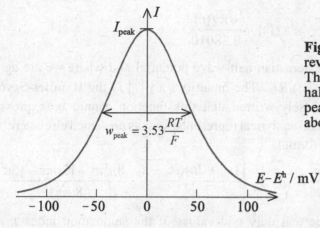

Figure 12-13 Anatomy of a reversible voltammetric peak. The peak width is measured half-way to the summit. The peak has bilateral symmetry about the peak potential.

12:48
$$w_{\text{peak}} = \frac{4RT}{F}\ln\left\{1+\sqrt{2}\right\} = 90.6 \text{ mV at } 25°\text{ C}$$
reversible peak

Under conditions that are otherwise similar, voltammetric peaks measured under irreversible or quasireversible regimes have peaks that are *lower* and *wider* than reversible peaks, but which enclose the *same area*.

The third standard shape encountered in reversible voltammetry results from a linear-scan experiment (page 346) under reversible conditions. This shape, which lacks symmetry of any kind, has not acquired a distinctive name akin to "wave" or "peak"; here we shall call it a "hybrid", because it has properties intermediate between a wave and a peak, as evident in Figure 12-14.

The shape of the **reversible voltammetric hybrid** is described by the equation[1240]

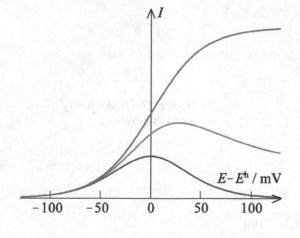

Figure 12-14 The reversible wave, the reversible hybrid, and the reversible peak.

[1240] The term λ{3/2} is the lambda number (see Glossary) of argument 3/2, equal to 1.6888.

12:49 $\qquad \dfrac{I(E)}{I(E^{h})}=\left(\dfrac{\sqrt{2}\,\pi}{\lambda\{3/2\}}\right)\sqrt{\pi}\chi\{\varepsilon\}=\dfrac{\sqrt{\pi}\chi\{\varepsilon\}}{0.38010} \qquad$ reversible hybrid

where $I(E^{h})$ is the current at the nernstian half-wave potential and where we are again using the abbreviation $\varepsilon = F(E-E^{h})/RT$. The quantity $\sqrt{\pi}\,\chi\{\ \}$ is the **Randles-Ševčik function**[1630]. It is sometimes falsely written that this function cannot be expressed analytically. In fact there are several analytical representations, as presented elsewhere[1241]. The most generally useful is the formula

12:50 $\quad \sqrt{\pi}\,\chi\{\varepsilon\}=\dfrac{149}{392}+\dfrac{54\varepsilon}{455}-\dfrac{13\varepsilon^{2}}{296}+\sqrt{\dfrac{\pi}{2}}\sum_{n=1,3}^{19}\left\{\dfrac{(\varepsilon_{n}+2\varepsilon)\sqrt{\varepsilon_{n}-\varepsilon}}{\varepsilon_{n}^{3}}-\dfrac{8n^{2}\pi^{2}+12n\pi\varepsilon-15\varepsilon^{2}}{8(n\pi)^{7/2}}\right\}$

where $\varepsilon_{n}=\sqrt{\varepsilon^{2}+n^{2}\pi^{2}}$. Notice that only odd values of the summation index n are employed in this summation.

As illustrated in Figure 12-15, the reversible hybrid has a blunt peak at a potential more positive (by $1.1090\,RT/F$ or 28.5 mV at 25°C) than E^{h} and is higher than $I(E^{h})$ by a factor of 1.1726. The peak height is easily measured, but the corresponding potential is less easily located, because the peak is broad and flat. Accordingly, the **half-peak potential**

12:51 $\qquad E^{h}=E_{pk/2}+1.0934\dfrac{RT}{F}=28.1$ mV at 25°C $\qquad$ reversible hybrid

provides a more accurate route to the nernstian E^{h}. The peak width being inconveniently large, the criterion of reversibility for the hybrid, corresponding to the Tomeš criterion 12:43 for the wave and the peak width 12:48 for the peak, is more often taken as the voltage separating the peak from the half peak $E_{pk/2}$:

12:52 $\qquad E_{pk}-E_{pk/2}=2.2024\dfrac{RT}{F}=56.6$ mV at 25°C $\qquad$ reversible hybrid

The effect of nonreversibilty on a hybrid curve is to make the hump lower and wider.

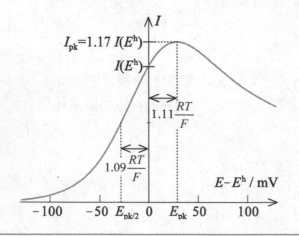

Figure 12-15 Anatomy of a reversible hybrid.

[1241] These are cited, along with commentary, in Web#1241.

Each of the three shapes – **wave**, **peak**, and **hybrid** – may be converted to either of the others by an operation of the calculus, as the chart below explains, though electrochemists seldom make use of these convenient conversions.

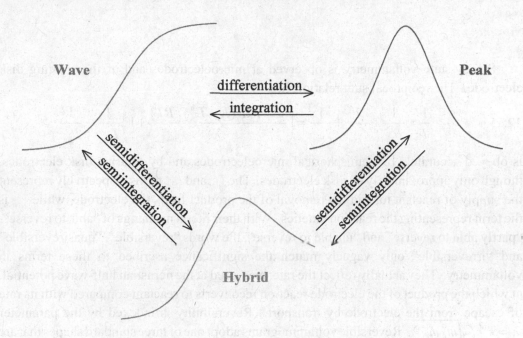

The operation of semiintegration and semidifferentiation may be unfamiliar, but conceptually they are, as their names imply, just half-way houses in carrying out ordinary integration or differentiation. In electrochemistry the operations are generally carried out with respect to time t, though E is an alternative[1241]. The symbolism

12:53
$$\frac{d^{1/2}}{dt^{1/2}} I(t) \qquad \text{current semiderivative}$$

represents the semiderivative of a time-dependent current. Such an operation would, for example, convert a reversible hybrid into a reversible peak. Similarly, the symbol

12:54
$$\frac{d^{-1/2}}{dt^{-1/2}} I(t) \qquad \text{current semiintegral}$$

depicts the semiintegral of a time-dependent current. Semioperations may be carried out by a variety of techniques: computationally by computer, mathematically[1242], or electronically[1243]. For data in the form of equally spaced time series, convenient algorithms

---

[1242] See Web#1242 for mathematical definitions and further information.

[1243] A transmission line[173] will semiintegrate the current passed through it or semidifferentiate a potential applied across it. In practical devices, the transmission line is replaced by a finite collection of resistors and capacitors.

for semiintegration and semidifferentiation are presented elsewhere[1244]. They can be used to interconvert the various voltammetric shapes[1245], whether reversible or not.

Summary

Steady-state voltammetry is observed at microelectrodes and at the rotating disk electrode. The reciprocal sum relationship

12:55
$$\frac{1}{I} = \frac{1}{I_{kin}} + \frac{1}{I_{rem}} + \frac{1}{I_{lim}} = \left(\frac{\exp\{\alpha F(E - E^h)/RT\}}{\lambda} + 1 \right) \frac{1}{I_{rem}} + \frac{1}{I_{lim}}$$

is obeyed accurately by hemispherical microelectrodes and by rotating disk electrodes, though only approximately by disk electrodes. The I_{lim} and I_{rem} terms respectively represent the supply of reactant to, and the removal of the product from, the electrode, while I_{kin} is the term representing the reaction kinetics. With their literal meanings of "able to reverse", "partly able to reverse" and "unable to reverse", the words "reversible", "quasireversible" and "irreversible" only vaguely match the significance ascribed to these terms in voltammetry. They actually reflect the rate, measured at the nernstian half-wave potential, at which the product of the electrode reaction reconverts to reactant compared with its rate of escape from the electrode by transport. Reversibility is indexed by the parameter $\lambda = k^{o\prime}/\sqrt{m_R^\alpha m_O^{1-\alpha}}$. Reversible voltammograms adopt one of three standard shapes that are closely interrelated.

[1244] See Web#1244 for several algorithms.

[1245] Use the Excel® spreadsheet algorithm provided in Web#1245 to explore the conversions illustrated on page 257. Instructions for these, and more generalized, conversions are included in the Web.

13

The Electrode Interface

Up to this point we have regarded the electrode as nothing more than a surface. On one side of this surface there exists a homogeneous electronic conductor; on the other side there is an ionic conductor, usually a solution. Although we have not insisted that the solution be homogeneous, the tacit assumption has been made that gradients of concentration and potential extrapolate smoothly from the bulk to the surface. In reality, the electrode interface is much more than a passive junction between two phases. In this chapter, several of the ways in which the electrode interface impinges on the properties of electrochemical cells will be explored.

Double Layers: three models of capacitance

On pages 15-17 we introduced the idea of a double layer, formed when an electric field is applied to a totally polarized electrode. The model presented there – a layer containing ions[1301] on the ionic-conductor side of the junction confronting a layer of equal and opposite electronic charges on the electronic-conductor side – is that of Helmholtz[1302]. The two layers of charge constitute a capacitor and, because the separation of the layers – the "*L*" in equation 1:18 – is of atomically small dimensions, the capacitance of a double layer should be huge, as indeed it is[1303]. The relationship

$$13:1 \qquad \frac{C_H}{A} = \frac{dq}{dE} = \frac{\varepsilon_H}{x_H} \qquad \text{Helmholtz model}$$

represents the capacitance according to this simple model. Here x_H is the thickness of the

---

[1301] At a metal | aqueous solution interface with a typical charge density of 0.10 C m^{-2}, a univalent ion would share the surface with about 15 water molecules. Confirm this estimate, or see Web#1301.

[1302] Hermann Ludwig Ferdinand von Helmholtz, 1821–1894, renowned German physicist.

[1303] Values measured at the Hg | *aq* interface are typically in the range $0.2-0.4$ F m^{-2}, depending on the applied potential, and the nature and concentration of the ions.

Electrochemical Science and Technology: Fundamentals and Applications, First Edition. Keith B. Oldham, Jan C. Myland, Alan M. Bond.
© 2012 John Wiley & Sons, Ltd. Published 2012 by John Wiley & Sons, Ltd.

layer, the **Helmholtz layer**, and ε_H is its permittivity, but the precise significance that attaches to each of these quantities separately is open to interpretation.

The most numerous and successful experimental studies[1304] that have been made in attempting to understand the double layer have focused on the junction between aqueous ionic solutions and mercury. There are three reasons for the choice of mercury, all stemming from its liquidity. As for stripping analysis (pages 177 – 182) and transient voltammetry (Chapter 16), this metal is chosen for the ease with which its surface may be renewed and thereby kept clean. The second reason for choosing mercury is that its surface area may be increased easily, for example by flowing mercury into the droplet[1305] shown in Figure 13-1.

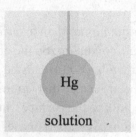

Figure 13-1 A mercury droplet, shown here in cross section, steadily grows as mercury slowly flows from the supporting capillary tube.

The third reason arises from the structureless state of liquid mercury compared with the crystalinity of solid metals. A sample of a solid metal is usually a polycrystalline mosaic which presents regions of different crystallographic indices to the electrolyte solution, as well as intergranular regions of chaotic structure (page 216). Thus measurements made on solid metals usually provide only average values, the precise composition of the average being dependent on the metallurgical treatment that the sample received in its fabrication. Some very exacting experiments have been made on metal surfaces carefully prepared to have only a single crystallographic face exposed. Those experiments reveal that, indeed, the electrochemical properties change from one crystallographic plane to another.

The interface between any two phases possesses an energy – the **interfacial energy** – that is proportional to its area A and equal to the product σA, where σ is the **surface tension** of the interface[1306]. It is the goal of minimizing the interfacial energy that causes small volumes of liquids to adopt their characteristic near-spherical shapes. For a charged mercury | ionic-solution interface, this tendency is opposed by the lateral repulsion between the charges. Accordingly, the surface tension is dependent on the charge density on the

---

[1304] Especially those of David C. Grahame, 1912 – 1958, a skilful and meticulous U.S. physical chemist. Consult his writings, or one of the many textbooks that report his results, for further information.

[1305] Growing mercury drops, formed as in Figure 13-1, provide the working electrodes in the technique known as **polarography**, which was the predecessor of voltammetry. Polarography is not discussed in this book, but a précis of the method will be found at Web#1305.

[1306] γ often replaces σ. The surface tension of an air | water interface is 0.07198 N m^{-1} (or J m^{-2}).

electrode in accordance with the **Lippmann equation**[1307]

13:2 $$\frac{d\sigma}{dE} = -q \qquad \text{Lippmann equation}$$

where E is the electrode potential. In this discussion, q is the charge density on the surface of the *electronic* conductor; electroneutrality mandates that there be an equal and opposite charge on the ionic conductor and close to the surface. On differentiating the Lippmann equation, an expression, namely

13:3 $$\frac{d^2\sigma}{dE^2} = -\frac{dq}{dE} = -\frac{C}{A}$$

is obtained for the capacitance.

If the capacitance were a constant, a graph of surface tension versus potential would give a parabola, as in Figure 13-2. The parabola, which has the equation[1308]

13:4 $$\sigma = \sigma_{zc} - \frac{C}{2A}(E_{zc} - E)^2 \qquad \text{electrocapillary relationship}$$

peaks at the **potential of zero charge**, E_{zc}. Such graphs, called **electrocapillary curves**[1309], are, indeed, found experimentally to be very close to parabolic in many interfaces, such as between mercury and a 100 mM aqueous solution of sodium fluoride. In those cases the coordinates of the maximum in the electrocapillary curve are

13:5 $$E_{zc} \approx -230 \text{ mV versus SHE}; \qquad \sigma_{zc} \approx 0.43 \text{ N m}^{-1}$$

There are three distinct ways of measuring the electrode capacitance:

(a) doubly differentiating the electrocapillary curve, as evident from equation 13:3;

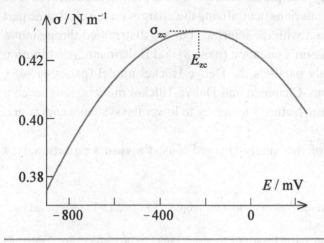

Figure 13-2 Simple electro-capillary curve at a mercury electrode. The simplicity of such a curve is often impaired by the presence of adsorbing species.

[1307] Gabriel Jonas Lippmann, 1845–1921, Luxembourger physicist who worked in both France and Germany; Nobel laureate 1908 for his development of interferometry. A derivation of equation 13:2 will be found at Web#1307.

[1308] Perform two differentiations of equation 13:4 to confirm 13:3. See Web#1308.

[1309] The name originates from a method of measuring σ by locating the liquid's meniscus in a capillary tube.

(b) electrically, using a.c. as described in Chapter 15 or by less satisfactory d.c. methods;

(c) by measuring the current flowing to an expanding electrode held at a constant potential E. Because method (c) implies a constant charge density q, it follows that[1310]

$$13:6 \qquad I = \frac{d}{dt}Q = \frac{d}{dt}(qA) = q\frac{dA}{dt} = \frac{C}{A}(E - E_{zc})\frac{dA}{dt} = C(E - E_{zc})\frac{d}{dt}\ln\{A\}$$

There is broad agreement between results from the three methods but unless both conductors – electronic and ionic – are liquids, only method (b) is applicable.

Double-layer capacitances generally behave nonideally, displaying dependence on the applied voltage. Changes by a factor of 2 or more over a one-volt range are not uncommon. In this circumstance one must distinguish two measures of capacitance: the **differential capacitance**

$$13:7 \qquad\qquad\qquad C(\text{differential}) = \frac{dQ}{dE}$$

and the **integral capacitance**

$$13:8 \qquad\qquad C(\text{integral}) = \frac{Q(E)}{E - E_{zc}} = \frac{1}{E - E_{zc}} \int_{E_{zc}}^{E} C(\text{differential})\, dE$$

It is the differential capacitance that is the more important in the present context and that is henceforth implied, unless we state otherwise.

Except perhaps at the highest charge densities, the Helmholtz model of the double layer at a metal|(electrolyte solution) interface implies that the capacitance should not depend on potential or on the bulk ionic concentration, whereas both these dependences are observed. Another model, the **Gouy-Chapman model**[1311], does predict such effects. The premise of this model is that the counterions neutralizing the charges on the electronic part of the double layer are not in contact with the interface but are distributed throughout a **diffuse zone**[1312]. In employing Poisson's equation (page 8) and Boltzmann's distribution law (page 38), the treatment closely parallels the Debye-Hückel model (pages 41–45), which it predated. In both the Gouy-Chapman and Debye-Hückel models, ions reach a compromise in a competition between Nature's tendency to lower the system's energy and her entropic urge towards disorder.

For an electrolyte consisting of two singly-charged ions, **Poisson's equation** 1:14

[1310] What value has the $d(\ln\{A\})/dt$ term if mercury flows into the droplet of Figure 13-1 at a constant rate? See Web#1310.

[1311] Louis Georges Gouy, 1854–1924, and David Leonard Chapman, 1869–1958, respectively French and English physical chemists. Gouy addressed not only solutions, such as those of NaF and MgSO$_4$, with equal populations of ions of anions and cations, but also the more difficult problem in which there is a two-to-one ratio, as for K$_2$CO$_3$ or CaCl$_2$. Chapman's rederivation was restricted to the former class.

[1312] The word "diffuse" means "spread out" or "dispersed". Though both share the same etymology, this adjective has no connection here to the verb "diffuse", which refers to a particular dispersion mechanism.

relates the second spatial derivative of the local potential ϕ to the local charge density ρ, and hence to the difference between the cation and anion concentrations, through the relation

$$13:9 \qquad \frac{d^2\phi}{dx^2} = \frac{-\rho(x)}{\varepsilon} = \frac{-F}{\varepsilon}\left[c_C(x) - c_A(x)\right] \qquad \text{Poisson's equation}$$

Electroneutrality demands that the total net charge in solution be equal and opposite to the charge on the electronic conductor, so that

$$13:10 \qquad F\int_0^\infty \left[c_C(x) - c_A(x)\right]dx = -q$$

where x is the rectilinear coordinate measured from the interface into the solution. By comparing 13:10 with the integral of equation 13:9, one concludes that the field at the surface of the solution is

$$13:11 \qquad -\left(\frac{d\phi}{dx}\right)^s = \frac{q}{\varepsilon} \qquad \begin{array}{l}\text{field at the}\\ \text{interface}\end{array}$$

If the charge density q on the surface of the electronic conductor is positive, anions will outnumber cations in the diffuse zone, which will acquire a positive potential compared with the bulk potential.

Boltzmann's distribution law, equation 2:26, provides a link between the equilibrium concentrations of an ion at two sites in a solution and the work needed to carry the ion from one site to the other. Choosing the departure site as at x, and the destination site as "at infinity", where ϕ is defined as zero, the law predicts that

$$13:12 \qquad \frac{c_i(\infty)}{c_i(x)} = \exp\left\{\frac{-N_A}{RT}w_i^{x\to\infty}\right\} = \exp\left\{\frac{-N_A}{RT}\left(-z_iQ_0\phi(x)\right)\right\} = \exp\left\{\frac{z_iF}{RT}\phi(x)\right\}$$

Making use of equations 13:9, 13:11, and 13:12, the Gouy-Chapman model predicts the capacitance of the double layer to be given by

$$13:13 \qquad \frac{C_{GC}}{A} = \sqrt{\frac{2F^2\varepsilon c}{RT}}\cosh\left\{\frac{F}{2RT}(E - E_{zc})\right\} \qquad \text{Gouy-Chapman model}$$

The derivation of this exact result is presented elsewhere[1313].

The Gouy-Chapman model predicts that the capacitance will increase with the bulk ionic concentration, as generally observed. A catenary-shaped[1314] dependence of capacitance on potential is predicted by the model, with its minimum at the potential of zero charge, as in the green curve of Figure 13-3 overleaf and, indeed, such a minimum is observed experimentally. However, at potentials well removed from E_{zc}, capacitance values predicted by equation 13:13 grossly exceed measured values.

[1313] See Web#1313. The derivation is restricted to the $z_C = 1$, $z_A = -1$ case, but other values are also tractable.

[1314] A rope or chain, sagging under gravity, adopts a catenary shape, described by a hyperbolic cosine.

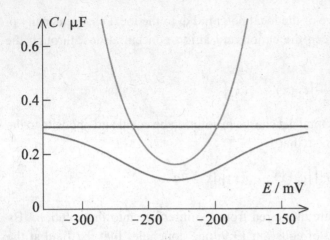

Figure 13-3 The relationship of capacitance to electrode potential according to three models of the double layer: those of **Helmholtz, Gouy and Chapman**, and **Stern**. The following parameters were adopted: $T = T°$, $c = 10$ mM, $x_H = 0.2$ nm, $\varepsilon_H = 6 \times 10^{-11}$ F m^{-1}, $\varepsilon = 7 \times 10^{-10}$ F m^{-1}, $A = 1$ mm^2, and $E_{zc} = -230$ mV versus SHE.

Stern[1315] recognized that the experimental capacitance results were better matched by a melding of the Helmholtz and Gouy-Chapman models, some of the counterions being in a diffuse zone and some in a **compact layer**[1316] immediately adjacent to the interface. Capacitances arising from the two regions would be in series and so the overall capacitance[158] would be calculable from the formula

13:14 $$\frac{1}{C_S} = \frac{1}{C_H} + \frac{1}{C_{GC}} = \frac{x_H}{A\varepsilon_H} + \frac{\sqrt{RT/2F^2\varepsilon c}}{A\cosh\{F(E - E_{zc})/RT\}} \qquad \text{Stern model}$$

The red curve of Figure 13-3 is predicted by the Stern model. The permittivity enters equation 13:14 twice, but its significance differs between the two instances. The permittivity ε in the disordered diffuse zone will be close to that of pure water, about $79\varepsilon_0$. However, the permittivity ε_H in the compact layer is that of water-plus-ions confined in a narrow region, compressed, and ordered by an intense local field; estimates suggest that the permittivity here is much lower, perhaps as small as $5\varepsilon_0$. A refinement to the Stern model allows for different values of x_H on either side of the capacity minimum, a narrower compact layer being hypothesized when occupied by unhydrated anions, whereas the strongly hydrated cations are believed to occupy a wider layer, as pictured in Figure 13-4.

We have yet to address the question of how the electrical parameters vary within the double layer; that is, how do the potential ϕ, the field X, and the charge density ρ in the solution depend on the distance from the electrode. The mathematics to establish these profiles according to the Stern model is quite elaborate and has been relegated[1317];

[1315] Otto Stern, 1889 – 1969, German physicist who emigrated to the U.S. in 1933; Nobel laureate 1943 for researches on molecular beams.

[1316] Also known as a **Helmholtz layer** or a **Stern layer**.

[1317] to Web#1317 where, not only are the profiles of ϕ, X, and ρ derived, but those of c_A and c_C also. Draw a diagram showing the concentration profiles of the ions under the conditions listed for Figure 13-5.

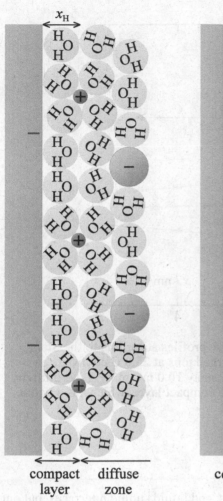
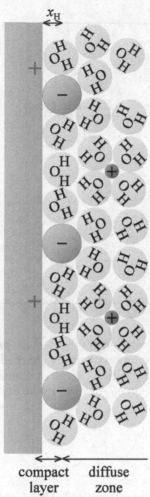

Figure 13-4 Cartoon of the supposed structure of the interfacial region of a double layer when the electronic conductor is charged negatively (left) or positively (right). Only a small portion of the diffuse zone is shown.

compact layer — diffuse zone   compact layer — diffuse zone

nevertheless, Figure 13-5 overleaf shows examples of the profiles. That the potential, field, and charge density decay to values close to their bulk values within about two Debye lengths (page 42), is evident from that figure.

Capacitance measurements[1304], and other experiments at the interface between mercury and aqueous solutions of inorganic ions, suggest that the Stern model, elaborated where necessary by the adsorption-induced effects described later, captures the essentials of double layer structure. While the model provides understanding of the $Hg \mid aq$ interface, the current state of theory is far from being of predictive quality: capacitance and other properties of the electrode interface must be measured; reliable values cannot be calculated. Sadly, the interfaces between aqueous solutions and *solid* metals are even less well understood, as are those at electrodes in which the ionic conductor is a nonaqueous solution or an ionic liquid.

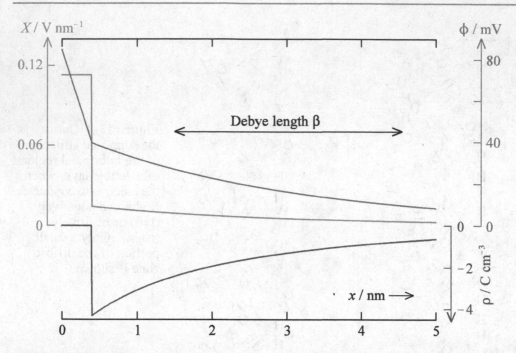

Figure 13-5 The potential, field, and charge density profiles according to the Stern model. A 10.0 mM aqueous solution of singly charged ions at 25.0°C confronts an electronic conductor bearing a positive charge of density 10.0 mC m^{-2}. A permittivity of $\varepsilon = 78.54\varepsilon_0$ is assumed, except $\varepsilon_H = 10.0\varepsilon_0$ in the compact layer, which has a width of 0.400 nm.

Adsorption: invasion of the interface

Adsorption occurs also at liquid | gas, solid | gas, and liquid | liquid interfaces, but our attention here is confined to those interfaces of the solid | ionic-solution or mercury | ionic-solution types that function as electrodes. If a solute i, molecular or ionic, has a lower energy when located on the electrode surface there is a tendency for it to occupy such a site, temporarily or permanently, rather than remaining dissolved. This phenomenon is called **adsorption**[1318]. When a solute greatly prefers the dissolved state, it may shun the surface region to some extent, leading to "negative adsorption".

Studies of adsorption have often been made at mercury electrodes, for reasons with which readers of the previous section will be familiar. Many ions, particularly anions[1319], are adsorbed at electrode interfaces, entering the compact portion of the double layer. For

---

[1318] Sometimes the phrase "specific adsorption" is used, but the adjective is redundant. Unless one regards, as we do not, ions driven to the interface by electrical forces as being "adsorbed" then all adsorption is specific.

[1319] It appears that all anions, except OH$^-$(aq) and F$^-$(aq), adsorb to some extent at a mercury surface whereas no inorganic cations, except Tl$^+$(aq), do so.

example, at the junction of a mercury electrode with an aqueous solution of potassium bromide, the bromide anions are adsorbed so that, even at the potential of zero charge on the mercury, there are many Br^- anions in the compact layer. This brings K^+ cation partners into the diffuse zone, along with some extra Br^- anions. The maximum in the electrocapillary curve shifts in a negative direction, as illustrated here.

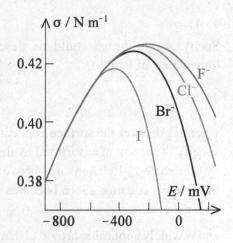

Neutral, and especially organic, molecules generally adsorb over a range of potentials that includes the potential of zero charge. Their effect, exemplified in Figure 13-6, is twofold. Firstly, to depress the differential capacitance[1320] over that range, because the content of the compact layer is replaced by larger molecules that endow a lower permittivity. Secondly, to enhance C, sometimes dramatically, at the edges of the adsorption range, because a small change in potential may cause a larger change in double-layer composition in these borderline regions. At extreme potentials of either sign, the adsorbed molecules are desorbed to make room for ions driven by strong coulombic forces to the interface.

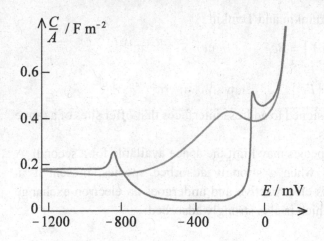

Figure 13-6 Capacitance-versus-potential graphs, measured by a.c., for the Hg | 1000 mM KCl(*aq*) interface without and with 10 mM pentanol, revealing evidence of strong adsorption of the $CH_3(CH_2)_4OH$ molecules in the range $-0.7\ V < E < -0.1\ V$. At lower frequencies the sharp peaks become higher, while at high frequencies they disappear because the changing electric field outruns the adsorption/desorption speeds.

Presence on a two-dimensional surface is a more ordered situation than the freedom enjoyed in three-dimensional space, so the adsorption of a solute is entropically opposed, even if energetically favored. An equilibrium generally exists between a dissolved solute and its adsorbed counterpart

---

[1320] Notice that an average value of C/A is about $0.3\ F\ m^{-2}$. Thus a disk electrode of radius 1.8 mm would have an interfacial capacitance of close to 3 microfarads. This is a value of C that we use in calculations elsewhere in this book.

13:15 $i(soln) \rightleftarrows i(ads)$

Such an equilibrium could be described in terms of an equilibrium constant (with dimensions of length)

13:16 $$\frac{\Gamma_i}{c_i} = K'$$

where Γ_i denotes the surface concentration (mol m^{-2}) of i. However, it is more usual to specify the extent of adsorption by the fraction θ_i of the available surface that is occupied by adsorbate i. Relationships between θ and c are called **adsorption isotherms**[1321]. In these terms equation 13:16 becomes

13:17 $\theta_i = Bc_i$ Henry isotherm[1322]

and is widely applicable provided that θ is small. B is known as an **adsorption coefficient**. As the coverage increases, the adsorbed entities interact repulsively with each other, leading to departure from proportionality to the solute concentration. If there is a limit to the extent of adsorption, as in the common case in which only a **unimolecular film** may form, then θ will approach unity at high values of c. A number of isotherms have been proposed to address this behavior, the simplest being

13:18 $\dfrac{1}{\theta_i} = 1 + \dfrac{1}{Bc_i}$ Langmuir isotherm[1323]

Others, including the isotherms of Frumkin and Temkin

13:19 $\dfrac{\theta_i}{1 - \theta_i} \exp\{g\theta_i\} = Bc_i$ Frumkin isotherm[1324]

13:20 $\theta_i = \alpha \ln\{Bc_i\}$ Tempkin isotherm[1324]

incorporate additional parameters designed to address interfaces that offer sites of a range of attractiveness to the solute species.

The presence of one adsorbed species may limit the space available for a second. A particularly simple situation arises when a strongly adsorbed species, present as a unimolecular film on an electrode, is electroactive and undergoes an electron-exchange reaction to form a second species, which is also strongly adsorbed:

[1321] The unexpected term "isotherm" is a carryover from adsorption studies with gases in which the extent of adsorption is measured as a function of temperature and pressure. When pressure alone was changed, the relationships were termed "isotherms" because they were conducted at constant temperature.

[1322] So called by analogy with Henry's Law of gas solubility (William Henry, 1775 – 1836, English physician and chemist).

[1323] Irving Langmuir, 1881–1957, multitalented U.S. scientist; Nobel laureate 1932 for his work on unimolecular films.

[1324] Alexander Naumovich Frumkin, 1895 –1976, preeminent Russian electrochemist; and his colleague Mikhail Isaakovich Temkin, 1908–1991, Russian physical chemist.

13:21
$$R^{z_R}(ads) \rightleftharpoons e^- + O^{z_O}(ads)$$

Such a process is described as a **surface-confined reaction**. Only one of R and O need be an ion, but both could be. Associated with any unimolecular film is a quantity Γ_{max} that represents the surface concentration (mol m^{-2}) of adsorption sites. The actual charge density of the film in our example is

13:22 $\quad q = \Gamma_{max} F \left[z_R \theta_R + z_O \theta_O \right] = \Gamma_{max} F \left[z_R \theta_R + (z_R + 1)(1 - \theta_R) \right] = \Gamma_{max} F \left[z_R + 1 - \theta_R \right]$

Any change in the composition of the film will be accompanied by the flow of faradaic current, the current being

13:23
$$I = Ai = \frac{dq}{dt} = -\Gamma_{max} AF \frac{d\theta_R}{dt}$$

Nernst's law will apply, almost certainly when such a cell is at rest and perhaps under transient conditions too. From equation 6:20 the electrode potential will be

13:24
$$E = E^o + \frac{RT}{F} \ln \left\{ \frac{a_O}{a_R} \right\} \qquad \text{Nernst's law}$$

and in accord with the activity assignment noted in approximation 2:15, this means that

13:25
$$E = E^{o\prime} + \frac{RT}{F} \ln \left\{ \frac{\theta_O}{\theta_R} \right\} = E^{o\prime} + \frac{RT}{F} \ln \left\{ \frac{1 - \theta_R}{\theta_R} \right\}$$

The second equality in 13:25 is valid if the entire surface of the electrode is covered by a unimolecular layer that is an assemblage of equally sized R(*ads*) and O(*ads*) species. Inversion of equation 13:25 leads to

13:26 $\quad \theta_R = \dfrac{1}{1 + \exp\left\{ F(E - E^{o\prime})/RT \right\}} = \dfrac{1}{2} - \dfrac{1}{2} \tanh\left\{ \dfrac{F}{2RT}(E - E^{o\prime}) \right\} \qquad$ coverage

What voltammetry results from this cell? We have a situation, almost unique in voltammetry, in which neither electroactive species requires transport[1325]. Consider that the potential, initially sufficiently negative that the adsorbate is entirely in the R state, is then scanned positively at a rate v in the manner typical of linear-potential-scan voltammetry (page 346). Then

13:27
$$\theta_R = \frac{1}{2} - \frac{1}{2} \tanh\left\{ \frac{F}{2RT}(E_{initial} + vt - E^{o\prime}) \right\} \qquad \begin{array}{l} \text{linear scan} \\ \text{reversible} \end{array}$$

This expression gives the fractional coverage of the electrode by species R, but note that it is the derivative of this quantity that equation 13:23 shows to be proportional to the faradaic current density. It follows that

[1325] Of course, transportation is required for the counterions, but these would generally be in excess, and cause no polarization.

$$I = -FA\Gamma_{max} \frac{d}{dt}\left[\frac{1}{2} - \frac{1}{2}\tanh\left\{\frac{F}{2RT}\left(E_{initial} + vt - E^{o'}\right)\right\}\right]$$

$$13:28 \qquad = \frac{F^2 A\Gamma_{max} v}{4RT}\,\text{sech}^2\left\{\frac{F}{2RT}\left(E_{initial} + vt - E^{o'}\right)\right\}$$

$$= \frac{F^2 A\Gamma_{max} v}{4RT}\,\text{sech}^2\left\{\frac{F}{2RT}\left(E - E^{o'}\right)\right\}$$

You will recognize this $\text{sech}^2\{F(E-E^{o'})/RT\}$ term from equation 12:45 as describing the reversible peaked curves illustrated in Figure 12-13. Thus the voltammogram is described by

$$13:29 \quad I = \frac{F^2 Av\Gamma_{max}}{4RT}\,\text{sech}^2\left\{\frac{F}{2RT}(E - E_{peak})\right\} \qquad \begin{array}{l}\text{surface-confined reversible}\\\text{voltammetry; linear-potential scan}\end{array}$$

Of course, other degrees of reversibility are also encountered. A practical example is shown in Figure 13-7.

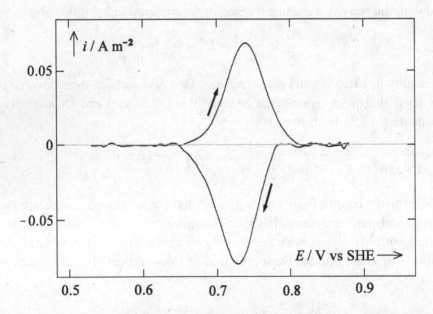

Figure 13-7 A linear potential-scan voltammogram for the oxidation of cytochrome-c peroxidase[1326], a surface-confined reaction. The data relate to a monolayer coverage of the enzyme on a graphite electrode, scanned at a rate of 20 mV s^{-1} in a phosphate buffer. The figure represents only the faradaic portion of the voltammogram. The much larger non-faradaic current has been subtracted.

[1326] Cytochrome-c is an electron-transporting protein of the mitochondria, owing its logistic power to changes in the oxidation state of an iron atom located in a heme group. Cytochrome-c peroxidase is an enzyme based on the cytochrome-c structure and able to reduce hydrogen peroxide to water.

Adsorption is an essential precursor to, or component of, many electrode reactions; for example, in the reactions discussed on pages 78–79. It also plays a role in preventing many reactions that would otherwise occur, as in offering protection from corrosion (Chapter 11). Adsorbed species playing a role in mechanisms may influence voltammetry. The next section discusses other ways in which interfacial properties affect voltammetry.

The Interface in Voltammetry: nonfaradaic current, Frumkin effects

In our discussion of surface-confined voltammetry in the previous section, the I in our equations represented the *faradaic* current arising from reaction 13:21. But, because the potential of the electrode was being changed, there was necessarily also a continuous current flow to recharge the capacitance of the double layer. The magnitude of this **nonfaradaic current** or **charging current** in that experiment, as in any experiment in which the potential is scanned linearly, is given by

13:30
$$I_{nf} = \frac{dQ}{dt} = \frac{dQ}{dE}\frac{dE}{dt} = Cv \qquad \text{charging current} \\ \text{scanned potential}$$

where v is the scan rate (volts per second) and C is the (differential) capacitance of the working electrode. This nonfaradaic current simply adds[1327] to the faradaic current I, only the total current being measurable:

13:31
$$I_{meas} = I + I_{nf}$$

This is an ever-present problem in transient voltammetry. Voltammetric interest is predominantly in the magnitude of the *faradaic* current as the potential changes. Yet the potential cannot be changed without eliciting an unwanted *nonfaradaic* current.

Unfortunately, the differential capacitance of the double-layer is by no means constant. This was demonstrated by the green curve in Figure 13-6 for the case of a mercury | (aqueous-solution) electrode, which is one of the "best behaved" interfaces. In addition to the capacitive current that flows across an interface, there are often additional currents that can be attributed to heterogeneous chemical reactions, not involving the species of voltammetric interest. These masquerade as capacitances inasmuch as they may yield a brief transient current when the potential is changed. Such reactions are not generally classified as "faradaic", though in reality they are. They are components of the so-called **background current**, which is the current that flows across the electrode in the absence of any dissolved reactant species. The chart overleaf identifies four other possible contributors to the background current other than double-layer capacitance. Though the figure legend attributes it to "capacitance", the sharply rising current towards the right-

---

[1327]   though it does not necessarily follow that the nonfaradaic current is unchanged by the concomitant reaction.

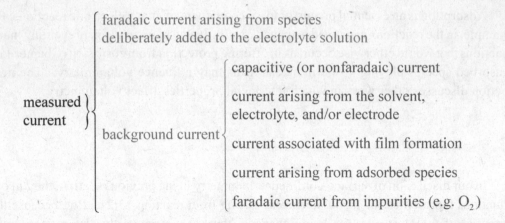

margin of Figure 13-6 is more likely due to a very slight oxidation reaction[1328] forming Hg_2Cl_2; such contributions to the background current occur close to the edges of the polarization "window" (page 130). Currents due to film formation and to adsorbed species may be identified in Figure 13-8, which shows the current when a polycrystalline platinum electrode is slowly polarized from 0.0 V to 1.5 V and back to 0.0 V. There is only a small

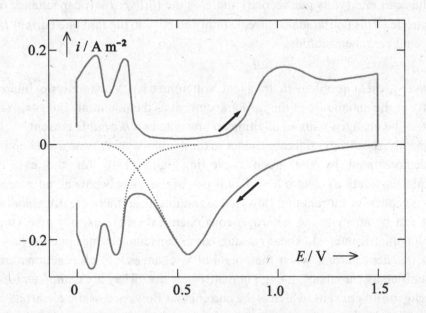

Figure 13-8 The current-voltage curve when a platinum electrode is subjected to a slow triangular potential perturbation over the range $0.0 \text{ V} \leq E(t) \leq 1.5 \text{ V}$ in sulfuric acid. The graph is illustrative only, because the detail depends on the scan rate, the acid concentration, the reversal potentials and particularly on the metallurgical and electrochemical pretreatment of the platinum.

[1328] Write its equation or see Web#1328.

range (from about 0.3 V to 0.7 V on the positive-going scan) in which the platinum behaves as a totally polarized electrode, with a small current that is wholly capacitive. At more positive potentials a current attributable to the reaction

13:33 $$Pt(s) + 3H_2O(\ell) \rightarrow 2e^- + PtO(s) + 2H_3O^+(aq)$$

is observed. After the scan direction reverses[1329] at 1.5 V, a cathodic current gradually builds, corresponding to the removal of the oxide layer by the reverse of reaction 13:32. The cathodic peak at 0.6 V represents the conflict between the increasing rate of reduction of the oxide, as the potential becomes less positive, and its decreasing amount. Most of the PtO has gone by 0.3 V, but a new cathodic reaction, the formation of adsorbed hydrogen

13:34 $$H_3O^+(aq) + e^- \rightarrow H(ads) + H_2O(\ell)$$

commences. After potential reversal at 0.1 V, H(*ads*) is reoxidized. Notice the near-mirror-image relationship of the anodic and cathodic current in the 0.1 V – 0.3 V range, suggesting that reaction 13:33 is behaving reversibly. This contrasts with the offset in the PtO peaks, which implies severe irreversibility of reaction 13:32. When the experiment is repeated with single-crystal samples of platinum metal, only one pair of adsorption peaks[1330] is found, strongly suggesting that the twin peaks arise from different crystal faces in the polycrystalline platinum experiment.

The H(*ads*) and PtO(*s*) deposits on the platinum electrode do not impede that electrode's ability to support faradaic processes. Indeed, the presence of such layers probably goes a long way to explaining why one "inert" metal will readily support a particular electrode reaction that requires a more extreme potential at another. Differences of this sort are responsible for the phenomenon of **electrocatalysis** (Chapter 4).

Much effort in designing and implementing transient voltammetry is aimed at lessening the contamination of the faradaic "signal" by the intrusive "background", or at least being cognizant of the magnitude of the interference. This may, or may not, involve measurement of one or more values of the double-layer capacitance C. Among the strategies that have evolved, in various contexts, in pursuit of this goal are:

(a) Systematically subtract the constant Cv where C is a measured or estimated value, from I. This is better than nothing, but it assumes the double-layer capacitance to be independent of potential, which it rarely is.

(b) Perform a separate "blank" experiment, identical in every respect to the voltammetric experiment of interest except that the electroactive species is absent. Then assume $I = I_{meas} - I_{blank}$ and make a point-by-point subtraction. This is a very common procedure and it certainly helps. However, the blank experiment is made on a totally polarized electrode and conditions may be significantly different in the real experiment. Different species are

---

[1329] The initial and reversal potentials are chosen to be well inside the "window", so as to avoid appreciable formation of O_2 or H_2.

[1330] In fact, the peaks at about 0.25 V correspond to the Pt(100) face, while those at about 0.10 V are from the 110 face. The "100" and "110" are Miller indices (see crystallographic texts).

then present, at varying concentrations, and the hope that these do not affect the composition, and therefore the capacitance, of the double layer is probably in vain.

(c) Often there is at least one region in a voltammogram where it is believed that no faradaic process occurs. The current trace in this (these) region(s) is therefore purely nonfaradaic and can be extrapolated (interpolated) into the faradaicly active region, providing a basis for subtraction. An example of a faradaic current badly contaminated by background is shown in Figure 13-9. Amazingly, the peak shown in Figure 13-7 was extracted from this figure by just such a process. Of course, interpolation procedures are not limited to surface-confined reactions.

(d) Comparison of equations 13:29 and 13:30 shows that, in the case of surface-confined voltammetry, both I and I_{nf} are proportional to the scan rate v. This is unusual however; more often the faradaic current is proportional to the square root of the scan rate (page 346). It then follows that lowering the scan rate will lessen charging-current interference. There will usually, however, be a limit imposed by other considerations on how low the scan rate can be. Of course, in the limit of a *very* slow scan rate, we are back to near-steady-state voltammetry (Chapter 12), in which the nonfaradaic current is indeed negligible.

(e) Make use of the different scan-rate dependencies of I and I_{nf} by running two or more voltammograms at different scan rates. For only two scan rates, the formulas[1331]

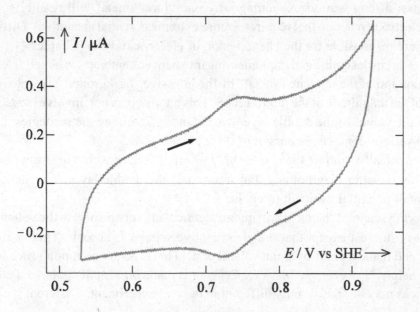

Figure 13-9 A faradaic current badly contaminated by background.

[1331] Derive these formulas; see Web#1331. Also suggest a procedure if many different scan rates were employed.

13:35 $\quad (I)_1 = \dfrac{v_2 (I_{\text{meas}})_1 - v_1 (I_{\text{meas}})_2}{v_2 - \sqrt{v_1 v_2}}$ and $\quad (I_{nf})_1 = \dfrac{v_1 (I_{\text{meas}})_2 - \sqrt{v_1 v_2}\,(I_{\text{meas}})_1}{v_2 - \sqrt{v_1 v_2}}$

permit the individual currents to be disentangled. Use of such formulas is based on the assumption that only the *height*, and not the *shape*, of the voltammogram and its nonfaradaic counterpart are affected by the scan rate. Thus it could be taken to apply to reversible voltammograms but not to the rising portion of a quasireversible voltammetric wave.

(f) In a similar way, and with similar reservations, the fact that the faradaic current is proportional to the bulk concentration of the electroactive species, whereas the charging current is expected to be electroactive-concentration-independent, may be exploited by running voltammograms at different bulk concentrations of the electroreactant. For only two concentrations[1332] the correction formulas are

13:36 $\quad (I)_1 = \dfrac{c_1^b [(I_{\text{meas}})_2 - (I_{\text{meas}})_1]}{c_2^b - c_1^b}$ and $\quad (I_{nf})_1 = \dfrac{c_2^b (I_{\text{meas}})_1 - c_1^b (I_{\text{meas}})_2}{c_2^b - c_1^b}$

(g) For a.c. voltammetry (pages 318–322) the different phase shift appropriate to faradaic and nonfaradaic currents provides a vehicle for the removal of charging current. The net a.c. current is analyzed in terms of the equivalent circuit shown to the right, in which R_u represents the uncompensated resistance (page 210). A subsidiary a.c. measurement is sometimes made to elucidate C and R_u, even when the prime investigatory tool is a d.c. technique.

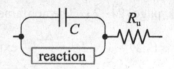

(h) For large-amplitude a.c. voltammetry (pages 322–327) capacitance does not contribute to the second and higher harmonics.

(i) For cyclic voltammetry (pages 347–354), point-by point addition of the currents from the **forward** and **backward** scans should remove the nonfaradaic component exactly. The resultant net faradaic current, shown as the red curve in the diagram on the right must then be compared with an unfamiliar theory that similarly melds the two currents.

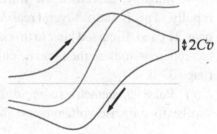

(j) A less drastic application to cyclic voltammetry is to use the current drop, $2Cv$, that occurs at the reversal potential (see Figure 13-8 for an example) to estimate the double-layer capacitance, though the presence of uncompensated resistance often corrupts the sharp drop the would otherwise occur.

[1332] How would you proceed to eliminate the nonfaradaic current if several concentrations were studied? See Web#1332.

(k) Replace the ramped voltage by a staircase in which the potential changes in many abrupt steps. Moreover, do not measure the current continuously, but only during a brief interval immediately prior to the next step, that is at a time of almost Δt_{step} after the previous step. Here Δt_{step} and ΔE_{step} are the "tread" and "riser"of the staircase, so that the quotient $\Delta E_{step}/\Delta t_{step}$ plays a role equivalent to the rate v of a scan. The motivation here is to provide an interval during which the potential does not change, so that the nonfaradaic current can decay to a negligible value.   The dE/dt derivative is theoretically infinite during the application of a potential step to a capacitor, and equation 13:30 therefore predicts a charging current that is instantaneous and infinite. Of course, infinite currents are never observed, in part because there is always some uncompensated resistance present in the cell. As the circuit shown on the previous page makes clear, the nonfaradaic current flows through a series combination of a capacitor and a resistor, as described on pages 21–23. The nonfaradaic component of the current[1333] at the measurement instant is

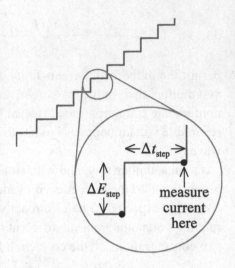

13:37 $$I_{nf} = \frac{\Delta E_{step}}{R_u} \exp\left\{ \frac{-\Delta t'_{step}}{R_u C} \right\}$$ nonfaradaic response to potential step

and this will generally be much smaller[1334] than the nonfaradaic current that would have been recorded in the equivalent ramped experiment. Reducing the uncompensated resistance, R_u *increases* the nonfaradaic current immediately after the potential step but, more valuably, *decreases* the **time constant**, R_uC, so that the interference dies away rapidly. This is one of several reasons for making R_u as small as possible, as advocated on page 211 and discussed later in this section. Of course, replacing a ramp by a staircase also has repercussions on the faradaic current, but fortunately these are much less dramatic[1335] (page 339).

(ℓ) Pulse voltammetries of various kinds (pages 339–346) employ a discriminatory tactic similar to staircase voltammetry: the current is measured only at instants that follow a

---

[1333] This equation serves our purpose here, but it is inexact inasmuch as it ignores the flow of *faradaic* current through the uncompensated resistance, a complicating factor.

[1334] As in Web#1334, use equations 13:37 and 13:30 to compare the nonfaradaic currents for a staircase and the equivalent ramp, for an experiment in which $\Delta E_{step} = 3.0$ mV, $\Delta t_{step} = 30$ ms, $C = 3.0$ nF and $R_u = 10$ kΩ, 5.0 kΩ, 2.0 kΩ, 1.0 kΩ, or 0.50 kΩ.

[1335] The faradaic distinction between a ramp and a staircase fades as ΔE_{step} and Δt_{step} decrease, their ratio being kept constant. In fact, many function generators that purport to output ramps, actually produce staircases with very small (about 0.1 mV) potential increments. Inferior instruments use larger ΔE_{step} values.

period of constant electrode potential, that period being long enough for the nonfaradaic current to have decayed almost to zero.

(m) The experimental voltammogram is accepted as having a substantial nonfaradaic component. An estimated value[1336] of the capacitance (perhaps of the uncompensated resistance, too) is built into the voltammetric model, which thereby seeks to predict the combined faradaic-plus-nonfaradaic voltammogram. A penalty paid for endorsing this approach is that theoretical predictions of the voltammetric behavior are no longer valid because the simplicity of the applied signal has been corrupted.

(n) An *unknown* value[1336] of the capacitance is built into the model. That is, C is regarded as one of the parameters of the model (others might be rate constants, transfer coefficients, etc). Subsequently, using a multiparameter fitting routine, the capacitance is evaluated by a comparison of experiment with the model.

Voltammetry is the search for a relationship between the potential E and the current I. It must be appreciated, however, that, unless the double-layer capacitance C and the uncompensated resistance R_u are negligible, neither E nor I is directly measurable. The voltammetrically pertinent potential is that applied to the boxed "reaction" element in the diagrammatic circuit shown in Figure 13-10. It is related to the applied voltage (measured by the potentiostat's voltmeter) by

13:38 $\qquad E = \Delta E_{appl} - E_{RE} - I_{meas} R_u \qquad$ voltammetrically pertinent potential

where I_{meas} is the *total* current – faradaic plus nonfaradaic – that is measured by the potentiostat. Likewise, the voltammetrically pertinent current is the faradaic current, the current flowing through the boxed element. Its relation to measurable quantities is

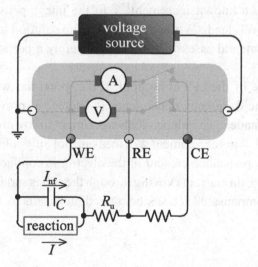

Figure 13-10 Neither I, the current through the faradaic element, nor $E-E_{RE}$, the voltage across it, is directly measurable. The potentiostat's ammeter measures $I_{meas} = I + I_{nf}$, while its voltmeter records the applied voltage $\Delta E_{appl} = E - E_{RE} - I_{meas} R_u$.

[1336] not necessarily a *single* value of C. The model might incorporate a formula that allows for potential dependence.

13:39 $\quad I = I_{meas} - C\dfrac{dE}{dt} = I_{meas} - C\dfrac{d\Delta E_{appl}}{dt} + R_u C\dfrac{dI_{meas}}{dt}$ $\quad$ voltammetrically pertinent current

Equation 13:37 shows that, in voltammetry, the true electrode potential E can be established with confidence only if the uncompensated resistance has a reliably known value, or is negligible. Similarly, equation 13:38 shows that uncompensated resistance is also a hindrance to the accurate determination of the faradaic current. The problem is particularly severe with many nonaqueous solvents, in which the conductivity cannot be increased as easily as in water. Just as with the capacitance, there are several strategies aimed at measuring, or reducing, the **uncompensated resistance**:

(a) Use a well designed Luggin capillary (page 210), closely spaced to the working electrode, to diminish the uncompensated resistance.

(b) Because R_u is inversely proportional to the conductivity κ of the cell solution, use as high a concentration of supporting electrolyte as is compatible with the goals of the experiment. A particular problem arises with the many organic solvents in which most salts dissolve only sparingly.

(c) Knowing the solution's conductivity and the geometry of the reference electrode with respect to the working electrode, calculate[1337] R_u. For example, equation 10:31 could be used for a large working electrode, while the equation

13:40 $\quad\quad\quad\quad R_u = \dfrac{\arctan\{x_{RE}/r_{disk}\}}{2\pi\kappa r_{disk}}$

is applicable to a small RE mounted above the center of a disk working electrode[1338].

(d) A useful diagnostic tool is to *add* a known increment[1339] to the uncompensated resistance by inserting a resistor into the WE lead. While this provides no solution to the problem, it does allow a valuable experimental assessment of how severely a particular resistance affects your voltammogram.

(e) Using a.c., measure the impedance of the cell at an applied d.c. potential where faradaic current is absent. Analysis of the frequency dependence (Chapter 15) provides both R_u and C. The measurement may be made in an independent experiment following the voltammetry, but with the cell undisturbed. This experiment will measure not only solution resistance that is uncompensated, but any resistance present in the electronic conductor.

(f) In the **current-interruption** method, the current flowing through the cell is suddenly interrupted. The voltage across the uncompensated resistance immediately falls to zero,

---

[1337] See Web#1337.

[1338] Show that the uncompensated resistance is 34 Ω when the RE is positioned 1.0 mm from the center of a disk WE of 1.0×10^{-5} m^2 area and the aqueous electrolyte solution is 0.1 molar KCl. See Web#1338.

[1339] by placing a resistor in the lead connecting the potentiostat to the working electrode.

but that across the other elements remains initially unchanged[1340]. Accordingly, there is an immediate diminution of magnitude IR_u in the voltage, from which the uncompensated resistance can be found. Adsorption can lead to false values, however.

(g) Another method[1341] of measuring R_u is available whenever ferrocene is used as an internal reference (page 353) in cyclic voltammetry.

(h) With R_u known, the potentiostat circuit may be modified[1342] to add a quantity $I_{meas}R_u$ to the applied voltage, thereby entirely compensating for the ohmic loss. However, such a **positive feedback** procedure, while feasible in concept, is difficult to execute successfully. If one tries to make a 100% correction in this way, minuscule time delays between the measurement of the current and the implementation of the compensatory voltage can throw the potentiostat circuitry into destructive oscillations. Partial compensation can be effective, however, and positive feedback is sometimes used to ameliorate, rather than remove, uncompensated resistance. Often this approach is used heuristically: compensating resistance is progressively added until oscillation is encountered, then backed off until stability is restored.

(i) As with double-layer capacitance [items (m) and (n), page 277] the uncompensated resistance may be treated as a component of the model and evaluated in a multiparameter fit.

Another way in which interfacial structure has a bearing on voltammetry is through the so-called **Frumkin effects**. Though our discussion of the double layer, earlier in this chapter, was based on a totally polarized electrode, there is no reason, in the absence of adsorption, to believe that double-layer structure is significantly different during the occurrence of an electrode reaction. Frumkin[1324] realized that there are two ways in which the presence of a double layer would modify the traditional theory of electrode kinetics (pages 130–136). These effects arise because, at the instant at which electron exchange occurs, the reactant is located within the double layer. For lack of a better measure, the reaction site may be identified at a point distant x_H, the width of the compact layer, from the geometric interface.

The first Frumkin effect represents an adjustment to the concentration of reactant R at the reaction site. This is an equilibrium effect; we are not here concerned with transport or kinetics. From Boltzmann's distribution law, the reactant concentration is

13:41 $$c_R^H = c_R^b \exp\left\{\frac{-z_R F}{RT}\phi_H\right\}$$ first Frumkin effect

at the site. Here, as before, we have assigned a value of zero to the potential in the bulk solution. If the reactant is an anion, its concentration is enriched, because ϕ_H will generally

---

[1340] For an explanation, see Web#1340.

[1341] Web#1341 contains details.

[1342] For the circuitry used, see A.J. Bard and L.R. Faulkner, *Electrochemical Methods*, 2nd edn, Wiley, 2001, Section 15.6.3.

be positive during an oxidation. The concentration[1343] of the product O will obey a formula similar to 13:40.

The second Frumkin effect addresses the influence that the double layer has on the kinetics of the electrode reaction. Recall that, at a fundamental level, the reason why increasing the electrode potential, say by δE, accelerates a one-electron electrooxidation reaction is that the Gibbs energy barrier is thereby lowered by $F\delta E$ and this appears as an increment of $(1-\alpha)F\delta E/RT$ in the logarithm of the forward rate constant together with a decrement of $\alpha F\delta E/RT$ in the logarithm of the backward rate constant. This is the story that the Butler-Volmer model tells. Frumkin, however, recognized that, when the reactant is sited at x_H, only a portion of the δE increment serves as a reaction accelerant, the remainder being "wasted" on the diffuse double layer. He postulated that, for example

$$13:42 \qquad \ln(k_{ox}) \propto \exp\left\{ \frac{(1-\alpha)F}{RT}\left(E - \phi_H - E^{o'}\right) \right\} \qquad \text{second Frumkin effect}$$

The quantitative repercussions of the second Frumkin effect are reported elsewhere[1344].

Dramatic evidence supporting the existence of Frumkin effects comes from voltammetric study of the irreversible reduction of such multicharged anions as the peroxydisulfate

$$13:43 \qquad\qquad\qquad S_2O_8^{2-}(aq) + 2e^- \rightarrow 2SO_4^{2-}(aq)$$

With low concentrations of supporting electrolyte, and in a range of potentials removed from the vicinity of E_{zc}, conditions are encountered in which the reaction rate actually *decreases* as the potential becomes more negative. This surprising result can be explained by the Frumkin effects. According to equation 13:40, the concentration of a doubly-charged anion[1345] at x_H is decreased tenfold by a 30 mV negative shift of ϕ_H, and this can more than offset the increasing electroreduction rate.

There is no doubt that the Frumkin effects are real and important. Nevertheless, electrochemists seldom make the appropriate **double-layer corrections** to calculate "true" values of the kinetic parameters $k^{o'}$ and α from the "apparent" values measured assuming Butler-Volmer kinetics. The reason for this is that data needed to make the corrections reliably are not available other than for mercury electrodes and, even when the kinetic experiments *are* performed at mercury, the conditions are seldom compatible with those required for double-layer studies. Because ϕ_H is smaller and more constant in solutions of high ionic strength, Frumkin effects are minimized by using large concentrations of supporting electrolyte, as earlier remarked upon (page 199).

[1343] If the potential is -5.9 mV at the Helmholtz plane, calculate the ionic concentrations there in a 100 mM solution of calcium chloride, $CaCl_2$. See Web#1343.

[1344] See Web#1344.

[1345] There is some evidence, however, that the reducible species is actually a *singly*-charged ion pair, such as $NaS_2O_8^-(aq)$.

Nucleation and Growth: bubbles and crystals

For the most part, electrochemical processes do not produce new phases. Exceptions include the creation of crystals of a metal on an electronic conductor not of that metal,

13:44 $$M^{z_M}(soln) + z_M\,e^- \rightarrow M(s)$$

and bubble formation at electrodes. Two stages are involved: **nucleation**, which is the creation of a minute sample of the new phase, and the subsequent **growth** of the nucleus into a macroscopic phase. When the new phase is crystalline and is formed electrochemically at an electrode, as in equation 13:44, the term **electrocrystallization** is used.

Electrocrystallization has much in common with such phenomena as the boiling of superheated liquids, precipitation from supersaturated solutions, and droplet formation from supercooled vapors. In all these cases, the process involves a **metastable system**: one in which progress to the final state is favored by a substantial negative Gibbs energy change, but in which the process has difficulty getting started, because the initial steps along the road toward equilibrium involve an *increase* in G, as illustrated in Figure 13-11. Perhaps the most familiar illustration is the supersaturated carbon dioxide solution formed by removal of pressurization from a carbonated beverage. Thermodynamically, the process $CO_2(soln) \rightarrow CO_2(g)$ should occur readily[1346], but this would require microbubbles of gas to form within the solution. The pressure within a bubble is augmented by the term

13:45 $$\Delta p = \frac{2\sigma}{r_{bubble}}$$ the Laplace[162] bubble-pressure equation

where σ is the surface tension. This equation predicts huge pressures when the bubble

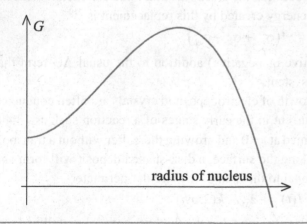

Figure 13-11 Variation of Gibbs energy with the size of the nucleus. Though this diagram actually describes the formation of a liquid from a supercooled vapor[1347], the same principles apply to all metastable systems.

[1346] As it does when ebullient victors shake champagne bottles, thereby introducing many tiny bubbles to serve as nucleation sites.

[1347] The curve obeys the equation $G = 4\pi\sigma r^2 - (4\pi RT\rho/3M)r^3 \ln\{p/p^\circ\}$, for water vapor at 0°C, where M/ρ is the molar volume of liquid water and p/p°, here equal to 4.0, is the ratio of the prevailing pressure to the vapor pressure. G is the free energy of the droplet with respect to the vapor from which it is formed. At the peak in the curve, corresponding to the critical nucleus size, about 90 water molecules are present in the droplet.

radius is small[1348], implying that small bubbles are unstable; the gas inside them "wants" to redissolve. Big bubbles, once formed, are stable and will increase in size by capturing dissolved gas. But if small bubbles won't endure, how can they ever grow? In the absence of impurities or blemishes that foster nucleation, some metastable systems do retain their metastability almost indefinitely[1349], but in the carbonated beverage example there are usually enough **nucleation sites** on the walls of the bottle to foster the growth of a few small bubbles to a size at which spontaneous growth becomes possible. These nucleation sites are suitably shaped blemishes: perhaps a small pit within which gas can reside without having a curved pressure-inducing boundary with the beverage.

Similar, though less easily quantified, considerations apply to electrocrystallization. There is a limited number of nucleation sites on the electrode surface, often locations where the atomic packing of the electrode material has gone awry. Moreover, as one might expect, the more negative the potential of the electrode, the more nucleation sites come into play. The dependence of the rate of nucleation on the applied overvoltage η is found to obey an equation of the form

13:46
$$\begin{pmatrix} \text{rate of nucleus} \\ \text{creation} \end{pmatrix} \propto \exp\left\{ \frac{-(\text{constant})}{\eta^2} \right\} \qquad \text{nucleation}$$

and there is theoretical support for a relationship of this shape from consideration of the statistical probability of unlikely configurations occurring spontaneously.

The formation of a new phase of area A at an electrode creates two surfaces – one between the electronic conductor and the new crystal (with a surface tension σ_{ec}), and another between the ionic conductor and the new crystal (surface tension σ_{ic}) – but destroys part of the preexisting surface – that between the electronic and ionic conductors (surface tension σ_{ei}). The net surface Gibbs energy created by this replacement is[1350]

13:47
$$A\left(\sigma_{ec} + \sigma_{ic} - \sigma_{ei}\right)$$

and this may be a significant (positive or negative) addition to the usual ΔG term that determines the null potential of the system.

Studies of the nucleation and growth of electrodeposited crystals are often conducted by studying the time-evolution of current in the early stages of a reaction such as 13:44. Firstly consider a single nucleus, formed at $t = 0$ and growing thereafter without a transport impediment. If the growth occurs along the surface, a disk-shaped deposit will form and grow by accretion at a rate proportional to the area of the circular perimeter

13:48
$$I(t) = -z_M Fck(2\pi rh)$$

where c is the concentration of $M^{z_M+}(soln)$ ions, h is the thickness of the crystalline layer

---

[1348] Calculate the pressure within a gas bubble in water containing 1000 gas molecules, on the basis of Laplace's equation and the gas law $pV = nRT$. See Web#1348.

[1349] Very pure liquid water has been cooled to $-40°C$ without freezing!

[1350] The formula is appropriate only for a two-dimensional crystal sheet.

and k is a potential-dependent rate constant with the m s^{-1} unit. The total charge needed to produce a disk-shaped deposit of radius r is

$$13{:}49 \qquad Q(t) = \frac{-\pi r^2 h z_M F \rho_M}{M_M}$$

where M_M/ρ_M is the molar volume of the crystalline layer. The time-derivative of 13:49 must equal expression 13:48, from which it follows that

$$13{:}50 \qquad \frac{dr}{dt} = \frac{cM_M k}{\rho_M}$$

Integration produces $r = cM_M kt/\rho_M$ and when this result is substituted back into 13:48, one finds

$$13{:}51 \qquad I(t) = \frac{-2\pi z_M F k^2 M_M}{\rho_M} c^2 t \qquad \begin{array}{l}\text{single nucleus, surface growth} \\ \text{no transport restriction}\end{array}$$

with the important conclusion that the current should increase linearly with time and depend on the square of the ion concentration. On the other hand, if growth of the nucleus can occur in three dimensions, a hemispherical deposit will develop with the rate of accretion being proportional to the growing hemisphere's area. Reasoning similar to that just presented leads to the prediction that[1351]

$$13{:}52 \qquad I(t) = \frac{-2\pi z_M F k^3 M_M^2}{\rho_M^2} c^3 t^2 \qquad \begin{array}{l}\text{single nucleus, 3-D growth} \\ \text{no transport restriction}\end{array}$$

These equations assume that the supply of metal ions poses no limitation, but if the rate of crystallization is governed by the diffusion of those ions to a growing hemisphere, the current obeys a relation that is approximated by

$$13{:}53 \qquad I(t) \approx \frac{-\pi z_M F}{3} \sqrt{\frac{8D^3 M_M c^3 t}{\rho_M}} \qquad \begin{array}{l}\text{single nucleus, 3-D growth} \\ \text{diffusive transport control}\end{array}$$

One seldom finds a solitary nucleus in nature. Nevertheless, it is possible to study **instantaneous nucleation**. A small and constant number of nuclei is formed by briefly pulsing the electrode to a large negative potential that fosters nucleus creation, then studying their growth at a lesser overpotential at which new nuclei are unlikely to form. Another technique is to study a system in which new nuclei are forming at a slow but constant rate. This is termed **progressive nucleation**. The measured current is then the sum of the currents from all the growing nuclei, some old, some new:

$$13{:}54 \qquad I(t) = \sum_n I_n(t - t_n) \qquad 0 \le t_n \le t$$

Here t_n is the time of birth of the nth nucleus and I_n is the current that it contributes. If nuclei are being born frequently and at a constant rate v_{nuc}, then the summation may be

---

[1351] Derive equation 13:52 or see Web#1351.

replaced by an integral

13:55
$$I(t) = v_{nuc} \int_0^t I_{single}(t)\,dt$$

For example, if individual nuclei obey equation 13:53, the net current will be

13:56 $I(t) \approx \dfrac{-4\pi z_M F r_{nuc}}{9}\sqrt{\dfrac{2D^3 M_M c^3 t^3}{\rho_M}}$ progressive nucleation, 3-D growth diffusion control

Ultimately, of course, the growing nuclei compete for space and begin to overlap. Even this may be taken into account mathematically[1352], but we shall not pursue that complication.

A metal M electrocrystallizing on a foreign substrate will adopt a crystal habit that seeks to minimize the Gibbs energy term in expression 13:47. This may not be a configuration conducive to further crystallization of M atoms. That is, a monolayer of M may form at a potential less than that predicted by the Nernst equation for the deposition of bulk metal. This **underpotential deposition** yields a metal layer with properties distinct from the bulk metal: such layers have found applications in electronics and catalysis.

Multilayer crystallization of a salt is eventually limited by the exhaustion of one of the ions. The ion that is not exhausted will often adsorb on the crystal, forming a partial layer as illustrated in Figure 13-12. Because the extra ions fit so perfectly onto the preexisting lattice, this tends to be a rather specific effect, of which advantage is taken in a particular type of sensor. The adsorbed ions create a field within the salt crystal and this field can be detected by a field-effect transistor[1353] placed behind a thin layer of the crystal. Such sensors are called **isfets** or **chemfets**[1354].

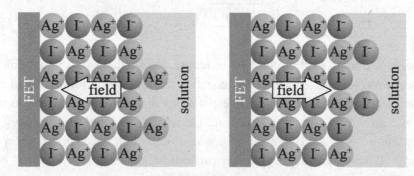

Figure 13-12 Adsorption of "extra" ions causes a field that is sensed by the nearby field-effect transistor.

[1352] through the statistical theory associated with the names of Kolmagarov and Avrami.

[1353] a semiconductor device that uses the presence of a field on a p-type "gate" to control the flow of electrons between two n-type regions.

[1354] acronyms for "ion-selective field-effect transistor" and "chemical sensing field-effect transistor".

Summary

Many features of the electrode interface affect electrochemical reactions. The double layer that always exists at the junction between an ionic and an electronic conductor has a large capacitance and gives rise to nonfaradaic currents whenever the electrode potential is changed. Frumkin effects allow the double layer to perturb both the concentration of ions close to the electrode and the rate of the electron-transfer reaction. The interface is also the site of adsorption, which frequently plays a role in electrochemical reaction mechanisms. New phases are often reluctant to form at electrode interfaces because nucleus formation is statistically infrequent and small nuclei may be stillborn.

14

Other Interfaces

In the previous chapter, we addressed phenomena at an electrode; that is, at the interface between an ionic conductor and an electronic conductor. There are, however, several other types of interface that are of interest to electrochemists, and three of these will be discussed in the present chapter. The first is the interface between an ionic conductor and a semiconductor. Inasmuch as electrons cross this interface, it earns the name "electrode". However, in the other two interfaces to be addressed here, there is no direct role played by electrons: the interfaces are *not* electrodes. One of these electron-free interfaces is formed at the junction between two distinct ionic conductors. The most common manifestations of this type of interface are those formed when two immiscible liquids, each containing a dissolved electrolyte, meet. The final type of interface discussed in this chapter is that between an ionic conductor and an insulator; in particular, the junction of an aqueous solution and glass or silica.

> ionic conductor
> electronic conductor

> ionic conductor
> semiconductor

> one ionic conductor
> a second ionic conductor

> ionic conductor
> insulator

Semiconductor Electrodes: capturing the energy of light with photochemistry

When, in the previous chapter, we were discussing the diffuse distribution of charges in the ionic conductor at a polarized electrode, did you wonder why there was no diffuse layer on the electronic-conductor side? The answer is that metals are such good conductors of electricity that any such layer would be of subatomic size. Electrodes having poorly conducting electronic conductors do, however, support diffuse charges as part of a double-layer interface. The phenomenon is particularly important in the case of semiconductor electrodes, in which the conductivities of the electronic and ionic conductors may be

Electrochemical Science and Technology: Fundamentals and Applications, First Edition. Keith B. Oldham, Jan C. Myland, Alan M. Bond.
© 2012 John Wiley & Sons, Ltd. Published 2012 by John Wiley & Sons, Ltd.

comparable. However, you will not find the effect discussed in the literature in terms of a diffuse layer. In view of the band structure[124] associated with semiconductors, it is referred to as **band bending**.

An important concept in solid-state physics is that of the **Fermi**[1401] **energy**, which is the energy that an electron would have in the **Fermi level**. We write "would have" because the Fermi level is not actually occupied; it represents a weighted average of the energies of the electrons occupying the valence and conduction bands. Apart from its sign, the Fermi energy[1402] $\Delta\varepsilon$ is essentially the **work function** of the solid, which is the energy required to remove (to infinity[1403]) an electron from a phase; it represents the average energy of an outer electron. In semiconductors the Fermi level lies somewhere between the energy levels of the conduction and valence bands, depending on the occupancy of those bands. In an intrinsic semiconductor, the Fermi level lies midway between the two bands because the population of electrons in the conduction band equals the population of holes in the valence band. In an n-type semiconductor, the level is just below the level of the conduction band, because the carriers – electrons – are predominantly in that band. Conversely, the Fermi level is just above the energy level of the valence band, in a doped semiconductor of the p-type. The Fermi level of a semiconductor reflects the electron activity in a semiconductor in a way analogous to that in which the standard potential[1404] of a metallic electrode governs the activity of its electrons.

When the working electrode of an electrochemical cell is metallic, there is always an abundance of electrons to take part in an electrochemical reaction. This is not necessarily the case, however, when the electrode is a semiconductor. For example, reductions readily occur at an n-type semiconductor, because there are plenty of conduction-band electrons, whereas oxidations are disfavored, because of the paucity of holes to accept electrons into the valence band. Of course, the opposite is true of semiconductors of the p-type, for which their abundance of holes facilitates the oxidation

$$14:1 \qquad\qquad R(aq) + h^+(sc) \rightleftarrows O(aq)$$

In fact, the rate of oxidation can be regarded as proportional, not only to the concentration of R at the interface, but also the concentration of valence-band holes there. One can observe limiting currents that represent hole depletion. At more positive potentials, the more conventional reaction

---

[1401]   Enrico Fermi, 1901–1954, Italian physicist. Following his 1938 Nobel Prize, he emigrated to the U.S.A., where he worked on nuclear transformations.

[1402]   When, as in solid-state physics, energy is measured in electron volts, the distinction between a difference in potential and a difference in energy becomes largely one of units and, accordingly, the symbol E is often used for both potential and energy. Here we make a distinction and use ΔE for potential difference and $\Delta\varepsilon$ for an energy difference. For electrons, the relation between the two is $\Delta\varepsilon = Q_0\Delta E$.

[1403]   An electron infinitely remote and in free space in said to occupy the **vacuum level**.

[1404]   Attempts to quantitatively relate the electrode potential to the Fermi level rely on the concept of the **absolute electrode potential**, based on models such as that of Kanevskii[602].

14:2 $$R(aq) \rightleftarrows e^-(sc) + O(aq)$$

comes into play, with the electrons entering the conduction band.

Notwithstanding that the term "double layer" is not used in the semiconductor literature, a space charge, structured akin to a Gouy-Chapman diffuse region, generally forms inside the semiconductor when it is brought into contact with an aqueous solution. Typically, it is several nanometers wide. Unlike a double layer in solution, however, the diffuse region appears not to be accompanied by any compact layer. The space charge[1405] forms by a redistribution of the charge carriers brought about by the intense field caused by the juxtaposition of the ionic conductor, as illustrated in Figure 14-1. In consequence, the conduction and valence bands "bend"; that is, the energy levels of the bands near the

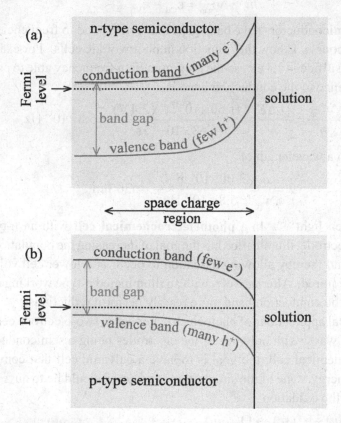

Figure 14-1 Energy levels in semiconductors polarized by contact with an ionic conductor. (a) In this n-type semiconductor, a positive space charge has formed, bending the conduction and valence bands upwards. The positive charge is provided by a depletion of electrons from the **conduction band** and an enrichment of the (otherwise sparse) complement of holes in the **valence band**. (b) The converse situation occurs when a p-type semiconductor acquires a negative space charge.

[1405] which may go by the name **depletion layer** in semiconductor parlance.

interface come to differ in energy from those deep inside the semiconductor. If the aqueous solution contains a redox pair, so that the electrode potential is adjustable, the bands within the semiconductor can be "tuned" to some extent. In particular, it may be possible to inhibit band bending. The potential needed to accomplish this is spoken of as the **flat-band potential**; it is analogous to the potential of zero charge in a traditional electrode.

Seldom are semiconductor│ionic-conductor junctions as efficient as other types of electrode in promoting chemical reactions. Interest in semiconductor electrodes is primarily motivated by a property that other electrodes do not share: their ability to absorb light energy. Semiconductors can capture photons when the energy of the photon matches the band gap of the semiconductor. Because photon energies are equal to Planck's constant[840] h multiplied by the light's frequency, light can be absorbed if

14:3 $$h\nu \geq \Delta\varepsilon_{gap} = \varepsilon_{CB} - \varepsilon_{VB}$$

Fortunately, many semiconductors have band gaps that correspond to frequencies present in sunlight. That, of course, is how they function in photovoltaic cells. For example, the band gap in cadmium sulfide is 2.4 eV, so the minimum light frequency able to promote an electron from the valence to the conduction band is

14:4 $$\nu = \frac{\Delta\varepsilon_{gap}}{h} = \frac{Q_0\Delta E}{h} = \frac{(1.60\times10^{-19}\,C)(2.4\,V)}{6.63\times10^{-34}\,J\,s} = 5.8\times10^{14}\,Hz$$

which is equivalent to a wavelength of

14:5 $$\lambda = \frac{c}{\nu} = \frac{3.00\times10^8\,m\,s^{-1}}{5.8\times10^{14}\,Hz} = 520\,nm$$

corresponding to green light[1406]. In a **photoelectrochemical cell** with an n-type semiconductor working electrode, illumination has the goal of increasing the population of holes in the valence band and thereby allowing oxidation to occur at a lower cell voltage than would otherwise be required. Alternatively, with an illuminated p-type working electrode, electrons populating the conduction band more readily perform reductions.

To pursue practical applications of photoelectrochemistry, two-electrode cells may be configured in several ways, with only one of the electrodes being a semiconductor. In a photovoltaic electrochemical cell, the goal is to have a galvanic cell that converts light energy to electrical energy. One of the simplest arrangements would be to have an n-type photoanode at which the oxidation

14:6 $$R(soln) + h^+(sc) \rightarrow O(soln) \qquad \text{semiconductor photoanode}$$

coupled with a conventional cathode

14:7 $$O(soln) + e^-(metal) \rightarrow R(soln) \qquad \text{conventional cathode}$$

both utilize the same redox couple. Figure 14-2 illustrates the processes involved in such

---

[1406] Rutile, a semiconducting form of TiO_2, absorbs light only of wavelengths < 420 nm. Calculate the band gap. See Web#1406.

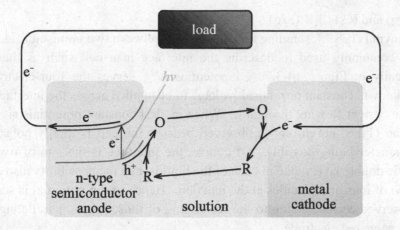

Figure 14-2 Schematic diagram of a galvanic photoelectrochemical cell.

a cell. An alternative to this direct way of operating a galvanic photocell, is to use a **mediator** which more readily reacts with holes. Dyes often fill this role, being converted to an excited state at the semiconductor interface and subsequently reacting with the O/R couple. Another configuration has electrosynthesis as its objective. It is exemplified by the **photosplitting** of water, in which the reactions are

14:8 $\quad 6H_2O(\,soln) + 4h^+(sc) \rightarrow O_2(g) + 4H_3O^+(aq)$ $\qquad$ semiconductor photoanode

and

14:9 $\quad 4H_3O^+(aq) + 4e^-(metal) \rightarrow 2H_2(g) + 4H_2O(\ell)$ $\qquad$ conventional cathode

This process has not been realized as a self-sustaining operation, but **photoassisted electrolysis** can reduce the cell voltage below that required in the dark.

Phenomena at Liquid│Liquid Interfaces: transfers across "ITIES"

Though (aqueous solution)│(ionic liquid) interfaces have been investigated, most electrochemical studies of the junction between two ionic conductors are made with two immiscible liquids, each containing dissolved electrolyte, meeting at a plane[1407]. One of the liquids is usually water and, to ensure immiscibility, the other is a hydrophobic organic solvent, such as nitrobenzene, $C_6H_5NO_2$, or 1,2-dichloroethane, $(ClCH_2)_2$. Whereas most simple salts, lithium chloride for example, dissolve readily in water to provide ions, such as $Li^+(aq)$ and $Cl^-(aq)$, that confer conductivity, only bulky ions with lipophilic groups are sufficiently soluble in organic liquids to provide an adequately conductive path. Tetra-butylammonium tetraphenylborate is an example of a salt that yields such ions, namely

---

[1407] Though planar to macroscopic viewing, such junctions are surprisingly irregular at a molecular level.

$(C_4H_9)_4N^+(org)$ and $(C_6H_5)_4B^-(org)$.

The acronym ITIES[1408], standing for the interface between two immiscible electrolyte solutions, is commonly used to describe the interface in a cell such as that shown diagrammatically in Figure 14-3. A bipotentiostat[1233] serves the four-electrode cell permitting a known, constant or ramped, voltage to be applied across the interface, while measuring the current flowing through it. When modest constant potentials are applied across such an ITIES, no current is observed because the cell is totally polarized, no reaction or transfer being possible. Of course, the interface is the site of two Gouy-Chapman-style double layers, one in each of the liquids, and the possibility also exists of the adsorption of ions or molecules at the junction. Hence, if the potential is scanned, a current is observed corresponding to the recharging of those layers, paralleling similar behavior at a polarized electrode.

If a larger d.c. voltage (about 200 mV) is applied, the aqueous layer being positive, some of the tetraphenylborate anions will overcome their reluctance to enter the aqueous layer,

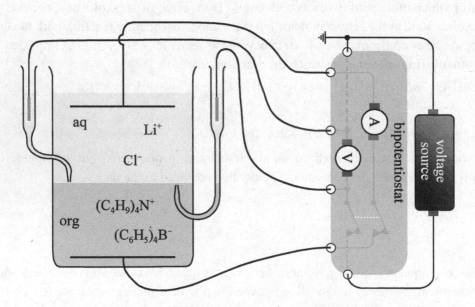

Figure 14-3 Diagram of an apparatus for the electrochemical study of the interface between an aqueous solution of lithium chloride and a solution of tetrabutyl-ammonium tetraphenylborate in a hydrophobic organic solvent. Both reference electrodes could be of the usual Ag│AgCl│Cl⁻ (aq) variety, with the one serving the organic layer dipping into a solution of tetrabutylammonium chloride.

[1408] Pioneering studies were carried out by Ernst Hermann Riesenfeld (German physical chemist, 1877–1957, who worked in Sweden during the Nazi era); he was Nernst's brother-in-law. For a review of ITIES, see H.H. Girault, *Electrochemistry at Liquid-Liquid Interfaces* in: A.J. Bard and C.G. Zoski (Eds.), *Electroanalytical Chemistry: A Series of Advances, Vol. 23*, CRC Press, Boca Raton, FL, 2010.

14:10
$$(C_6H_5)_4B^-(org) \rightleftarrows (C_6H_5)_4B^-(aq)$$

causing an **ion transfer current**[1409]. At a somewhat more positive potential, some lithium ions will desert their preferred aqueous environment and leak into the organic phase

14:11
$$Li^+(aq) \rightleftarrows Li^+(org)$$

further enhancing the current. If the polarity is reversed, it is the cations that travel from the organic phase into the aqueous phase

14:12
$$(C_4H_9)_4N^+(org) \rightleftarrows (C_4H_9)_4N^+(aq)$$

and/or the anions that make the converse journey.

14:13
$$Cl^-(aq) \rightleftarrows Cl^-(org)$$

prompting a flow of current in the opposite direction.

Electrochemical measurements in aqueous systems are based on assigning a potential of zero to the SHE, as explained on page 106. This scale can serve the aqueous chamber of the Figure 14-3 cell, but some convention is needed to provide an interpretation of voltages measured across an ITIES. Because ferrocene, $(C_5H_5)_2Fe$ (frequently abbreviated to Fc), is soluble in many liquid solvents, and because it easily establishes an electron-transfer equilibrium with the ferrocenium cation

14:14
$$Fc(soln) \rightleftarrows e^- + Fc^+(soln)$$

in most of those solvents, the potential of this couple has become the accepted way of intercomparing electrode potentials measured in diverse solvents. It is appropriate, therefore, to adopt the same convention in seeking to interpret intersolvent potentials in ITIES cells. That is, the standard potential of reaction 14:14 in one of the two ITIES solvents is equated to that in the other. A vindication of this approach comes from the observation that, if decamethylferrocene $((CH_3C)_5)_2Fe$ (abbreviated dmFc) is used as an alternative calibrant, exactly the same result is obtained. Figure 14-4 shows the joint scales and how several redox pairs fit into the scheme.

If a particular ion i is present on both sides of an ITIES at equilibrium, a version of Nernst's law exists in the form

14:15
$$\Delta E = \phi^w - \phi^o = \frac{\Delta G_i^{w \to o}}{z_i F} + \frac{RT}{z_i F} \ln\left\{ \frac{a_i^o}{a_i^w} \right\}$$

where $\Delta G_i^{w \to o}$ is the standard molar Gibbs energy change accompanying the transfer[1410] of an ion from water (superscript ^{w}) into the organic liquid (superscript ^{o}). Studies of ITIES

[1409] Such currents may be considered "faradaic", though not in the usual sense of that term.

[1410] Of course, one inevitably deals with the simultaneous transfer of two ions (one via the reference electrodes). As in other contexts, the hydrogen ion is taken as the reference standard and is assumed to have the same standard Gibbs energy in all solvents. As in Web#1410, predict the interfacial potential difference if the chloride ion, Cl^-, exists in 100-fold excess on the aqueous side of a water│nitrobenzene ITIES. Which side is positive? Explain the polarity.

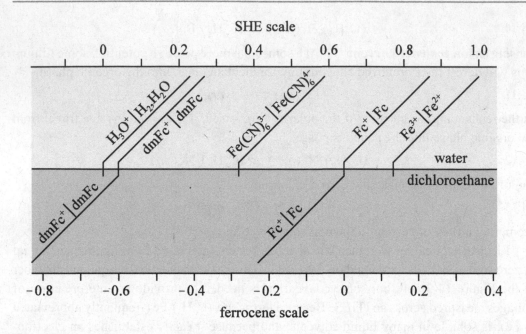

Figure 14-4 Reconciliation of the potential scales for a water│dichloroethane junction.

can thereby be used to determine, for example, that the standard Gibbs energy change is 43.9 kJ mol^{-1} for the transfer of a chloride ion, Cl$^-$, from water to nitrobenzene. Moreover, the application of a voltage across an ITIES can be employed to alter the surface composition of one or both of the phases. For example, protons can be "pumped" out of the aqueous phase into an organic phase containing a lipophilic base B to effect the transfers illustrated in Figure 14-5. Thereby the local pH of the aqueous phase is increased, while the acidity of the organic phase is enhanced locally by a change in the B/BH$^+$ buffer ratio in the organic phase.

However, equilibrium across an ITIES may be established only slowly. Investigations of the rate of transfer of the organic cation acetylcholine[1411] from dichloroethane across its

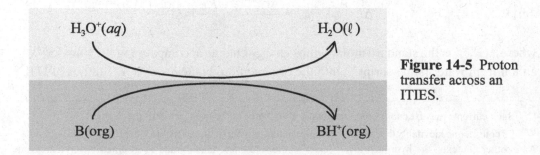

Figure 14-5 Proton transfer across an ITIES.

[1411] See pages 191– 192 for the formula of acetylcholine and to read of its role as a neurotransmitter.

interface with an aqueous solution, and vice versa, show transfer rates proportional to $\exp\{\mp \alpha F\Delta E/RT\}$, with α values close to 0.5. In this respect, the transfers show strong parallels with electron-transfer kinetics, as interpreted through the rival treatments of Butler-Volmer and Marcus-Hush (Chapter 7).

With great relevance to the mechanism of ion-selective electrodes (Chapter 6) are studies of **facilitated transport** across ITIES. An example of a facilitator is provided by the ionophore **valinomycin** (page 172) which is able to assist the potassium K^+ ion in crossing an ITIES into the organic phase. Without the facilitator, the process

$$14:16 \qquad K^+(aq) \rightleftarrows K^+(org)$$

is slow, but equilibrium is much more rapidly approached in the presence of valinomycin, which will be represented by Vm in the equations that follow. One can envisage several mechanisms whereby this facilitation might occur. Some of the following processes

$$14:17 \qquad Vm(org) \rightleftarrows Vm(aq)$$

$$14:18 \qquad K^+(aq) + Vm(aq) \rightleftarrows KVm^+(aq)$$

$$14:19 \qquad KVm^+(aq) \rightleftarrows KVm^+(org)$$

and

$$14:20 \qquad K^+(org) + Vm(org) \rightleftarrows KVm^+(org)$$

could be involved[1412]. Experiments suggest that, at least in this case, the complexation of the ion occurs wholly on the organic side of the interface, the ionophore's role being to accelerate the dispersal of the ion out of the double layer there. That is, reactions 14:16 and 14:20 are the important ones. Of course, the ultimate equilibrium state is independent of the details of the mechanism and must satisfy the laws governing the various equilibria.

Figure 14-5, on the facing page, illustrates a proton transfer taking place at an ITIES, but electrons may be transferred in a similar fashion, as pictured in Figure 14-6 overleaf. Evidently a reduction, of hexacyanoferrate(III) to hexacyanoferrate(II), occurring wholly within the aqueous phase, couples with an oxidation, occurring wholly in the organic phase, of ferrocene to the ferrocenium cation, with an electron crossing the interface. This has been called a **biphasic electron-transfer reaction**. The analogy of this ITIES to an electrode is apparent. But note that whereas an electrode can bring about *either* an oxidation *or* a reduction, *both* of these processes occur simultaneously at an ITIES. Notice, moreover, that a process with this result needs not necessarily occur by a purely heterogeneous route. Indeed, at least for certain concentrations of the four iron compounds, it has been demonstrated that the pictured reaction incorporates the three processes[1413]

[1412] Write equilibrium constants for the processes 14:16–14:20. The five constants are not independent; find a relationship between them. Consult Web#1412.

[1413] Which of these three is affected by changing the voltage across the ITIES? How might a positive change in that voltage affect the overall reaction? See Web#1413.

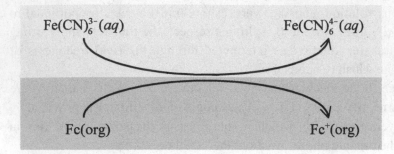

Figure 14-6 Electron transfer across an ITIES.

14:21 $$Fc(org) \rightleftarrows Fc(aq)$$

14:22 $$Fc(aq) + Fe(CN)_6^{3-}(aq) \rightleftarrows Fc^+(aq) + Fe(CN)_6^{4-}(aq)$$

and

14:23 $$Fc^+(aq) \rightleftarrows Fc^+(org)$$

at the water | nitrobenzene interface.

If the interfacial potential of the ITIES cell discussed in the previous paragraph is adjusted until no current flows, then a null condition exists, but not necessarily an equilibrium state. Transfer 14:21 generates no current, and could occur unnoticed. Moreover, if there exist trans-ITIES flux densities of the three ions that satisfy the condition

14:24 $$j_{Fc^+}^{o \rightarrow w} = j_{Fe(CN)_6^{3-}}^{w \rightarrow o} - j_{Fe(CN)_6^{4-}}^{w \rightarrow o}$$

then there is no net current, but no equilibrium either. A situation akin to an electro-chemical "mixed potential" (page 217) would exist. Thus, caution is needed in recognizing the existence of equilibrium. However, if it is confidently known that no trans-interface transfers occur, then Nernst's law may be taken to hold in the form

14:25 $$\Delta E_n = (constant) + \frac{RT}{F} \ln \left\{ \frac{a_{Fc^+}^o \, a_{Fe(CN)_6^{4-}}^w}{a_{Fc}^o \, a_{Fe(CN)_6^{3-}}^w} \right\}$$

The "constant" could be regarded as the standard potential for the ITIES cell.

Joint proton and electron transfer across an ITIES, shown in Figure 14-7, has also been demonstrated. Oxygen in the aqueous phase may be reduced by the oxidation of a hydroquinone in the organic phase[1414]. A potential difference applied across the ITIES, using the arrangement diagrammed in Figure 14-3, may radically modify processes such as this. Thus, the applied voltage in the pictured transfer can determine whether the product of oxygen reduction is water or hydrogen peroxide.

The expectation that certain ions and molecules will adsorb at an ITIES is well

---

[1414] Write balanced equations to demonstrate that both protons and electrons are transferred across the ITIES in the processes pictured in Figure14-7. Is there necessarily an accompanying current flow? See Web#1414.

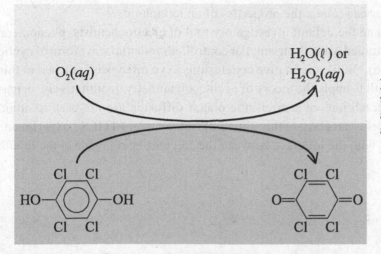

O₂(aq)

H₂O(ℓ) or
H₂O₂(aq)

Figure 14-7 Joint transfer of protons and electrons across an ITIES.

realized. Adsorption, particularly of surfactants, can lead to the mechanical instability of an ITIES: the gravitational stabilization provided by the density differential between the two liquids may be unable to counter a tendency of the surface to buckle in an attempt to increase the contact area. Establishing an association with the material discussed on page 187, is the observation that phospholipids (which are zwitterions[1415] at *pH* values of biological interest) strongly adsorb at an ITIES, forming a monolayer, rather than the bilayer that is the structure of overwhelming biochemical importance. Monolayer formation arises from the hydrophilicity of one end of the phospholipid molecule contrasted with the lipophilicity of the other end. Evidently there are pores in the phospholipid monolayer, because the small tetramethylammonium cations, $(H_3C)_4N^+$, are able to transfer through such an ITIES film whether or not the phospholipid monolayer is present, though with a smaller current in its presence. In contrast, an enhanced current has been observed for the transfer of Na^+ and K^+, unexpectedly showing that an unaugmented

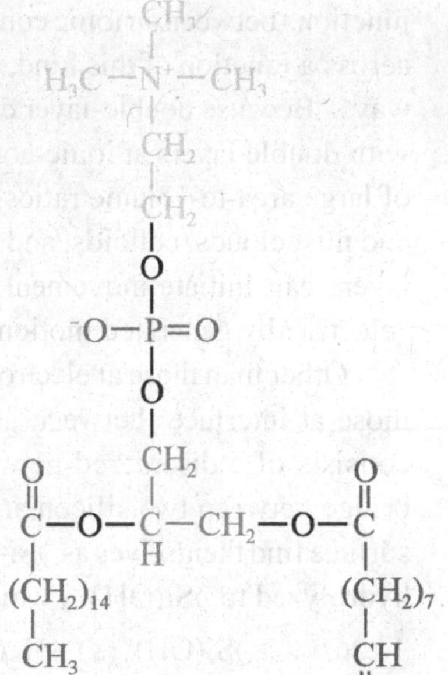

Figure 14-8 A typical phospholipid[1416].

[1415] A **zwitterion** is a species containing both positively and negatively charged groups.

[1416] All phospholipids incorporate a **phosphate** unit, and most have a **glycerol** unit, as shown in Figure 14-8. Various bases are incorporated, the **choline** option being illustrated. There is variety also in the two fatty acid units that give rise to the long hydrocarbon chains: those shown in Figure 14-8 arise from **palmitic** and **oleic** acids. Ester linkages bond the units together.

phospholipid monolayer can mimic the properties of an ionophore.

Because it has become the default investigatory tool of electrochemists, phenomena at ITIES are commonly studied using the triangular controlled-potential waveform of cyclic voltammetry (Chapter 16). While qualitative conclusions have often been reached in this way, it has proved difficult to apply the theory of cyclic voltammetry quantitatively, in part because the necessary conditions of semiinfinite planar diffusion are not easy to attain reproducibly. In an effort to ameliorate this situation, some "micro-ITIES" experiments have been conducted[1408] with the interface between the solvents being made at the mouth of a micropipette.

Electrokinetic Phenomena: the zeta potential

In recent sections we have encountered instances of double layers formed at junctions between an ionic conductor and another conductor. However, double layers also form at junctions between an ionic conductor and an insulator. Clearly, no electric current can flow across a junction of this kind, but the double layer may make its presence known in other ways. Because double-layer effects scale with the interfacial area, phenomena associated with double layers at ionic-conductor|insulator junctions are most noticeable in systems of large area-to-volume ratios, such as porous materials, suspensions, nanoparticles, dust and mist clouds, colloids, and so on, as well as in tubes of narrow bore. Electrical double layers can initiate movement and this is the source of the adjective **electrokinetic** [≡ "electrically generated motion"] used to describe these phenomena[1417].

Other than those at electrodes, the most extensively studied double layers are probably those at interfaces between glass or silica and an ionic aqueous solution. Fused **silica** consists of a disordered network of silicon and oxygen atoms, each oxygen forming a bridge between two silicon atoms. Figure 14-9 illustrates how the oxygen atoms on the surface find themselves as $\rangle Si{=}O$ groups. In an aqueous environment, these groups become hydrolyzed to $\rangle Si(OH)_2$, which is weakly acidic and ionizes as follows

14:26 $\rangle Si(OH)_2(s) + H_2O(\ell) \rightleftarrows \rangle Si(OH)O^-(s) + H_3O^+(aq)$ $K \approx 3 \times 10^{-6}$

Glasses have structures similar to silica, but some of their oxygen atoms are already ionized in the dry state, their charges being countered by those of metal cations. In either case, the surface of the solid becomes the negative member of a double-layer, with compensatory cations occupying a diffuse zone within an aqueous solution in contact with the insulator. The intensity of charge on the solid surface reflects the pH, being most negative at high pH values. Occasionally, the adsorption of highly-charged cations, for example

14:27 $\rangle Si(OH)O^-(s) + H_2O(\ell) + Al^{3+}(aq) \rightleftarrows \rangle SiOAlO^+(s) + H_3O^+(aq)$

[1417] There is no direct relationship to "electrode kinetics", despite the similar name.

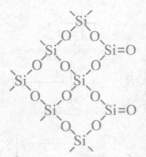

Figure 14-9 The structure of fused silica is three-dimensional and much more disordered than the representation shown here.

will reverse the polarity of the double layer, but generally the insulator surface is negative.

Because this geometry enhances the area/volume ratio, studies of electrokinetic phenomena are often conducted in capillary tubes, especially those of glass or silica. Figure 14-10 illustrates such a tube connecting two chambers, each equipped with an electrode (possibly Ag|AgCl) and filled with an ionic solution, as is the capillary tube. A double layer exists at the solution|solid interface, with the latter usually being negatively charged, so that a solution layer enriched in cations occupies a hollow cylinder or "sleeve" adjacent to the capillary wall. Of course, the thickness of this layer is very much narrower than the bore of the capillary.

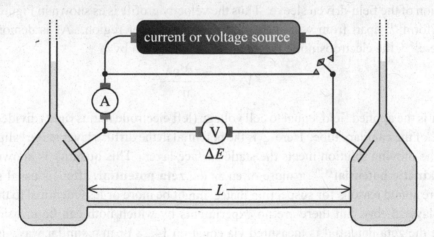

Figure 14-10 Apparatus to illustrate a variety of electrokinetic phenomena. The three-position switch can allow the imposition of a voltage, or current, or neither, on the cell.

Now imagine that the electrodes are used to apply an electric field along the capillary tube, acting in the direction diagramed in Figure 14-11 overleaf. The effect of this field on a positively charged sleeve is to promote its motion in the rightwards direction. The motion is not just that of the sleeve, the entire capillary contents within the sleeve will be dragged to the right in a phenomenon known as **electroosmosis** or **electroosmotic flow**. If the flow is allowed to continue, a head will build up in the right-hand vertical tube, opposing the

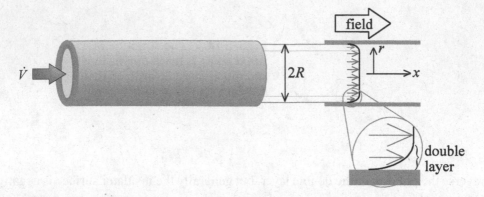

Figure 14-11 A sleeve of positively charged solution lines the solid tube. The electrostatic force on the sleeve causes motion of the entire tube content, save for a motionless molecular layer at the tube wall.

flow, until eventually an **electroosmotic pressure** is attained, inhibiting further motion.

Electroosmotic flow is laminar, but it does not have the quadratic dependence of velocity on radius, as described on pages 161–162, that is seen in Poiseuille flow. This is because electroosmotic flow is not pressure-driven; instead, it is induced by friction from the motion of the field-driven sleeve. Thus the velocity profile is as shown in Figure 14-11: it is uniform[1418] apart from within the narrow double-layer region. As is demonstrated elsewhere[1419], the electroosmotic flow rate (m³ s⁻¹) is given by

14:28
$$\dot{V}_{osm} = \frac{-\pi R^2 \varepsilon X}{\eta} \varsigma$$

where X is the applied field, equal to cell voltage (left electrode minus right) divided by the length L of the capillary tube. Here ς is the potential in the diffuse layer at the "slip plane" where the moving solution meets the static surface layer. This quantity is known as the **electrokinetic potential**[1420] or more often as the **zeta potential**, after its usual symbol. There are sound reasons for suspecting that ς might be more or less identical to the ϕ_H of double-layer theory, but there are no experiments by which both can be measured. In practice the zeta potential is measured via equation 14:28 or in a similar way. In dilute neutral aqueous solutions containing only singly charged ions at 25°C, the zeta potential is close to –150 mV at a glass surface, decreasing slightly with concentration, or about half that value at silica.

Equation 14:28 may be rewritten as

[1418] This uniformity is essential for the success of **electrophoresis**, the technique described on page 153.

[1419] See Web#1419 for the derivation of 14:28 and for a definition of **electroosmotic mobility**.

[1420] though it is actually a potential difference, measured with respect to the solution in the center of the tube at the same value of x.

14:29 $$\dot{V}_{osm} = \frac{A\varepsilon\varsigma}{\eta L}\Delta E$$ electroosmotic flow caused by voltage

where A is the cross-sectional area of the capillary tube and ΔE is the applied cell voltage. If the electrodes are depolarized, the applied voltage will not only cause the electroosmosis, it will also generate a current, of magnitude $\kappa A \Delta E / L$. One may therefore write 14:29 as

14:30 $$\dot{V}_{osm} = \frac{\varepsilon\varsigma}{\eta\kappa} I$$ electroosmotic flow caused by current

where κ is the conductivity of the solution. It is just as valid to think of this current as the cause of the electroosmotic flow, as it is to regard the applied voltage as the causal agent. Indeed, by replacing the potential source in Figure 14-10 by a current source, one can drive a current through the cell, and observe exactly the flow described by 14:30 and a potential drop described by 14:29. In a similar way, when flow has ceased because of the buildup of a reverse head, the electroosmotic pressure may be considered to be generated by the voltage or the current

14:31 $$\Delta p_{osm} = \frac{-8\pi\varepsilon\varsigma}{A}\Delta E = \frac{-8\pi L\varepsilon\varsigma}{A^2\kappa}I$$ electroosmotic pressure caused by voltage or current

In an alternative experiment, one may use the three-position switch to disconnect the source in Figure 14-10, and instead apply pressure (by a gravitational head or otherwise) so as to force solution through the capillary tube. Whereas, in the previous paragraph, an electrical stimulus (voltage or current) gave rise to a mechanical effect (flow or pressure), we now find the converse: mechanical stimuli produce an electrical effect. Thus, one has a choice of measuring a so-called **streaming potential**[1421] with the ammeter open-circuited or a **streaming current** with the ammeter shorting the voltmeter. Being pressure-driven, the flow through the capillary now obeys Poiseuille's law[842], which can be written as $\dot{V}_{Pois} = -A^2\Delta p_{osm}/(8\pi L\eta)$. The streaming potential may be considered to be caused by the forced flow of solution or by the pressure differential[1422]

14:32 $$\Delta E = \frac{-8\pi L\varepsilon\varsigma}{A^2\kappa}\dot{V} = \frac{\varepsilon\varsigma}{\eta\kappa}\Delta p$$ streaming potential caused by flow or pressure

and these alternative causes may likewise be considered to engender the streaming current

14:33 $$I = \frac{-8\pi\varepsilon\varsigma}{A}\dot{V} = \frac{A\varepsilon\varsigma}{L\eta}\Delta p$$ streaming current caused by flow or pressure

We have described a total of eight experiments, but they are all closely related. In fact, when the appropriate sign is selected, they are all incorporated into the universal

---

[1421] This is the usual name but it is actually the *difference* in potential of two electrodes.

[1422] By comparing the *SI* units of its three members, confirm that equation 14:32 is dimensionally homogeneous. See Web#1422.

dimensionless relationship

14:34
$$\left\{ \frac{L}{\kappa \varsigma A} I \quad \text{or} \quad \frac{1}{\varsigma} \Delta E \right\} = \pm \left\{ \frac{8\pi L \varepsilon}{A^2 \kappa} \dot{V} \quad \text{or} \quad \frac{\varepsilon}{\eta \kappa} \Delta p \right\}$$

The remarkable inclusiveness of this relationship is explained by the branch of physics known as **irreversible thermodynamics** and pioneered by Onsager[819].

Another important phenomenon, closely related to streaming potential, is **sedimentation potential**[1421]. If colloidal particles, of a density different from that of the solution in which they are suspended, are subjected to a gravitational or centrifugal force, their motion can lead to the loss of some of their diffuse double layer. In consequence an electric field (and therefore a potential difference) develops, restraining the motion. This is a hindrance if the object is to separate the particles from the medium and is commonly countered in **ultracentrifugation** by increasing the ionic strength of the medium, thereby reducing the zeta potential.

Summary

Three rather diverse junctions have been discussed in this chapter, though all arise when an ionic solution, and primarily an aqueous solution of an electrolyte, meets another phase. In all cases a double layer is created. In the case of the interface with an insulator, the presence of the double layer causes motion or pressure in a direction parallel to the surface. In the case of the interface with a semiconductor, there is a region of charge imbalance, not only in the ionic conductor, but also in the semiconductor phase, caused by a gradient of electron and/or hole concentrations; electrode reactions may occur and these may be fostered by the capture of photons. In the case of an ITIES, there are two double layers and their junction may allow the interphase transfer of molecules, ions or electrons, or may be the site of adsorption, with or without the passage of current.

15

Electrochemistry with Periodic Signals

When a periodic voltage signal is applied to an electrochemical cell, the current may settle, after a few cycles, into a repetitive pattern. The cell is then in a periodic state[1501], as defined on page 231. A periodic state is always preceded by transience[1502], but here we are concerned only with the state achieved after the initial transient has decayed to a negligible magnitude.

A voltage signal of period P is one in which the potential of the working electrode obeys the requirement that

15:1 $$E(t + P) = E(t) \qquad \text{periodicity}$$

at all times t within a wide window. For the most part, our discussion will concern sinusoidal voltages

15:2 $$\Delta E(t) = |E| \sin\{\omega t + \varphi_E\}$$

and currents

15:3 $$I(t) = |I| \sin\{\omega t + \varphi_I\}$$

As in Chapter 1, $|E|$ and $|I|$ are the amplitudes of the voltage and current, ω is the angular frequency and the φ's are phase angles. Much of this chapter is devoted to examining the electrochemical effects of these signals and the methodologies that employ them.

Also addressed in this chapter are two techniques that might be termed **near-periodic voltammetries** in which a slowly changing d.c. potential is applied, as well as the periodic excitation. The responding current has both d.c. and a.c. components, but analysis of the experiment is generally restricted to the periodic response. If the amplitude of the a.c. component is small, the response is linear – that is, the a.c. current is proportional to the a.c. voltage amplitude – then the behavior can be described in terms of an equivalent circuit. Larger amplitudes destroy linearity.

[1501] The misleading term "**stationary state**" is also in use.

[1502] As demonstrated in Web#1502, when the signal $|E|\sin\{\omega t\}$ is applied across a resistor R in series with a capacitor C, there is a transient current of magnitude $-|E|C\omega\exp\{-t/RC\}/[1+\omega^2R^2C^2]$ that must be allowed to decay to insignificance before the current becomes wholly sinusoidal. Other loads lead to similar transients.

Electrochemical Science and Technology: Fundamentals and Applications, First Edition. Keith B. Oldham, Jan C. Myland, Alan M. Bond.
© 2012 John Wiley & Sons, Ltd. Published 2012 by John Wiley & Sons, Ltd.

Nonfaradaic Effects of A.C.: measuring conductance and capacitance

Before turning to the major topic of this chapter, which concerns how electrochemical reactions respond to alternating electricity, the effect of a.c. on a totally polarized cell (page 193) will be addressed. No d.c. current can flow through such a cell, because there is no electrode reaction available at one or more of the electrodes. However, the capacitance that exists at all (electronic conductor)|(ionic conductor) interfaces, by virtue of the double layer there, provides a route for a.c. to cross each polarized electrode; the conductance of the ionic conductor completes the a.c. circuit through the cell.

Simple two-electrode cells similar to that illustrated in Figure 15-1 are generally used to measure the conductance of aqueous solutions of electrolytes and other liquid ionic conductors. Such a **conductivity cell** has the equivalent circuit shown here[1503], with each electrode represented by a capacitor and with the intervening conductor represented by a resistor. When a.c. current flows, there is a voltage drop across each of the three elements. In a conductivity measurement, one seeks to minimize the effect of the capacitors and this requires using a high frequency[1504] and having electrodes of large area. Often "platinized" platinum electrodes are used; these are coated with a finely granular layer of platinum, greatly enhancing their effective area. This often enables the capacitances to be completely ignored, so that the conductivity is computed from

15:4
$$\kappa = \frac{L}{AR} = \frac{L|I|}{A|E|}$$

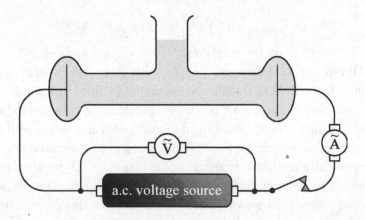

Figure 15-1 A conductivity cell.

[1503] or simply a resistor in series with *one* capacitor of magnitude equal to one-half of C. Explain why or see Web#1503.

[1504] Typically in the kilohertz range. Recall that the impedance of a resistor is frequency-independent, but that of a capacitor is inversely proportional to ω.

Here A is the nominal cross-sectional area of the cell and L is the effective interelectrode distance[1505]. Alternative to using the unmodified equation 15:4, some **conductivity meters** further discriminate against capacitance by measuring only that component of the current that is in phase with the applied voltage.

For highest precision, conductivities are determined by comparing the outcome of a conductivity measurement, in a particular apparatus, with a similar measurement of the conductivity of a standard solution. For example, a solution made by dissolving 7.43344 g of KCl in exactly one kilogram of water has a molarity of 0.1 and an accurately measured conductivity of 1.2886 S m^{-1} at 25.00°C, and this fact can be used to calibrate the cell.

Alternating current is also employed in the measurement of the capacitance of an electrode under conditions of total polarization. The criteria are now reversed; one needs to minimize the effect of the resistance component, and accentuate the importance of the capacitance. Thus a potentiostat-controlled three-electrode cell is generally used, thereby reducing the resistance to a small "uncompensated" fraction, R_u. The equivalent circuit is now the simple one shown to the right. If the current flowing through this series combination is $I(t) = |I|\sin\{\omega t + \varphi_I\}$, then the voltage generated by each component is

$$15{:}5 \qquad E_R(t) = R_u I(t) = R_u |I|\sin\{\omega t + \varphi_I\}$$

by Ohm's law and

$$15{:}6 \quad E_C(t) = \frac{Q(t)}{C} = \frac{1}{C}\int |I|\sin\{\omega t + \varphi_I\}\,dt = \frac{|I|}{\omega C}\cos\{\omega t + \varphi_I\} = \frac{|I|}{\omega C}\sin\left\{\omega t + \varphi_I + \frac{\pi}{2}\right\}$$

by relationship 1:19. Notice that the phases of the two voltages differ by $\pi/2$, and that the amplitude of the component of the voltage that is out of phase[1506] with the current is $|I|/\omega C$, from which the interfacial capacitance is calculable. The results of measurements such as this are often described in terms of impedance, as later in this chapter.

Faradaic Effects of A.C.: impedance, harmonics, rectification

A periodic variation in the potential of a working electrode does not necessarily result in a periodic current response from a faradaic process. Essential to ensuring current periodicity are the requirements that the working electrode *not* behave irreversibly and that there be an ample concentration of *each* member of the redox pair. The latter condition

---

[1505] Conductivity cells are rarely of perfectly cylindrical shape. The quantity L/A is known as the **cell constant**. The resistance of a conductivity cell filled with the standard solution described above is 35.73 Ω. What is the cell constant? What is the conductivity of a solution that, in the same cell, provides a resistance of 83.67 Ω? See Web#1505.

[1506] For one sinusoid to be "out of phase" with another means that their phase angles differ by $\pi/2$, as do a sine and a cosine of the same argument.

may be met either by having both members present in the bulk or by applying an unchanging d.c. potential, in addition to the periodic signal, to generate the member that is absent from the bulk.

The concept of impedance was introduced in Chapter 1. It may be measured by applying an a.c. voltage and measuring the resulting current, or by applying an a.c. current and measuring the voltage. If the voltage and current are respectively $|E|\sin\{\omega t + \varphi_E\}$ and $|I|\sin\{\omega t + \varphi_I\}$, then the **impedance** is defined as

15:7 $$Z = \frac{|E|}{|I|} \qquad \text{definition of impedance}$$

Its analog in d.c. electricity is resistance, with which impedance shares the ohm unit[1507]. In addition to the impedance, however, a second quantity, the phase shift $\varphi_I - \varphi_E$ may be measured. The table on page 26 lists values of the impedance and phase shifts for several circuit elements and other loads. In a sense, the chemistry occurring at an electrode interface introduces other "elements". These are additional to the nonfaradaic elements – the double-layer capacitance C and uncompensated resistance R_u – that are present even when an electrochemical reaction is absent.

Our first task in this section is to elucidate the *faradaic* effect of imposing a sinusoidal signal of frequency ω:

15:8 $$E(t) = |E|\sin\{\omega t + \varphi_E\} \qquad \text{applied a.c. signal}$$

on an electrode at which the simple one-electron reaction

15:9 $$R(soln) \rightleftarrows e^- + O(soln)$$

occurs. We seek to find how the amplitude $|I|$ and phase angle φ_I of the current

15:10 $$I(t) = |I|\sin\{\omega t + \varphi_I\} \qquad \text{assumed a.c. response}$$

are related to the corresponding properties of the applied signal. If there are ample concentrations of R and O present in the solution, then one might expect this current to elicit a sinusoidal response in the concentrations of the two species at the electrode surface

15:11 $$c_i^s(t) = c_i^b + |c_i^s|\sin\{\omega t + \varphi_i^s\} \qquad i = R,O$$

These expectations prove to be correct, as will now be demonstrated, provided that the amplitude of the applied signal is sufficiently small.

We assume typical voltammetric conditions: transport solely by semiinfinite diffusion to an electrode that is flat (or that can be treated as if it were) and of sufficient area A that edge effects are negligible. The Laplace transformation approach to modeling is inappropriate in this scenario and, as an alternative, we turn to the **Faraday-Fick relation** (derived elsewhere[1508] and reproduced in equation 15:13) that links the current to the

---

[1507] likewise **admittance**, defined as $|I|/|E|$, is analogous to conductance, both sharing the siemens unit.

[1508] See Web#1508. This formula incorporates the **contradiffusion relation** (see Web#1213).

concentrations of R and O at an electrode surface. The linkage operates via the semiintegral[1509] $M(t)$ of the current, and so the first step is to semiintegrate equation 15:10. One finds that

15:12 $\qquad M(t) = \dfrac{d^{-1/2}}{dt^{-1/2}} |I| \sin\{\omega t + \varphi_I\} \Big|_{-\infty} = \dfrac{|I|}{\sqrt{\omega}} \sin\left\{\omega t + \varphi_I - \dfrac{\pi}{4}\right\}$ current semiintegral

and, because the semiintegral is linearly related to the surface concentrations through the expressions

15:13 $\qquad \dfrac{M(t)}{FA} = \sqrt{D_R}\,[c_R^b - c_R^s(t)] = \sqrt{D_O}\,[c_O^s(t) - c_O^b]$ Faraday-Fick relation

one is led immediately to the results

15:14 $\qquad c_R^s(t) = c_R^b - \dfrac{|I| \sin\{\omega t + \varphi_I - \pi/4\}}{FA\sqrt{D_R\omega}};\qquad c_O^s(t) = c_O^b + \dfrac{|I| \sin\{\omega t + \varphi_I - \pi/4\}}{FA\sqrt{D_O\omega}}$

These equations match prediction 15:11 and provide expressions for the amplitudes and phase angles of the concentration excursions. The concentration of O is seen to lead[1510] the current by $\pi/4$ or 45°, whereas the concentration of R lags the current by[1511] $3\pi/4$. For future convenience it is useful to abbreviate these latest equations as

15:15 $\qquad \dfrac{c_R^s(t)}{c_R^b} = 1 - \dfrac{\left|c_R^s\right|}{c_R^b}\sin\{\omega t + \varphi_I - \pi/4\};\qquad \dfrac{c_O^s(t)}{c_O^b} = 1 + \dfrac{\left|c_O^s\right|}{c_O^b}\sin\{\omega t + \varphi_I - \pi/4\}$

where $\left|c_R^s\right|$, equal to $|I|/FA\sqrt{D_R\omega}$, is the amplitude of the concentration excursion of R at the electrode, and $\left|c_O^s\right|$ is similarly defined. Note that each of these excursions cannot exceed the corresponding bulk concentrations, as otherwise the equation would imply a nonsensical negative concentration at some point in the cycle. In this derivation, however, we shall later impose a much stronger restriction that ensures that the concentration excursions are small in absolute terms. Figure 15-2 overleaf illustrates that the concentration excursions differ in phase by π or 180° from each other, the total concentration $c_R^s(t) + c_O^s(t)$ being almost constant.

The question that must now be asked is whether the current postulated in 15:10 and the consequential concentration excursions expressed in equations 15:15 are consistent with the applied potential given in expression 15:8. To answer this we turn to the Butler-Volmer equation. Written in the style of 7:27, it states that

15:16 $\qquad \dfrac{I(t)}{Ai_n} = \dfrac{c_R^s(t)}{c_R^b}\exp\left\{\dfrac{-\alpha F}{RT}[E(t) - E_n]\right\} - \dfrac{c_O^s(t)}{c_O^b}\exp\left\{\dfrac{(1-\alpha)F}{RT}[E(t) - E_n]\right\}$ Butler-Volmer equation

---

[1509] See Web#1242 and especially entry (206).

[1510] Because the phase angle of the sinusoidally varying surface concentration of R is equal to $\varphi_I - \pi/4$, each minimum in the R's concentration precedes ("leads") that of the current (see Figure 15-2).

[1511] Because $-\sin\{\omega t + \varphi_I - \pi/4\} = \sin\{\omega t + \varphi_I + 3\pi/4\}$.

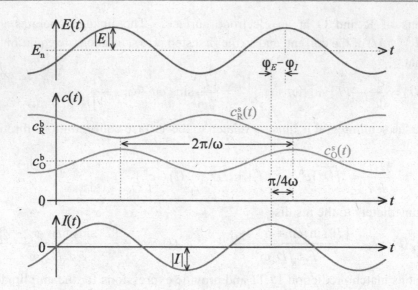

Figure 15-2 The sinusoidal behaviors of the current, the concentration of O at the electrode surface, the concentration of R at the electrode surface, and the electrode potential.

where i_n is the exchange current density, equal to $Fk^{o'}(c_R^b)^\alpha (c_O^b)^{1-\alpha}$. Because of the stipulation that the a.c. perturbation be of small magnitude, the potential $E(t)$ will differ only modestly from its null value. Therefore only minor errors are introduced on replacing each exponential function in equation 15:16 by the first two terms in its expansion. Thus

$$15:17 \qquad \frac{|I|\sin\{\omega t + \varphi_I\}}{Ai_n} \approx \frac{c_R^s(t)}{c_R^b}\left[1 - \frac{\alpha F}{RT}[E(t) - E_n]\right] - \frac{c_O^s(t)}{c_O^b}\left[1 + \frac{(1-\alpha)F}{RT}[E(t) - E_n]\right]$$

Notice that $I(t)$ has also been substituted from equation 15:10. When the two concentration ratios in this expression are replaced from the expressions in 15:15, some welcome cancellations occur and there remains

$$15:18 \qquad \frac{F}{RT}[E(t) - E_n] = \frac{|I|\sin\{\omega t + \varphi_I\}}{Ai_n} + \left(\frac{|c_R^s|}{c_R^b} + \frac{|c_O^s|}{c_O^b}\right)\sin\left\{\omega t + \varphi_I - \frac{\pi}{4}\right\} + \begin{array}{l} \textit{much smaller} \\ \textit{terms that will} \\ \textit{be ignored} \end{array}$$

after a regrouping of terms. The potential excursion $E(t) - E_n$ is seen to be the sum of two sinusoidal terms and is therefore itself sinusoidal and able to be represented as $|E|\sin\{\omega t + \varphi_E\}$, as predicted. This allows equation 15:18 to be reorganized into

$$|E|\sin\{\omega t + \varphi_E\} = |I|\left[\frac{RT}{FAi_n}\right]\sin\{\omega t + \varphi_I\}$$

15:19

$$+ \frac{|I|}{\sqrt{\omega}}\left[\frac{RT\sqrt{\omega}}{F|I|}\left(\frac{|c_R^s|}{c_R^b} + \frac{|c_O^s|}{c_O^b}\right)\right]\sin\left\{\omega t + \varphi_I - \frac{\pi}{4}\right\}$$

where the terms in square brackets are constants. The first one of these constants will be recognized from page 202 as the **charge-transfer resistance** R_{ct}:

15:20 $$\frac{RT}{FAi_n} = \frac{RT}{F^2 Ak^{o\prime}(c_R^b)^\alpha (c_O^b)^{1-\alpha}} = R_{ct} \qquad \text{charge-transfer resistor}$$

and incorporates the kinetic parameters of the electrode reaction[1512]. The second square-bracketed term in equation 15:19 is given the symbol W[1513]

15:21 $$\frac{RT\sqrt{\omega}}{F|I|}\left(\frac{|c_R^s|}{c_R^b} + \frac{|c_O^s|}{c_O^b}\right) = \frac{RT}{F^2 A}\left(\frac{1}{c_R^b\sqrt{D_R}} + \frac{1}{c_O^b\sqrt{D_O}}\right) = W \qquad \text{Warburg element}$$

This term reflects transport to and from the electrode surface, but not the electron transfer itself. With these notations introduced, equation 15:18 becomes

15:22 $$|E|\sin\{\omega t + \varphi_E\} = |I| R_{ct}\sin\{\omega t + \varphi_I\} + \frac{|I|W}{\sqrt{\omega}}\sin\left\{\omega t + \varphi_I - \frac{\pi}{4}\right\}$$

The electrode potential is seen to equal the sum of two sinusoidal terms, each proportional to the current amplitude. This exactly parallels the behavior of an a.c. current passing through a series combination of two "elements". Thus the voltage across an electrode carrying an alternating faradaic current of sufficiently small amplitude is analogous to two elements in series, one of which is characterized by R_{ct} and involves kinetic factors, whereas the second, characterized by W, involves transport terms[1514]. Because the first right-hand term in 15:22 shows a response with an amplitude that is independent of frequency and with a phase angle identical to that of the current, it emulates a resistor, as its R_{ct} symbol implies. The second right-hand term in 15:22 shows a potential response with an amplitude that is inversely proportional to the square-root of the frequency, and a phase angle leading the current by $\pi/4$. Reference to the table on page 26 shows that these are exactly the properties of a **Warburg element**[172]. In other words, the second element W behaves exactly like a transmission line in which the ratio of the resistance to capacitance is given by $R/C = W^2$.

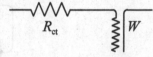

If the reaction behaves reversibly, $k^{o\prime}$ is so large that the charge-transfer resistance is

---

[1512] Demonstrate, as in Web#1512, that if R and O each have a bulk concentration of 1.0 mM and the formal rate constant $k^{o\prime}$ equals 1.0×10^{-4} m s^{-1}, then the charge-transfer resistance at an electrode of area 1.0×10^{-5} m^2 is 270 Ω.

[1513] Elsewhere you may find the symbol $\sqrt{2}\sigma$ or $\sqrt{2}\theta$ replacing W. A typical value is $W \approx 1.7$ kΩ s$^{-1/2}$ for a disk electrode of radius 1.8 mm. Confirm this by taking R and O each to have millimolar concentrations and diffusivities of 1.0×10^{-9} m^2 s^{-1}. See Web#1513.

[1514] $|I|R_{ct}$ and $|I|W/\sqrt{\omega}$ are analogous to the kinetic and transport overvoltages of Chapter 10.

negligibly small and the electrode is represented by the Warburg element alone[1515]. As for steady-state voltammetry on page 242, and rotating-disk voltammetry on page 249, it is useful to allocate a reversibility index, which for the present experiment is

$$15:23 \qquad \lambda = \frac{W}{R_{ct}\sqrt{\omega}} = \frac{k°}{\sqrt{\omega}}\left[\frac{1}{\sqrt{D_R}}\left(\frac{c_O^b}{c_R^b}\right)^{1-\alpha} + \frac{1}{\sqrt{D_O}}\left(\frac{c_R^b}{c_O^b}\right)^{\alpha}\right] \qquad \begin{array}{l}\text{reversibility index}\\ \text{for a.c. impedance}\end{array}$$

If $\lambda \gg 1$, the reaction is reversible and, as noted earlier, the Warburg element is then the dominant member of the faradaic impedance. If λ is in the neighborhood of unity, the experiment is quasireversible and both the charge-transfer resistance and the magnitude of the Warburg element can, in principle, be measured by a.c. experiments. Notice that the reversibility index includes the frequency so that it is possible to "tune" the experiment into quasireversibility and, moreover, to use the frequency to distinguish between the impedances of the two elements. One might think that, with $\lambda \ll 1$, irreversibility would cause the impedance to be purely that of the charge-transfer resistance, but recall that faradaic periodicity cannot be attained when an electrode behaves irreversibly.

Recognize that, provided the excursions are sufficiently small, one can apply the sinusoidal current 15:10 and measure the resulting potential response as in 15:8, with

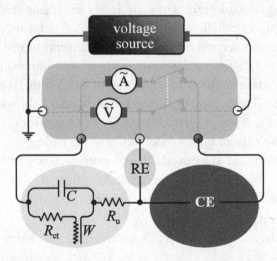

Figure 15-3 Elements of the equivalent circuitry for measurement of electrode impedance by a potentiostat-controlled three-electrode cell. The impedance of the "RE" element is of no major concern because no current flows through it. The "CE" element (which includes the majority of the cell solution) is likewise of little concern because its impedance is "compensated" by the potentiostat. The faradaic pathway has been represented by a resistor in series with a transmission line, as appropriate if the a.c. signal is of small amplitude.

[1515] Sometimes a Warburg element is regarded as a resistor and a capacitor in series, each of the same impedance. Such a representation, however, fails to capture the frequency dependence of the element.

precisely the same result as applying the voltage and measuring the current. Though the derivation is then simpler, the latter is the far more common option experimentally.

Appreciate that what is measured in practice is not the faradaic response alone, but its parallel combination with the non-faradaic response of the double-layer capacitance and their serial combination with the uncompensated resistance. The four elements that represent the overall load are included in Figure 15-3, which shows the essential circuitry for a.c. electrochemistry. The outcome of such an experiment is commonly reported in terms of impedance, as elaborated in the next section. Potentiostats specially designed for a.c. and capable of providing data over wide ranges of frequency[1516] are known as **impedance spectrometers** or **frequency-response analyzers** and they often provide graphical data output in a variety of formats, for example as the Sluyters plots discussed in the next section.

The approximation made in equation 15:17, namely that exponential functions may be replaced by the first two terms in their expansions if their arguments are small, is valid if $|E|$ is small in comparison with the quantity RT/F, which equals about 26 mV at room temperature. But what happens if an a.c. voltage of larger amplitude is applied? There are then three faradaic repercussions:

15:24 $$\left.\begin{array}{c} \text{imposition of the voltage} \\ E(t) = E_n + |E|\sin\{\omega t + \varphi_E\} \\ \text{on the working electrode when} \\ \text{the amplitude}|E| \text{ is not small} \end{array}\right\} \text{produces} \left\{\begin{array}{l} \text{fundamental response} \\ \text{harmonic responses} \\ \text{faradaic rectification} \end{array}\right.$$

The **fundamental response** is the current $|I|\sin\{\omega t + \varphi_I\}$, where $|I|$ remains similar to, though somewhat smaller than, the value predicted by the small-signal theory. The **harmonic responses** are currents of frequencies that are multiples of the frequency ω. The component that has a frequency of $n\omega$ is referred to as the nth **harmonic**[174]. The higher harmonic frequencies are generally of smaller magnitudes that are not linearly proportional to the applied amplitude $|E|$. In electrical technology, the conversion of a.c. into d.c. is known as "rectification", and so the term faradaic rectification in 15:24 implies that, in addition to sinusoidal currents of various frequencies, a d.c. current is also generated when an alternating potential of sizeable magnitude is applied to an electrode[1517].

Let us address the questions of why faradaic rectification occurs and how harmonics are generated. To capture the concepts in a nonrigorous fashion, it is useful to think of the a.c. properties of an electrode as arising from the change $\Delta\eta$ in the overvoltage $\eta = E(t) - E_n$ eliciting a sinusoidal change $I = |I|\sin\{\omega t + \varphi_I\}$ in the current. Pictorially, one can imagine

---

[1516] In practice, many of these instruments apply a large number of frequencies simultaneously rather than sequentially. It is essential the phases of the multiple frequencies be adjusted to avoid the total amplitude exceeding the "small amplitude" limit.

[1517] Correspondingly, a faradaic rectification voltage results from imposing an a.c. current of large amplitude.

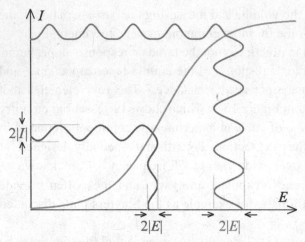

Figure 15-4 Heuristically illustrated in this diagram is the origin of harmonic frequencies and faradaic rectification. As the sinusoidal voltage excursions encounter the polarization curve, their oscillations are converted into oscillations of the current. The generated response is almost sinusoidal in the blue case, but markedly distorted in the red case. Phase shifts have been ignored.

the sinusoidal voltage repeatedly sliding up and down a small portion of the polarization curve and thereby generating a periodic current. Especially if the polarization curve is more-or-less linear, a small sinusoidal variation in voltage will generate a sinusoidal current, as in the blue scenario in Figure 15-4. A larger amplitude in the voltage will be more likely to encounter a curved portion of the polarization curve and generate distortion; the periodic current will no longer be sinusoidal, it will resemble the red scenario in the figure. The distorted current is the source of the harmonics and of rectification.

To see more quantitatively how these effects arise, return to equation 15:17 and note the effect of adding one more term in the expansion of each exponential. This introduces several new terms into the right-hand member of equation 15:18, the major one of which will usually be the quantity shown on the left of equation 15:25 below. Trigonometric identities allow conversion to the quantity on the right

$$15:25 \quad \frac{\left[\alpha^2 - (1-\alpha)^2\right]F^2|E|^2}{R^2T^2}\sin^2\{\omega t + \varphi_E\} = \frac{(2\alpha - 1)F^2|E|^2}{2R^2T^2}\left[1 + \sin\left\{2\omega t + 2\varphi_E - \frac{\pi}{2}\right\}\right]$$

Thus one sees that a second harmonic (one of frequency 2ω) is generated[1518], as well as an aperiodic potential (from the "1"). Both these are of a magnitude proportional to the square of the potential amplitude. Of course, adding further terms in the expansions of the two exponential terms leads to additional harmonics and to extra contributions to the rectification voltage. Fourier transformation (pages 26 and 323) is useful in establishing the magnitudes of harmonics and phase angles. More information on faradaic rectification will be found elsewhere[1519].

[1518] unless $\alpha = \frac{1}{2}$. Third harmonics and accompanying rectification are, however, generated irrespective of α.

[1519] See Web#1519.

Equivalent Circuits: deciphering the impedance

When a large-amplitude a.c. signal is applied to an electrode, the nonlinear phenomena of rectification and harmonic generation come into play, and the possibility of representing the impedance of an electrochemical cell by a network of elements ceases to be useful. In this section, therefore, we return to a consideration of small a.c. signals and investigate how impedance measurements[1520] can help to provide information about the working electrode.

Through equation 15:22, we were able to identify the response of a faradaic interface to a sinusoidal voltage of small amplitude as equivalent to a resistance R_{ct} and a Warburg element of magnitude W. It is said that a resistor in series with a Warburg element is the **equivalent circuit** of the faradaic process at an electrode under the conditions being discussed. One might be tempted to say that the impedance was "$R_{ct} + W/\sqrt{\omega}$", but that would be erroneous. Because a.c. properties incorporate a phase angle, impedances are vectors that require two quantities to be cited in order that the impedance may be specified unambiguously. A useful system for taking phase angle diversity into account is to use complex algebra and represent the impedance as the sum of two terms

15:26 $\qquad Z = Z' + jZ'' \quad \text{where} \quad j = \sqrt{-1} \qquad$ complex notation

The two components go by a variety of names. The Z' term is called the **resistive component**, the **in-phase component** or the **real component**. The Z'' term is variously named the **reactive component**, the **out-of-phase component**, the **imaginary component**[1521] or the **quadrature component**. Taking the square root of the product of the impedance and its complex conjugate[1522] yields the impedance

15:27 $\qquad \sqrt{(Z' + jZ'') \times (Z' - jZ'')} = \sqrt{(Z')^2 + (Z'')^2} = Z \qquad$ impedance

while the rule

15:28 $\qquad \arctan\left\{\dfrac{Z''}{Z'}\right\} = \varphi_I - \varphi_E \qquad$ phase shift

relates the phase angle of the current flowing through the impedance to that of the voltage across it. The impedance components of four common elements are listed in the table here. Though seldom encountered in electrochemistry, an inductor possesses the property of **inductance** L, present in coiled wires, which is measured in the henry (H) unit[1523].

Element	Z'	Z''
resistor R	R	0
inductor L	0	ωL
capacitor C	0	$\dfrac{-1}{\omega C}$
Warburg element W	$\dfrac{W}{\sqrt{2\omega}}$	$\dfrac{-W}{\sqrt{2\omega}}$

[1520] See U. Retter and H. Lohse in: F. Scholz (Ed.), *Electroanalytical Methods*, 2nd edn, Springer, 2010, pages 159–177, for details.

[1521] This is misleading terminology! Z'' is a real quantity; it is jZ'' that is imaginary.

[1522] The **complex conjugate** of the quantity $x + jy$ is $x - jy$.

[1523] Joseph Henry, 1797–1878, U.S. physicist and founding director of Washington's Smithsonian Institute.

When two elements are connected in series, the impedance of the pair is related to the individual impedances by the very simple formulas

15:29 $\qquad Z' = Z_1' + Z_2' \quad$ and $\quad Z'' = Z_1'' + Z_2'' \qquad$ elements 1 and 2 in series

The formulas for impedances in parallel are considerably more complicated[1524]:

15:30
$$\begin{cases} Z' = \dfrac{Z_1'Z_2'(Z_1'+Z_2') + Z_1'(Z_2'')^2 + Z_2'(Z_1'')^2}{(Z_1'+Z_2')^2 + (Z_1''+Z_2'')^2} \\[4mm] Z'' = \dfrac{Z_1''Z_2''(Z_1''+Z_2'') + Z_1''(Z_2')^2 + Z_2''(Z_1')^2}{(Z_1'+Z_2')^2 + (Z_1''+Z_2'')^2} \end{cases} \qquad \text{elements 1 and 2 in parallel}$$

In Figure 15-3, the four-element equivalent circuit that represents the working electrode is often known as the **Randles-Ershler circuit**[1630,1525]. With circuits of such complexity as this, a useful pictorial device is to draw the circuit diagram with the elements having lengths proportional to the magnitudes Z of their impedances, and at angles that correspond to the phase angle differences. Showing the four elements involved, Figure 15-5 is such a diagram for a working electrode exposed to a small-amplitude a.c. voltage. Such a diagram allows trigonometric analysis of the behavior of the network.

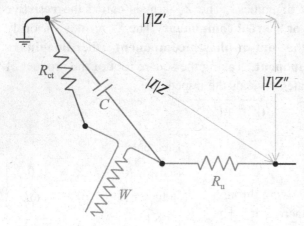

Figure 15-5 Equivalent circuit of a working electrode exposed to a small-amplitude a.c. voltage, drawn to represent also a vector diagram. The magnitudes of the impedances correspond to the lengths of the four colored symbols. The angle that each element makes to the horizontal duplicates the phase angle (with respect to the phase of the overall voltage) of the current passing though that element.

Alternatively, one may use the formulas in 15:29 and 15:30 to build up expressions for the overall component impedances of the Randles-Ershler circuit. Firstly, using 15:29, one derives the impedances of the series combination of the charge-transfer resistor and the Warburg element as

15:31 $\qquad\qquad Z' = R_{ct} + \dfrac{W}{\sqrt{2\omega}} \quad$ and $\quad Z'' = \dfrac{-W}{\sqrt{2\omega}}$

[1524] As in Web#1524, derive equations 15:30 by appreciating that, for a parallel arrangement of elements, it is the **admittances** (Y, equal to $1/Z$) that add in a simple manner analogous to equation 15:29.

[1525] Boris Vladimirovitch Ershler, 1908–1978, Russian electrochemist and radiation chemist.

Then, using the rules in 15:30, one can add the double-layer capacitance in parallel with the earlier pair, to discover[1526] that

15:32
$$\begin{cases} Z' = \dfrac{\sqrt{2\omega}\left(\sqrt{2\omega}R_{ct}+W\right)}{\left(\sqrt{2\omega}+\omega WC\right)^2 + \omega^2 C^2\left(\sqrt{2\omega}R_{ct}+W\right)^2} \\[3em] Z'' = \dfrac{-W\left(\sqrt{2\omega}+\omega WC\right)-\omega C\left(\sqrt{2\omega}R_{ct}+W\right)^2}{\left(\sqrt{2\omega}+\omega WC\right)^2 + \omega^2 C^2\left(\sqrt{2\omega}R_{ct}+W\right)^2} \end{cases}$$

Finally, the uncompensated resistance may be added in via formula 15:29. This leaves the out-of-phase impedance unchanged, but adds R_u to Z'. Thus

15:33
$$\begin{cases} Z' = R_u + \dfrac{\sqrt{2\omega}\left(\sqrt{2\omega}R_{ct}+W\right)}{\left(\sqrt{2\omega}+\omega WC\right)^2 + \omega^2 C^2\left(\sqrt{2\omega}R_{ct}+W\right)^2} \\[3em] Z'' = \dfrac{-W\left(\sqrt{2\omega}+\omega WC\right)-\omega C\left(\sqrt{2\omega}R_{ct}+W\right)^2}{\left(\sqrt{2\omega}+\omega WC\right)^2 + \omega^2 C^2\left(\sqrt{2\omega}R_{ct}+W\right)^2} \end{cases}$$

With this addition, we have now analyzed the impedance "seen" by the potentiostat[1527].

The object of measuring an electrode impedance is, of course, to determine the magnitudes of the four component elements, R_{ct} being particularly interesting because it opens the way to measuring electrode kinetics. It may seem a daunting task to compute four quantities from only two measurements: those of Z' and Z''. However, the experimenter has a number of tools. One is the ability to employ a very wide range of frequencies; values of ω ranging from 10^{-1} to 10^5 Hz being available in well equipped laboratories. The values of the double-layer capacitance and the uncompensated resistance, C and R_u, are sometimes treated as "known". They may be determined from another experiment[1528] without the electroactive couple, but otherwise similar, though there is no guarantee that the change in conditions does not cause a significant perturbation in the value of the double-layer capacitance. Another strategy is to vary the bulk concentrations: this has an inversely proportional effect (see equations 15:20 and 15:21) on both the resistance R_{ct} of charge transfer, and the magnitude W of the diffusional Warburg element,

---

[1526] Derive equations 15:32 or see Web#1526.

[1527] For a frequency of $\omega = 1.25$ kHz, calculate the in-phase and out-of-phase impedances for a Randles-Ershler circuit with the following elements: $C = 5.0$ μF, $W = 2.0$ kΩ s$^{-1/2}$, $R_{ct} = 200$ Ω, and $R_u = 50$ Ω. Hence find the amplitude and phase angle of the a.c. current that is passed by the circuit when interrogated by an a.c. voltage of 5.00 mV amplitude. See Web#1527.

[1528] or from the same experiment, but at a d.c. potential at which the reaction does not occur.

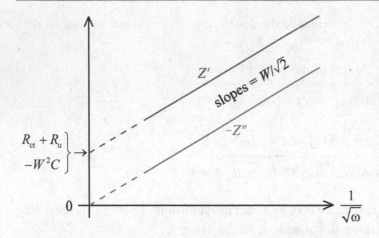

Figure 15-6 At low frequencies, Z' and $-Z''$ form parallel straight lines when plotted versus the reciprocal of the square root of frequency.

but is without major effect on the capacitance C of the double layer or the resistance R_u that remains uncompensated by the potentiostat.

When the frequency is sufficiently small, equations 15:33 simplify[1529] and show that both components are linear functions of $1/\sqrt{\omega}$, as illustrated in Figure 15-6. The lines should have the same slope, from which the magnitude W of the Warburg element may be measured. Other useful information may be accessed from the intercept.

Commonly the results of a set of impedance measurements, made over a range of frequencies, are plotted as an **Argand diagram**[1530]; that is, as a graph of the out-of-phase impedance as a function of the in-phase impedance. In such a plot, the frequency varies steadily as one proceeds along the graph line. In electrochemistry, Argand diagrams may go by the name of **Sluyters plots**[1531]. When the electrode reaction behaves reversibly, so that R_{ct} is zero, equations 15:33 simplify considerably and the in-phase and out-of-phase components of the impedance may be combined into

15:34 $\left[-Z'' - (Z' - R_u)\right]\left[(-Z'')^2 + (Z' - R_u)^2\right] = 2W^2C(Z' - R_u)^2$ reversible

by elimination of the frequency term. This may be rearranged to a cubic equation that can be solved[1532]. However, when an electrode behaves reversibly, the lion's share of the a.c. current passes through the faradaic branch of the Randles-Ershler circuit, permitting the capacitive branch to be ignored at all but the highest frequencies. In such circumstances, the Sluyters plot resembles Figure 15-7.

[1529] As in Web#1529, simplify equations 15:33 to show that the Z' and Z'' components are linear functions of $1/\sqrt{\omega}$. Then calculate the slopes and intercepts, at low frequency, of the lines in Figure 15-6 for elements of the values given in the caption of Figure 15-9.

[1530] Jean-Robert Argand, 1768–1822, French bookseller and amateur mathematician.

[1531] Jan H. Sluyters and Margaretha Sluyters-Rehbach, a productive team of electrochemists in The Netherlands. This method of data presentation is also called a **Cole-Cole plot** or, a **Nyquist**[1545] **plot**.

[1532] by rearrangement as a cubic equation in $(Z' - R_u)/Z''$ and then solved by the standard procedure detailed in K. Oldham, J. Myland, and J. Spanier, *An Atlas of Functions*, 2nd edn, Springer, 2009, page 142.

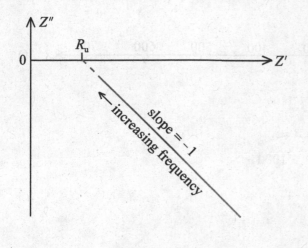

Figure 15-7 When the electrode reaction behaves reversibly, the Sluyters plot is linear at most frequencies.

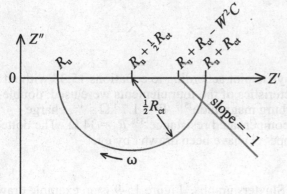

Figure 15-8 The semicircle, which obeys equation 15:35, is that expected for a Sluyters plot in which the magnitude W of the Warburg element is insignificant at all frequencies.

An analysis of the impedance of the Randles-Erschler circuit at low and high frequencies is given elsewhere[1533], along with the derivation of equation 15:35. Towards the other end of the reversibility spectrum, the charge-transfer resistance may often be so large that the Warburg element becomes irrelevant. If all terms involving W are omitted from equations 15:33, the residue may be manipulated[1533] into the equation

15:35 $$\left(Z' - R_u - \tfrac{1}{2}R_{ct}\right)^2 + (-Z'')^2 = \left(\tfrac{1}{2}R_{ct}\right)^2 \qquad W \text{ negligible}$$

This is the equation of a semicircular graph with the coordinates and characteristics shown in Figure 15-8. It is rare, however, for W to remain insignificant over a range of frequencies wide enough to delineate an entire semicircle. Instead, at the lowest frequencies, the relationship[1534]

15:36 $$-Z'' = Z' + R_u + R_{ct} - W^2 C \qquad \text{low frequency}$$

applies. This is the equation of the straight line shown on Figure 15-9 overleaf. In practice,

---

[1533] Equation 15:35 is derived in Web#1533.

[1534] See Web#1534 for the derivation of equation 15:36.

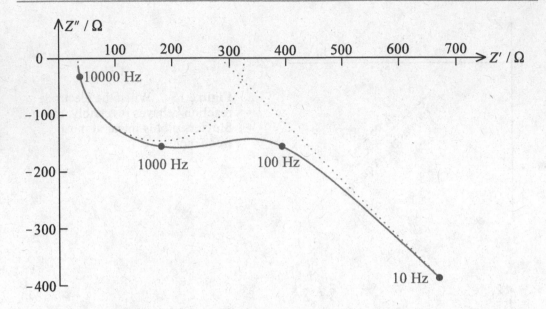

Figure 15-9 The full red line is a Sluyters plot according to equations 15:33 without approximation. The following characteristics of the four elements were used: double-layer capacitance[1320] $C = 3.0$ μF, Warburg magnitude[1513] $W = 1.7$ kΩ s$^{-1/2}$, charge-transfer resistance[1512] $R_{ct} = 270$ Ω, uncompensated resistance[1338] $R_u = 34$ Ω. The dotted semicircle and the straight line (of slope −1) have been drawn "by eye".

experiments often yield spoon-shaped Sluyters graphs. Figure 15-9 is an example drawn by using the data from the full equations 15:33 without approximation.

A.C. Voltammetry: discriminating against capacitive current

It was noted earlier that faradaic a.c. can cross an electrode interface only if both partners of a redox couple are present at ample concentrations. In a.c. voltammetry[1535], only one member, say R, is present in the bulk. Its partner, O, is generated by a d.c. current that accompanies the a.c. The experiment is carried out under typical voltammetric conditions (pages 233–234). Sophisticated instrumentation[1536] is needed to apply both a.c. and d.c., to separate the responses, and to measure such properties as the phase angle shift and the fundamental and harmonic responses.

The diagram in Figure 15-10 shows the potential waveform that is applied to the working electrode during a.c. voltammetry: it consists of a small (typically 10 mV)

[1535] invented by Grahame[1304].

[1536] such as a **lock-in amplifier** or **phase-sensitive analyzer**. Nowadays, both the input and output are handled digitally (page 27).

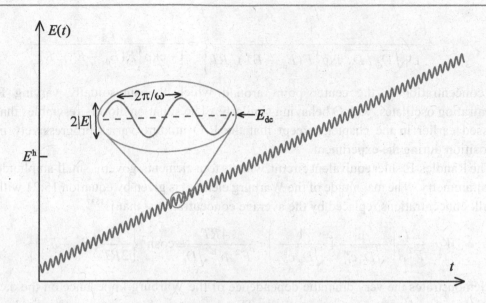

Figure 15-10 Applied signal in a.c. voltammetry. The diagram is not drawn to scale because, typically, there could be thousands of a.c. periods during the experiment.

amplitude sinusoid superimposed on a slow ramp:

15:37
$$E(t) = E_0 + vt + |E|\sin\{\omega t + \varphi_E\} = E_{dc} + |E|\sin\{\omega t + \varphi_E\}$$

For the one-electron oxidation, $R(soln) \rightleftarrows e^- + O(soln)$, the ramp would start at about 150 mV before (more negative than) the nernstian half-wave potential of the R/O couple and end close to 150 mV beyond E^h. What is measured is the amplitude $|I'|$ of that component of the a.c. current that is in phase with the applied potential and its dependence on the d.c. potential. The d.c. current is not constant and is of minor concern. The purpose of the d.c. is merely to adjust the ratio of the concentrations of R and O at the electrode surface.

Recognize that it is possible for a reaction to behave reversibly on the timescale of a slow scan, but not to respond reversibly to the rapid alternations of a high-frequency a.c. potential. It is to just such circumstances that a.c. voltammetry is directed. The Nernst equation then determines the ratio of the average concentrations at the electrode surface

15:38
$$\frac{c_O^{av}}{c_R^{av}} = \exp\left\{\frac{F}{RT}\left(E_{dc} - E^{o\prime}\right)\right\} \qquad \text{Nernst equation}$$

This, together with the linear relationship[1508.]

15:39
$$\sqrt{D_O}\,c_O^{av} = \sqrt{D_R}\,[c_R^b - c_R^{av}]$$

that links the two surface concentrations, enables the average concentration of R to be related to the d.c. potential by

$$15:40 \qquad c_R^{av} = \frac{c_R^b}{1 + \sqrt{D_O / D_R}\, \exp\left\{F(E_{dc} - E^{o\prime})/RT\right\}} = \frac{c_R^b}{1 + \exp\left\{F(E_{dc} - E^h)/RT\right\}}$$

This concentration is the center point around which the sinusoidally varying R concentration oscillates, with O behaving similarly. The situation closely resembles that discussed earlier in the chapter, except that the R/O mixture varies progressively in composition during the experiment.

The Randles-Ershler equivalent circuit, with its four elements, governs small-amplitude a.c. voltammetry. The magnitude of the Warburg element is given by equation 15:21 with the bulk concentrations replaced by the average concentrations; that is[1537]:

$$15:41 \qquad W = \frac{RT}{F^2 A}\left(\frac{1}{\sqrt{D_O}\, c_O^{av}} + \frac{1}{\sqrt{D_R}\, c_R^{av}}\right) = \frac{4RT}{F^2 A c_R^b \sqrt{D_R}} \cosh^2\left\{\frac{F}{2RT}\left(E_{dc} - E^h\right)\right\}$$

This demonstrates the very dramatic dependence of the Warburg impedance on the d.c. potential, with a deep minimum around E^h. The charge-transfer resistance also displays a minimum in a similar, though less pronounced, fashion. The general formulation of R_{ct} is elaborate[1538], so we shall only describe its behavior in the special, but nonetheless typical, case when $\alpha = \frac{1}{2}$; then

$$15:42 \qquad R_{ct} = \frac{RT}{F^2 A k^{o\prime} \sqrt{c_R^{av} c_O^{av}}} = \frac{2RT}{F^2 A c_R^b k^{o\prime}} \cosh\left\{\frac{F}{2RT}\left(E_{dc} - E^h\right)\right\} \qquad \alpha = \frac{1}{2}$$

It is the relative magnitudes of $W/\sqrt{\omega}$ and R_{ct} that determine the relative importance of kinetic and transport polarization and hence the reversibility of the experiment. As for other types of voltammetry, we can allocate a reversibility index as the ratio of these impedances at the nernstian potential E^h

$$15:43 \qquad \lambda = \left(\frac{W/\sqrt{\omega}}{R_{ct}}\right)_{E^h} = \frac{2k^{o\prime}}{\sqrt{D_R \omega}} \qquad \text{reversibility index}$$
$$\text{for a.c. voltammetry}$$

As always, a reversibility index much larger than unity implies reversibility and, in the case of a.c. voltammetry, that the Warburg element is the dominant impedance. When $\lambda \ll 1$ the charge-transfer resistance is dominant and the reaction behaves irreversibly. As for impedance spectroscopy, the presence of ω in the expression for the reversibility index opens the way to "tune" the experiment to suit different rate constant values.

What is measured in a.c. voltammetry is the current that is in phase with the applied a.c. voltage. As Figure 15-11 illustrates, this is made up of a faradaic and a nonfaradaic

---

[1537] See Web#1537 for an explanation of the second equality in 15:41.

[1538] In the general case an extra factor of $(D_O / D_R)^{(2\alpha - 1)/4} \exp\{(2\alpha - 1)F(E_{dc} - E^h)/2RT\}$ appears on the right-hand side of equation 15:42.

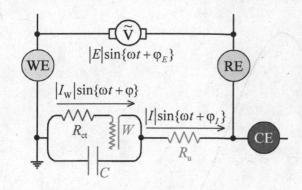

Figure 15-11 The Randles-Ershler circuit applies to a.c. voltammetry, though frequently the uncompensated resistance is neglibible.

component. It is demonstrated elsewhere[1539] that this in-phase current is given by

$$15:44 \quad |I'| = \frac{|E|\left[\omega R_{ct} + \sqrt{\omega/2}\,W\right]}{\sqrt{\omega R_{ct}^2 + \sqrt{2\omega}\,R_{ct}W + W^2}\left[\sqrt{\omega}R_u + \sqrt{\omega R_{ct}^2 + \sqrt{2\omega}\,R_{ct}W + W^2}\right]} \quad \begin{array}{l}\text{in-phase}\\\text{current}\end{array}$$

Frequently the uncompensated resistance is negligible[1540]. If, in addition, the system behaves reversibly, then equation 15:44 reduces to

$$15:45 \quad |I'| = \sqrt{\frac{\omega D_R}{2}}\,\frac{F^2 A c_R^b |E|}{4RT}\,\text{sech}^2\left\{\frac{F}{2RT}\left(E_{dc} - E^h\right)\right\} \quad \text{reversible}$$

after incorporating equation 15:41. The a.c. voltammogram then has the typical shape of a "reversible peak" as discussed on page 255. One such curve is included in Figure 15-12 overleaf. The peak occurs at E^h and its height is

$$15:46 \quad |I'|_{peak}^{rev} = \sqrt{\frac{\omega D_R}{2}}\,\frac{F^2 A c_R^b |E|}{4RT} \quad \begin{array}{l}\text{reversible}\\\text{peak current}\end{array}$$

Another curve in this figure is for irreversible conditions, for which

$$15:47 \quad |I'| = \frac{|E|}{R_{ct}} = \frac{F^2 A c_R^b k^{o'} |E|}{2RT}\,\text{sech}\left\{\frac{F}{2RT}\left(E_{dc} - E^h\right)\right\} \quad \text{irreversible}$$

and the other two are for intermediate, quasireversible cases. Notice that the hyperbolic secant term is squared in the reversible case, but not when the reaction behaves irreversibly.

Used with mercury drop electrodes, and especially in the skilled hands of Smith[1541], a.c. voltammetry provided some of the earliest accurate measurements of electrochemical rate constants.

[1539] See Web#1539.

[1540] Demonstrate the simplification that occurs if $R_u = 0$. Then, as in Web#1540, go on to derive equation 15:45.

[1541] Donald E. Smith, 1936–1985, American electrochemist.

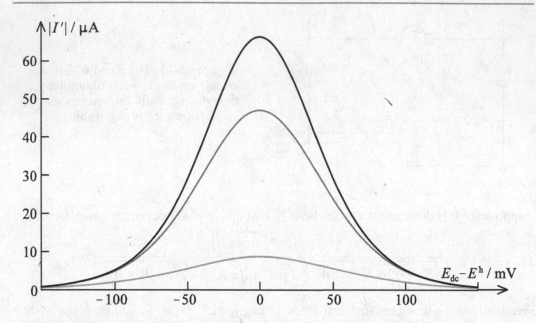

Figure 15-12 In a.c. voltammetry the amplitude of the component of the a.c. current that is in phase with the applied a.c. voltage is plotted versus the d.c. potential, leading to a peak in the vicinity of the nernstian potential. The following constants apply to all four examples of a.c. voltammograms: electrode area $A = 1.0 \times 10^{-5}$ m^2, frequency $\omega = 4000$ Hz, a.c. voltage amplitude $|E| = 5.0$ mV, transfer coefficient $\alpha = 0.5$, uncompensated resistance $R_u = 0$, bulk concentration $c_R^b = 1.0$ mM, diffusivity $D_1 = 1.0 \times 10^{-9}$ m^2 s^{-1}. The formal rate constants are $k^{o\prime} \geq \mathbf{10^{-1}}$ **m s^{-1}** (reversible), $k^{o\prime} = 1 \times 10^{-3}$ **m s^{-1}** (quasireversible), and $k^{o\prime} = 1 \times 10^{-4}$ m s^{-1} (irreversible).

Fourier-Transform Voltammetry: the harmonic response to an a.c. signal

Fourier[175] demonstrated that it is possible to represent *any* periodic signal of period P by a series (usually an infinite series) of cosines and sines of frequencies that are multiples of $\omega = 2\pi/P$. The case of the square-wave was exemplified on page 27. With per(t) representing any periodic signal, the general formula of a **Fourier series** is

15:48 $\qquad$ $\text{per}(t) = \dfrac{c_0}{2} + \sum\limits_{m=1}^{\infty} c_m \cos\{m\omega t\} + s_m \sin\{m\omega t\}$ $\qquad$ Fourier series

As this equation shows, an aperiodic term is often present; this constant may be regarded as a cosine of zero frequency. The coefficients in the Fourier series may be found from the formulas[1542]

[1542] These are often called Euler equations. Leonhard Euler, 1707–1783, gifted Swiss mathematician.

15:49 $\qquad c_m = \dfrac{\omega}{\pi} \displaystyle\int_0^{2\pi/\omega} \mathrm{per}(t)\cos\{m\omega t\}\,\mathrm{d}t \qquad m=0,1,2,\cdots$ $\qquad$ cosine coefficients

and

15:50 $\qquad s_m = \dfrac{\omega}{\pi} \displaystyle\int_0^{2\pi/\omega} \mathrm{per}(t)\sin\{m\omega t\}\,\mathrm{d}t \qquad m=1,2,3,\cdots$ $\qquad$ sine coefficients

The process of determining the Fourier series of a periodic function is known as **Fourier analysis**[1543]. The result of Fourier analysis is often presented in the form of a Fourier spectrum, as in Figure 1-21, or as a power spectrum in which the quantity[1544] $\sqrt{c_m^2 + s_m^2}$ is plotted versus m.

Though mathematicians define the operation differently, the procedure that has come to be popularly known as **Fourier transformation** is an adaptation of Fourier analysis designed to handle data available as a time series rather than as a continuous variable. The effect of the transformation is to convert a set of measurements, taken at evenly spaced instants, into a second set of digitized data that describes the measurements in terms of frequency. Usually, **inverse Fourier transformation** is employed subsequently, to regenerate those segments of the time series whose frequencies are of particular interest. Fourier transformation/inversion has been applied beneficially to many fields of science and technology. In most of these, including electrochemical applications, there are three stages in the overall procedure[1545]

15:51 $\qquad$ time-series data $\xrightarrow{\text{transformation}}$ transformed data (all frequencies) $\xrightarrow{\text{filtration}}$ the $m\omega$ segment of transform $\xrightarrow[\text{transformation}]{\text{inverse}}$ component of frequency $m\omega$ in original data

the second and third being repeated for each frequency $m\omega$ of interest. If the time-series data are ideally periodic, the transformed data have a comb-like structure, as in Figure 1-21, with a nonzero amplitude only at integer multiples of the fundamental frequency ω. For nearly-periodic data on the other hand, the transformed data are more like an optical spectrum, with peaks at all, or many, integer multiples of ω.

As in a.c. voltammetry, of which it is an adaptation, **Fourier-transform voltammetry** employs a d.c. modulated by a single[1546] sinusoidal a.c. voltage, as in Figure 15-10. One distinction between the two techniques is that, whereas the higher harmonics are an unwelcome complication in a.c. voltammetry, measuring them is the objective in the

---

[1543] Find the c_0, c_1, s_1, s_2, and s_3 coefficients of the square-wave shown in Figure 1-22 and formulated in 1:42. Compare your results with equation 1:43 and with Web#1543.

[1544] or sometimes twice this quantity or even its logarithm

[1545] For some further details on Fourier analysis, see Web#1545.

[1546] Multiple sinusoids may be employed, however, or even signals, such as square-waves and "white noise", that incorporate an infinity of harmonics.

Fourier-transform technique. An applied a.c. potential of frequency in the range 5–1000 Hz is generally used, with an amplitude, $|E|$, even as large as 100 mV. Typically a scan rate v of 50 mV s^{-1} is employed, with a d.c. potential range $E_{final}-E_0$ of about 500 mV, embracing the nernstian half-wave potential of the reaction under study. As a result, very many harmonics are present and those up to the 6th, or more, often occur at amplitudes that are well in excess of noise. Sometimes, the d.c. scan is reversed, as in cyclic voltammetry (page 347).

The effect of the signal $E(t) = E_0 + vt + |E|\sin\{\omega t\}$ on the reversibly behaved reaction $R(soln) \rightleftarrows e^- + O(soln)$ can be modeled via the **reversible pan-voltammetric relation**[1547]

$$15:52 \qquad \frac{M(t)}{FA} = \frac{c_R^b\sqrt{D_R}}{2}\left[1 + \tanh\left\{\frac{F[E(t)-E^h]}{2RT}\right\}\right] \qquad \text{the reversible pan-voltammetric relation}$$

This valuable relationship applies in Fourier-transform voltammetry, as it does in most manifestations of the typical voltammetric conditions detailed on pages 232–233. $M(t)$ is the semiintegral[1242] of the current and therefore, on semidifferentiation and incorporation of the signal, one arrives at the formula:

$$15:53 \qquad \frac{I(t)}{FA} = \frac{c_R^b\sqrt{D_R}}{2}\frac{d^{1/2}}{dt^{1/2}}\left[1 + \tanh\left\{\frac{E_0 + vt + |E|\sin\{\omega t\} - E^h}{2RT/F}\right\}\right] \qquad \begin{array}{l}\text{current in}\\ \text{reversible a.c.}\\ \text{voltammetry}\end{array}$$

that governs a.c. voltammetry, including its Fourier-transform variant. The expression in 15:53 does *not* describe a periodic function. However, it will be *nearly* periodic if the timescale associated with the d.c. scan, namely $(E_{final}-E_0)/v$, is much larger than the a.c. timescale, which is $2\pi/\omega$. This inequality is well satisfied in Fourier-transform voltammetry, as practiced, providing legitimacy to the Fourier transformation/inversion approach.

Except at the half-wave potential E^h of a reversible process, Fourier transformation through numerical procedures is the only practical avenue open to process equation 15:53. The procedure follows the strategy in scheme 15:51: the digitized current is Fourier transformed, the entire transform is set to zero except in the immediate vicinity of the frequency $m\omega$ of interest, and then inverse transformation is applied to the remnant. Figure 15-13 illustrates the second and third steps in the procedure. The net result is to split the current into a d.c. component (which is generally ignored), the fundamental, and higher harmonics. The result is often displayed, as here, in the form of a graph of the amplitude of the envelope of the absolute value of the sinusoids as a function of the d.c. potential (or time). Note how diverse the individual patterns are, in both size and shape, with the number of elements in each pattern equaling the harmonic number m.

[1547] The pan-voltammetric relation, and its special case – the reversible pan-voltammetric relation – are derived in Web#1547. See page 364 for the full equation.

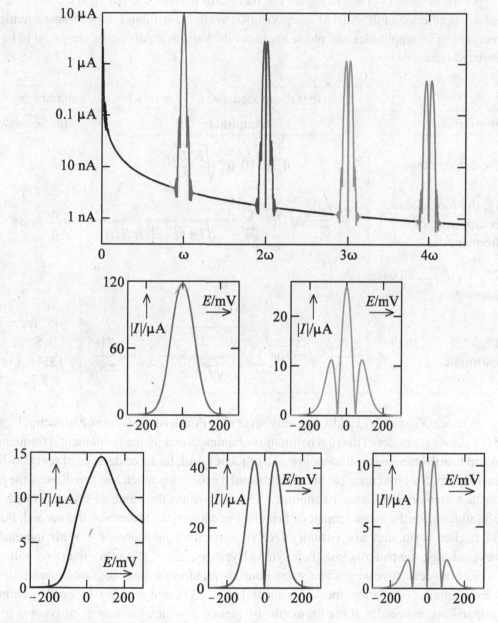

Figure 15-13 The upper curve, a **power spectrum**, logarithmically plots frequency-domain data resulting from Fourier transformation of measurements from a Fourier-transform voltammogram of an electrode reaction behaving reversibly. Each colored segment of this plot has been separately inverse-Fourier transformed to yield the time-domain data shown in the five smaller diagrams. The ordinate of the five plots is the amplitude of the envelope of the magnitudes of the current sinusoids for the 1st, 2nd, 3rd, and 4th harmonics, as well as the d.c. component. Data: $A = 10^{-5}$ m^2, $T = T^\circ$, $D_R = D_O = 10^{-9}$ m^2 s^{-1}, $c_R^b = 1$ mM, $E^h = 0$ V, $v = 50$ mV s^{-1}, $\omega = 9$ Hz, $|E| = 80$ mV.

At the instant $t = (E^h - E_0)/v$, as the d.c. ramp transits the half-wave potential, an analytic prediction of the current components is available without Fourier transformation/ inversion. The amplitudes and phase angles of the various components are found to be as tabulated here.

	Current components at E^h in reversible a.c. voltammetry	
Harmonic	Amplitude[1548]	Phase angle
the d.c. component	$0.38010 A c_R^b \sqrt{\dfrac{F^3 D_R v}{RT}}$	–
1st, the fundamental component, of frequency ω	$\dfrac{4RTAc_R^b}{|E|} \sqrt{D_R \omega} \sum_{j=1,3}^{\infty} 1 - \dfrac{1}{\sqrt{1 + (F|E|/j\pi RT)^2}}$	$\dfrac{\pi}{4}$ or $45°$
2nd, 4th, 6th, ..., all even harmonics	0	–
1st, 3rd, 5th, ..., the mth harmonic	$\dfrac{4FAc_R^b}{\pi \beta^m} \sqrt{m D_R \omega} \sum_{j=1,3}^{\infty} \dfrac{\left[\sqrt{j^2 + \beta^2} - j\right]^m}{\sqrt{j^2 + \beta^2}} \quad \beta = \dfrac{F|E|}{\pi RT}$	$+45°$ for $m =$ 1,5,9,... $-135°$ for $m =$ 3,7,...

Whereas Figure 15-13 addresses only reversible Fourier-transform voltammetry, Figure 15-14 shows examples of the first harmonic voltammograms (at the fundamental frequency, ω) and fourth harmonic voltammograms (frequency 4ω), for three degrees of reversibility. The magnitude of the current at the fundamental frequency is much less sensitive to the rate constant than are the higher harmonics. This underlines the value of Fourier-transform voltammetry for the measurement of fast electrode kinetics. Moreover, the second, third, and higher harmonics are virtually free of capacitive interference. With increasing irreversibility, the patterns lose their symmetry, especially if α is other than one-half.

The only effective way of analyzing Fourier-transform voltammograms quantitatively is by digitally simulating the experiment (page 333) and making theory/experiment comparisons, especially at the harmonic frequency at which the voltammogram is most responsive to the sought parameter. The simulation may incorporate such interferences as uncompensated resistance and background currents from capacitance and other effects. Electrode reactions that are more complicated than a single-step, one-electron transfer may

---

[1548] Note that the summations, which involve only odd j's, converge very rapidly. Appreciate that the "0" in this column for even harmonics does not mean that these harmonics are absent; only that the half-wave potential is a node in the harmonic pattern.

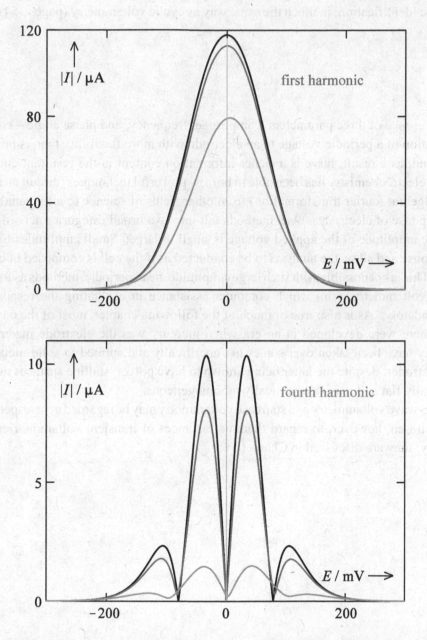

Figure 15-14 First and fourth harmonic components in Fourier-transform voltammetry, presented as the amplitude of the envelope of the sinusoidal current plotted versus d.c. potential, for a reversible, a quasireversible, and an irreversible oxidation. Data: $k^{o\prime}$ = **0.10**, 1.0×10^{-3}, and 1.0×10^{-4} m s^{-1}, α = 0.5, other values as in Figure 15-13.

be addressed similarly, permitting Fourier-transform voltammetry to be employed for mechanistic identification, in much the same way as cyclic voltammetry (pages 354–362) is used.

Summary

The presence of three parameters – amplitude, frequency, and phase angle – endows the application of a periodic voltage to an electrode with more flexibility than is possible with d.c. and, as a result, there is a richer information content in the resultant current. Moreover, electrochemistry has been able to borrow powerful techniques (circuit analysis, complex algebra, fourier transformation) from other fields of science to aid in analyzing the a.c. response of electrodes. A.c. methods fall into two broad categories according to whether the amplitude of the applied voltage is small or large. Small amplitudes foster a linear response and allow the analysis to be conducted as if the cell is composed of circuit elements. This is not possible with such large-amplitude, near-periodic, methods as fourier-transform voltammetry, with which computer assistance in evaluating the response is almost mandatory. As is also true of much of the following chapter, most of the methods discussed here were developed in an era when mercury was the electrode material of choice; they have been taken over somewhat uncritically and applied to solid metal and carbon electrodes, despite the latter being known to have polycrystalline surfaces that are not atomically flat and may be chemically inhomogeneous.

Square-wave voltammetry and staircase voltammetry may be regarded as near-periodic. We have chosen, however, to regard them as instances of transient voltammetries and, accordingly, they are discussed in Chapter 16.

16

Transient Voltammetry

Examination of the way in which a working electrode responds to an applied signal is one of the most fruitful ways of studying electrode reactions. The applied signal may be a current, but more usually it is a rather simple time-dependent potential program. The alternative name **potentiodynamic voltammetry** distinguishes experiments in which the potential changes from those in which it is mostly constant. Almost invariably the experiment is conducted under the control of a potentiostat and is subject to all the other standard voltammetric conditions itemized on pages 232–233. In Chapters 12 and 15 we describe voltammetry that leads, respectively, to a steady response and to a periodic (or nearly periodic) response. In the present chapter our concern is mostly for experiments that respond by producing a time-dependent current that is neither constant nor repetitive[1601]. As reported in Chapter 13, the presence of a parallel capacitive pathway for current, and the inevitable presence of a small uncompensated resistance are nuisances that practitioners of transient voltammetry must endure; we shall not here be unduly concerned with these interferences, which may be addressed by strategies akin to those discussed on pages 271–279. Usually electrodes are planar and large enough that edge effects may be ignored. Variety in transient voltammetry arises from different choices of signals designed to optimally elicit the information that the experimenter seeks, or to discriminate against unwanted effects. As in other realms of human activity, tradition is also a factor.

Modeling Transient Voltammetry: mathematics, algorithms, or simulations

It is pointless to carry out an experiment that produces a transient current unless one can interpret the size and shape of the response. Thus, predicting the outcome of a transient voltammetric experiment is a necessary adjunct to the experiment, and it is frequently a more demanding phase of the investigation than the experiment itself. The name **modeling**

[1601] Sometimes cyclic voltammetry is prolonged until the response becomes repetitive. Figure 13-9 is such a voltammogram, to which the name **ultimate cyclic voltammogram** has been given.

Electrochemical Science and Technology: Fundamentals and Applications, First Edition. Keith B. Oldham, Jan C. Myland, Alan M. Bond.
© 2012 John Wiley & Sons, Ltd. Published 2012 by John Wiley & Sons, Ltd.

is given to the predictive activity that creates and analyzes a mathematical model of the way in which we imagine the electrode to behave, in terms of various parameters and, on that basis, proceeds to predict the current-time relationship. If the model is realistic and flexible, it should be possible, by selecting appropriate parameter values, to bring the predictions into close registry with the experimental curve. If not, the model is faulty. If, with a credible set of parameters, the model agrees closely with experiment, the model is vindicated and the parameters are accepted as probably correct. There are three routes that can be identified as separate approaches to the prediction of the outcome of a transient voltammetric experiment:

16:1   voltammetric modeling $\begin{cases} \text{mathematical analysis} \\ \text{semi-analytical methods} \\ \text{digital simulation} \end{cases}$

though the distinctions among the three may not always be clear-cut.

More complex electrochemistry is addressed later. Here we shall be concerned only with the one-electron, one-step oxidation R($soln$) $\rightleftarrows$ e⁻ + O($soln$), the product being initially absent, and assume obedience to Butler-Volmer kinetics. In these circumstances there are nine parameters on which the current depends, namely:

16:2   input data $\begin{cases} \text{temperature } T; \text{ electrode area } A; \text{ bulk concentrations} c_R^b \text{ and } c_O^b; \\ \text{diffusivities } D_R \text{ and } D_O; \text{ formal potential } E^{o\prime} \text{of the reaction;} \\ \text{formal heterogeneous rate constant } k^{o\prime}; \text{ transfer coefficient } \alpha. \end{cases}$

though the two listed in orange are irrelevant when the reaction behaves reversibly. In addition to these nine, there is a need to specify the way in which the electrode potential $E(t)$ evolves with time. The modeler must assign known or supposed numerical values to all the input parameters in order to predict the voltammogram quantitatively, whichever of the three modeling approaches is followed.

The formal potential has been included in the 16:2 listing but an alternative reference potential is often used in voltammetric theory. This is the **nernstian half-wave potential**[1602]

16:3 $$E^h = E^{o\prime} - \frac{RT}{2F} \ln\left\{ \frac{D_O}{D_R} \right\}$$ nernstian half-wave potential

The Nernst equation may incorporate either reference potential, as in the formulations

---

[1602] Note the concordance between this name and the italicized passage in Web#1547. Definition 16:3 is consistent with that given on page 253. If $D_O = D_R$, the nernstian half-wave potential and the formal potential are identical. By what voltage do these potentials differ, at 25°C, if there is a 10% discrepancy between the diffusivities? See Web#1602.

16:4 $\quad \dfrac{\sqrt{D_O}\, c_O^s(t)}{\sqrt{D_R}\, c_R^s(t)} = \exp\left\{\dfrac{E(t) - E^h]}{RT/F}\right\} \qquad$ Nernst $\qquad \dfrac{c_O^s(t)}{c_R^s(t)} = \exp\left\{\dfrac{E(t) - E^{o\prime}}{RT/F}\right\}$

equations

but the Butler-Volmer equation is considerably simpler[1603] when referenced to the formal potential

16:5 $\quad \dfrac{I(t)}{FA} = \dfrac{k^{o\prime}}{\xi^\alpha(t)}\left[c_R^s(t)\xi(t) - c_O^s(t)\right] \quad$ where $\quad \xi(t) = \exp\left\{\dfrac{E(t) - E^{o\prime}}{RT/F}\right\} \quad$ Butler-Volmer

as is the pan-voltammetric relation[1547]. We shall mostly refer potentials to $E^{o\prime}$ in this chapter, though expressions for *reversible* voltammograms benefit from being redrafted in terms of E^h, and therefore we sometimes adopt that alternative.

Mathematical analysis is the most satisfactory of the three modeling approaches in 16:1, but sadly it is the method most likely to be inaccessible in all but the simplest cases. This approach is to combine the mathematical equations representing the various relationships – those describing the signal, the Butler-Volmer expression, Fick's laws, Faraday's law, and so on – into a single equation that shows explicitly how the measured output of the experiment, usually the current, is predicted to change with time. Incorporated into this equation will be the various input parameters of the experiment, as detailed above. Mathematical analysis is a pencil-and-paper operation; a computer is not needed, though one may be useful in converting the formula into a set of numbers for comparison with the experimental results.

Not infrequently, in attempting to model a voltammetric experiment mathematically, one is successful in creating an equation, or a set of equations, that contains all the pertinent parameters, including the time-dependent current, but which cannot be cast into the form of a single explicit equation. In these circumstances, a **semianalytical method** is often a second-best possibility. The equation(s) are solved with help from a **numerical algorithm**. Such techniques are often "progressive", by which is meant that the calculation of the current at one instant relies on values already calculated for previous instants in time. Such semianalytical procedures involve considerable arithmetic that nowadays is invariably carried out by computer, though the earliest applications predated today's computer ubiquity. A progressive numerical procedure of very wide applicability is provided by the rather elaborate formula[1604] given as equation 16:6 overleaf. This powerful voltammetric algorithm permits the calculation of the faradaic current that flows in response to any applied potential signal $E(t)$. It applies for any degree of reversibility of the single-electron

---

[1603] Demonstrate that this is so by constructing an analogue of the Butler-Volmer equation 16:5 that uses the abbreviation $\xi^h(t) = \exp\{F[E(t) - E^h]/RT\}$, or see Web#1603.

[1604] Web#1547 gives a derivation of the **pan-voltammetric relation**, from which algorithm 16:6 derives. Web#1604 includes a procedure for semiintegration and details of a **spreadsheet algorithm** by means of which algorithm 16:6 may be conveniently implemented.

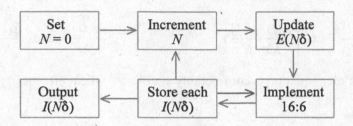

Figure 16-1 Flowsheet for the universal voltammetric algorithm 16:6. The **blue** arrow indicates the reuse of previously stored current values.

transfer and irrespective of whether R, or O, or both are present initially. Employing the $\xi(t)$ abbreviation defined in equation 16:5, the algorithm is

$$16{:}6 \quad I(t) = \frac{\dfrac{FA}{\sqrt{\delta}}\left[c_R^b\xi(t)-c_O^b\right]-\left[\dfrac{\xi(t)}{\sqrt{D_R}}+\dfrac{1}{\sqrt{D_O}}\right]\displaystyle\sum_{n=1}^{N-1}w_{N-n}I(n\delta)}{\dfrac{\xi^\alpha(t)}{k^{o\prime}\sqrt{\delta}}+\dfrac{\xi(t)}{\sqrt{D_R}}+\dfrac{1}{\sqrt{D_O}}} \quad \text{where} \begin{cases} N = t/\delta \\ w_0 = 1 \\ w_n = \left(1-\dfrac{1}{2n}\right)w_{n-1} \end{cases}$$

The weights may be precalculated, but notice that the nth current is weighted by w_{N-n}, not by w_n. All the input parameters listed in 16:2 are incorporated into this **spreadsheet algorithm**[1604]. In addition, a choice must be made of δ, a very small time interval[1605]. To exploit this, and similar, algorithms, one may follow the flowsheet shown in Figure 16-1. A crude value of $I(\delta)$ is first calculated using $E(\delta)$ together with the $N = 1$ instance of formula 16:6 [when $N = 1$, the sum is "empty" and contributes nothing to $I(\delta)$]. Then set $N = 2$ and, after resetting the potential to $E(2\delta)$, calculate $I(2\delta)$ [the sum now has one term, namely $I(\delta)/2$]. Next, set $N = 3$ and calculate $I(3\delta)$, after resetting the potential to $E(3\delta)$ [the sum now has two terms, namely $\{I(2\delta)/2\}+\{3I(\delta)/8\}$]; and so on. Though the earliest current values will be crude approximations, the error in subsequent outputs soon becomes insignificant. A spreadsheet is available[1604] to facilitate the procedure.

In applying mathematical analysis to, and devising semianalytical methods for, voltammetric problems, there are two techniques that have proved particularly fruitful, each lying somewhat outside the mathematical mainstream. These are Laplace transformation and the semicalculus. The particular feature of **Laplace transformation**[162] that is of value in solving problems in voltammetry is its ability to convert a partial differential equation, and particularly Fick's second law, into an ordinary differential equation, and thereby aid in the solution of such an equation. The great virtue of the **semioperators**[1242] is their ability to bypass the need to solve Fick's second law. In most applications of the semicalculus to

---

[1605] A one-thousandth of the total duration of the experiment, or thereabouts, would often be suitable. Nowadays, potentials are commonly applied, not as a continuously varying signal, but as succession of very short, constant-potential segments. In such cases δ could be the period of such segments, or a multiple thereof.

voltammetry, the only formulas needed are those based on the expression[1606]

$$16{:}7 \qquad \frac{\sqrt{D_R}\left[c_R^b - c_R^s(t)\right]}{FA} = \frac{\sqrt{D_O}\left[c_O^s(t) - c_O^b\right]}{FA} = M(t) = \frac{d^{-1/2}}{dt^{-1/2}}I(t) \qquad \text{Faraday-Fick relationships}$$

which, through the semiintegral $M(t)$, relates the faradaic current $I(t)$ directly to the concentrations of the redox pair at the electrode surface. The Faraday-Fick relations are derived[1508] in Chapter 15 and are the keys to the creation of the **pan-voltammetric relation**[1604] that will find frequent use in this chapter.

The computer is essential for **digital simulation**[1607], a powerful and versatile technique, which is different in concept from the other two modeling routes, although it shares with many semianalytical procedures the property of being progressive. Digital simulation, however, is not only progressive in time, but also in space. In a digital simulation, a specific memory location in the computer is allocated to represent a corresponding site in the diffusion field. Each computer location holds a number that corresponds to the concentration of a pertinent species. There may be several memory locations representing a single site, each keeping track of the concentration of a particular solute. With the passage of time, the stored numbers change repeatedly as each site interacts diffusionally with its neighbors, and perhaps chemically with other species at the same site. The memory locations that represent the electrode surface are especially important, these being the ones that hold numbers representing concentrations that change also on account of the electrochemical reaction. Though the geometry of an electrochemical cell is three-dimensional, symmetries generally permit the simulation to be conducted in one, or at most two, spatial dimensions.

In devising a digital simulation, one replaces the continuity of Fick's laws by discrete approximations that involve small, though not necessarily constant, intervals in space and time. In this way, a differential equation is replaced by a multitude of algebraic equations and thereby solved. There is a great art in devising such replacements[1608], so as to minimize computer time, while preserving adequate accuracy, and this activity has now evolved into a sophisticated vocation in which many experts are engaged. The creation of a successful simulation routine is an exacting and time-consuming process. So exacting, in fact, that electrochemists nowadays mostly rely on commercial simulation packages[1609].

[1606] As a simple application of the Faraday-Fick relation[1508], use the table in Web#1242, to derive Cottrell's equation (page 157) which describes the current response to the sudden imposition of a large positive potential (a "potential leap") on an electrode at which the reaction $R(soln) \rightarrow e^- + O(soln)$ can occur. See Web#1606.

[1607] Here we restrict attention to the **finite-difference** class of digital simulation. Another class, known as **finite-element** digital simulation is also in use and is particularly efficient in cases of complicated cell geometry.

[1608] See D. Britz, *Digital Simulation in Electrochemistry,* 3rd edn, *Lecture Notes in Physics*, Springer, 2005.

[1609] Dating from 1991, DigiSim® was the earliest to be widely used; it models cyclic voltammetry primarily. There are more recent alternatives, including the versatile, DigiElch®. See Appendix B of A.J. Bard and L.R. Faulkner, *Electrochemical Methods* 2nd edn, Wiley, 2001, for an outline of how simulation programs work.

"**Discretization**" means the treating of a continuous variable, such as time or a spatial coordinate, as if it varies discontinuously, in small discrete steps. The threefold classification of modeling approaches in 16:1 could be based on what is, and what is not, discretized, as the table demonstrates. Of course,

Modeling approach	Discretization?	
	space	time
mathematical analysis	no	no
semianalytical methods	no	yes
digital simulation	yes	yes

discretization is an error-inducing approximation, but if the steps are small, and if wise choices are made in designing the algorithms that execute the model, those errors can be made insignificant.

Potential-Step Voltammetry: single, double and multiple

The imposition on a working electrode of a sudden change in potential is the simplest voltammetric experiment. Here we consider the electrode to be initially held at some potential E_0, well negative of the standard potential, such that the reaction

16:8 $$R(soln) \rightleftarrows e^- + O(soln)$$

does not occur. At time $t = 0$, the electrode potential is suddenly changed to a more positive value E_1, and maintained at that value. Because the ensuing current $I(t)$ is measured as a function of time, this experiment is an example of **chronoamperometry**[1610]. Mathematically, the experiment is summarized by

16:9 $$E(t) = \begin{cases} E_0 & t < 0 \\ E_1 & t > 0 \end{cases} \xrightarrow[\text{experiment}]{\text{potential-step}} I(t) = \begin{cases} 0 & t < 0 \quad \text{potential-step} \\ ? & t > 0 \quad \text{chronoamperometry} \end{cases}$$

where the "?" represents the sought response. We shall assume the Butler-Volmer equation to hold; the version given in equation 16:5, namely

16:10 $$\frac{I(t)}{FA} = \frac{k^{o\prime}}{\xi_1^\alpha(t)}\left[c_R^s(t)\xi_1 - c_O^s(t)\right] \quad \text{where} \quad \xi_1 = \exp\left\{\frac{E_1 - E^{o\prime}}{RT/F}\right\} \quad \begin{matrix}\text{Butler-}\\ \text{Volmer}\end{matrix}$$

being appropriate, with a constant ξ_1 now replacing the variable ξ. **Potential-step chrono-amperometry** provides a rare example of a transient voltammetric experiment in which straightforward modeling by mathematical analysis[1611] is feasible, the result being

16:11 $$I(t) = FAc_R^b k^{o\prime}\xi_1^{1-\alpha}\exp\left\{b^2 t\right\}\text{erfc}\left\{b\sqrt{t}\right\} \quad \text{where} \quad b = k^{o\prime}\left(\frac{\xi_1^{1-\alpha}}{\sqrt{D_R}} + \frac{\xi_1^{-\alpha}}{\sqrt{D_O}}\right)$$

[1610] Measurement (-metry) of current (-ampero-) versus time (chrono-)

[1611] See Web#1611 for two routes to the solution. Also given there is a simplified formula for large $b^2 t$.

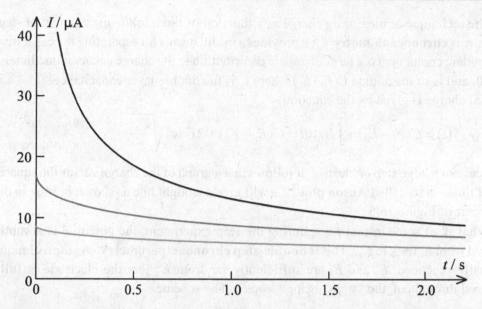

Figure 16-2 Potential-step chronoamperometry. Responses to a potential step for a **reversible**, quasireversible, and irreversible electrode reaction. Data: $E_1 = E^h + 30$ mV, $T = T^\circ$, $D_R = D_O = 1.0 \times 10^{-9}$ m^2 s^{-1}, $A = 1.0 \times 10^{-5}$ m^2, $c_R^b = 1.0$ mM, $\alpha = 0.5$, $k^{o\prime} = 10^{-2}$, 10^{-5}, and 5×10^{-6} m s^{-1}.

Figure 16-2 illustrates the responses for an irreversible, a quasireversible, and a reversible case. There is a sharp initial rise in current, followed by an unending decline. In the reversible case, characterized by a large $b\sqrt{t}$, the current obeys the simpler formula[1611]

$$16:12 \qquad I(t) = \frac{FAc_R^b k^{o\prime}\xi_1^\alpha}{b\sqrt{\pi t}} = \frac{FAc_R^b}{\sqrt{\pi t}\left[\dfrac{1}{\sqrt{D_R}} + \dfrac{1}{\xi_1\sqrt{D_O}}\right]} \qquad \begin{array}{l} \text{reversible} \\ \text{potential-step} \\ \text{chronoamperometry} \end{array}$$

with a spike that is theoretically of infinite height.

Equation 16:11 shows that the current immediately after the imposition of the step is

$$16:13 \qquad I(0) = FAk^{o\prime}\xi_1^{1-\alpha} = FAk^{o\prime}\exp\left\{\frac{(1-\alpha)F}{RT}\left[E_1 - E^{o\prime}\right]\right\} \qquad \begin{array}{l} \text{spike} \\ \text{height} \end{array}$$

and it might be thought that this is a good way to measure electrode kinetics. It is difficult, however, to accurately record currents immediately after a step and, moreover, the faradaic spike is then obscured by the large nonfaradaic current from double-layer charging. The inconvenience of a spike may be avoided either by semiintegrating the current[1612], or by integrating it to give the charge $Q(t)$.

[1612] See Web#1612 for a method of measuring the kinetics via the semiintegral $M(t)$. Also included is a recipe for carrying out semiintegration in such circumstances.

The technique of measuring charge as a function of time, following a potential step, is known as **chronocoulometry**[1613]; it provides a useful means for separating the capacitive and faradaic components. The nonfaradaic contribution to the charge occurs immediately, at $t = 0$, and is of magnitude $C(E_1 - E_0)$ where C is the double-layer capacitance[1614]. Thus the total charge is given by the equation

$$16{:}14 \quad Q_{\text{total}}(t) = C(E_1 - E_0) + \int_0^t I(t)\,\mathrm{d}t = C(E_1 - E_0) + 2FAc_R^b \sqrt{\frac{D_R t}{\pi}} \qquad \text{potential-leap[1615]}$$
$$\text{chronocoulometry}$$

in the case of a large step or "leap". It follows that a graph of the charge versus the square-root of time – a so-called **Anson plot**[1616] – will give a straight line as shown in **blue** in the upper part of Figure 16-3.

What if, at some instant $t = t_1$ during the leap experiment, the potential is abruptly changed from E_1 back to E_0? This is **double-step chronoamperometry**. As the technique is usually practiced, E_0 and E_1 are sufficiently far from E^h that the electrode is fully polarized throughout, the "steps" being "leaps". The scheme

$$16{:}15 \quad E(t) = \begin{cases} E_0 & t < 0 \\ E_1 & 0 < t < t_1 \\ E_0 & t > t_1 \end{cases} \xrightarrow[\text{experiment}]{\text{double-leap}} I(t) = \begin{cases} 0 & t < 0 \\ FAc_R^b\sqrt{\dfrac{D_R}{\pi t}} & 0 < t < t_1 \\ ? & t > t_1 \end{cases} \begin{array}{l} \cdot\text{double-leap} \\ \text{chrono-} \\ \text{amperometry} \end{array}$$

describes the experiment. It may be demonstrated[1617] that the (negative) faradaic current following this second potential leap is given by

$$16{:}16 \quad I(t > t_1) = -FAc_R^b\sqrt{\frac{D_R}{\pi}}\left(\frac{1}{\sqrt{t - t_1}} - \frac{1}{\sqrt{t}}\right) \qquad \begin{array}{l} \text{faradaic current following a} \\ \text{second (backwards) leap} \end{array}$$

and therefore, on integration

$$16{:}17 \quad Q(t > t_1) = Q(t_1) - 2FAc_R^b\sqrt{\frac{D_R}{\pi}}\left(\sqrt{t - t_1} - \sqrt{t} + \sqrt{t_1}\right) \qquad \begin{array}{l} \text{faradaic charge following} \\ \text{the second leap} \end{array}$$

as illustrated by the **blue** curve in Figure 16-4 (page 338). Formula 6:17 gives the faradaic charge. There will again be a capacitive contribution, equal to that produced by the first

---

[1613] For further information and references, see G. Inzelt in: F. Scholz (Ed.), *Electrochemical Methods*, 2nd edn, Springer, 2010, pages 147–158.

[1614] The *integral* double-layer capacitance; see page 262.

[1615] As before, we distinguish a "leap", which is to a potential extreme enough that diffusion alone governs the faradaic response, from a "step", that could be to any potential. The faradaic component of equation 16:14 is, of course, just the integral of the Cottrell equation (page 157).

[1616] Fred C. Anson, U.S. electrochemist.

[1617] Equations 16:16 and 16:17 are derived in Web#1617.

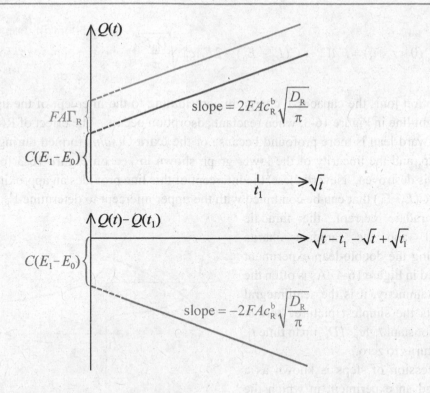

Figure 16-3 Chronocoulometry. In the upper segment of this Anson plot, the integrated current following a potential leap gives a **straight line** when plotted versus the square-root of time. The intercept measures the double-layer charge $C(E_1 - E_0)$. The lower **straight line** graph, which stems from a subsequent backwards leap, has an intercept and slope that differ from those in the upper half only in signs. The **green line** shows how the upper line is displaced by the presence of adsorption.

leap, but opposite in sign. A plot of $Q(t > t_1) - Q(t_1)$ versus $\sqrt{t - t_1} - \sqrt{t} + \sqrt{t_1}$ will generate the **red** straight line shown in the lower half of Figure 16-3.

Another event that may affect the intercept of an Anson plot is the possibility of **adsorption of the reactant**

$$
16{:}18 \qquad \begin{pmatrix} \mathrm{R}(soln) \\ \uparrow\downarrow \\ \mathrm{R}(ads) \end{pmatrix} \rightarrow \mathrm{e^-} + \mathrm{O}(soln) \qquad
\begin{array}{l} \text{co-oxidation of both solution-} \\ \text{phase and adsorbed reactant} \end{array}
$$

Because the adsorbed species requires no transport, reaction is immediate and this creates an additional charge of magnitude $FA\Gamma_R$ promptly at $t = 0$, where Γ_R is the surface concentration of the adsorbed reactant, thereby adding a third term to the equation describing the charge

16:19 $Q_{total}(0 < t < t_1) = FA\Gamma_R + C(E_1 - E_0) + 2FAc_R^b \sqrt{\dfrac{D_R t}{\pi}}$ three contributions to the charge during a potential leap[1618]

This adsorption joins the capacitive charge in contributing to the intercept of the upper green straight line in Figure 16-3, when reactant adsorption occurs. The effect of R(*ads*) on the backward leap is more profound because of the extra O(*soln*) formed during the forward leap, and the linearity of the lower graph shown in red on Figure 16-3 for no adsorption, is destroyed. Nevertheless, the intercept of this line provides an approximate estimate of $C(E_1 - E_0)$ that can be combined with the upper intercept to determine Γ_R.

The faradaic current, the faradaic semiintegral, and the faradaic charge accompanying the double-leap experiment are portrayed in Figure 16-4. As is often the case in voltammetry, it is the semiintegral that presents the simplest picture: here it equals the constant $FAc_R^b \sqrt{D_R}$ up to time t_1, and then returns to zero.

A succession of steps is known as a staircase and an experiment in which the potential program has this form is known as **staircase voltammetry**. If the ensuing current is measured late in the life of each step, a beneficial discrimination is provided against capacitive current, which decays much more rapidly, following each step,

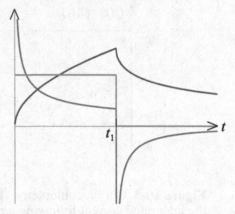

Figure 16-4 The changes in the faradaic current, semiintegral and charge during a double-leap experiment.

than does the faradaic current. The technique, illustrated on page 276, can deliver current samples that are almost purely faradaic. The theory of staircase voltammetry is rather complicated[1619] and the method is little used. When both ΔE_{step} and Δt_{step} are extremely small, the applied signal is close to a ramp in which the potential increases with time at a rate

16:20 $v = \dfrac{\Delta E_{step}}{\Delta t_{step}}$ staircase/ramp equivalence

As might be expected, the resulting current then becomes indistinguishable from that found

---

[1618]   By considering the *SI* units of all the terms, show that equation 16:19 is dimensionally homogeneous. See Web#1618. Also calculate the surface concentration of reactant that would make a contribution to the intercept equal to the capacitive contribution for a potential leap of 600 mV and a capacitance of 0.276 µF m^{-2}. Moreover, show that each adsorbing molecule occupies about 1.0 square micrometers of electrode area.

[1619]   See Web#1619 for the a general treatment of nernstian pulse voltammetry, including staircase voltammetry.

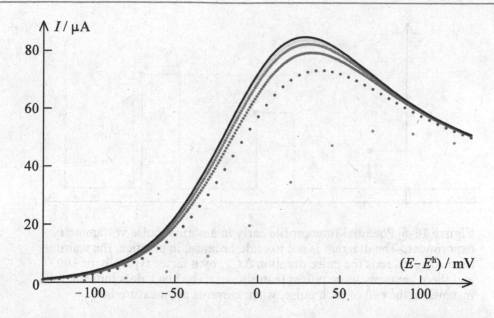

Figure 16-5 Reversible staircase voltammograms in which E and I relate to the potential and current at the end of each constant-potential segment. The ratio $\Delta E_{step}/\Delta t_{step}$ has a constant value of 1.0 V s^{-1} for all points, with $\Delta t = 25$, 5, 1, 0.2, and 0.04 ms. T, A, D, and c data as in Figure 16-2. The **black** curve is a linear-scan voltammogram under similar conditions.

in linear-potential-scan chronoamperometry (page 346). Figure 16-5 illustrates this. In fact, modern digital instruments which purport to generate ramp signals actually provide staircases with small steps (though often not of the submillivolt size that is desirable).

Pulse Voltammetries: normal, differential and square

A positive-going step soon followed by a negative-going step, or vice versa, is called a **pulse**. There are very many ways in which a succession of pulses may progressively attain evermore positive (or negative) potentials, and many of these pulse voltammetric possibilities have been explored by electrochemists. Here, however, only the three shown in Figures 16-6, 16-8, and 16-10 are addressed. An advantage of pulses, compared with the ramped voltammetry considered later, is that the current following a pulse soon loses most of its nonfaradaic component, whereas this interference is more-or-less constant with a ramp. A disadvantage is that the current must be measured intermittently, rather than continuously. As usual, our attention is directed to the one-step, one-electron, anodic reaction

$$\text{R}(soln) \rightleftharpoons \text{e}^- + \text{O}(soln)$$

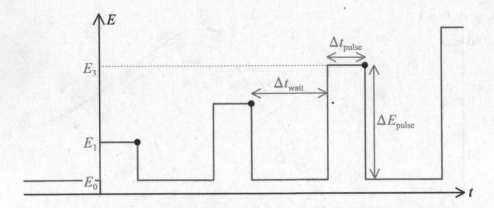

Figure 16-6 Potential-time profile early in a normal pulse voltammetry experiment. The diagram is not to scale because, in practice, the waiting time Δt_{wait} exceeds the pulse duration Δt_{pulse} by a factor typically of 100, and there are many more pulses than shown. The black dots indicate the instants, at the end of each pulse, when currents are measured.

taking place under the usual voltammetric conditions summarized on pages 232–233.

In **normal pulse voltammetry**[1620] the potential spends most of the time at a potential E_0 at which reaction 16:21 does not occur significantly, with brief excursions to increasingly more positive potentials, as illustrated in Figure 16-6. The idea is to allow the solution in the vicinity of the electrode to recover to its initial composition between pulses. Recovery is achieved, not only by diffusive replenishment of R from the bulk, but also by re-reduction of the O formed during the previous pulse[1621]. Thus the waiting time t_{wait} in normal pulse voltammetry far exceeds the pulse duration t_{pulse}. Current is measured immediately before the end of each pulse, by which time the capacitive current has decayed almost to zero. As will be clear from the discussion on page 276, the uncompensated resistance must be small to ensure a speedy decay of the nonfaradaic contribution to the current.

Because any one pulse is essentially independent of its predecessors, each pulse may be treated as a distinct step, and modeled as in the previous section. Thus, by analogy to equation 16:11, the sampled current obeys the formula

$$I = FAc_R^b k^{o'} \xi^{1-\alpha} \exp\{b^2 \Delta t_{pulse}\} \,\mathrm{erfc}\{b\sqrt{t_{pulse}}\}$$

16:22

$$\text{where} \quad b = k^{o'}\left(\frac{\xi^{1-\alpha}}{\sqrt{D_R}} + \frac{\xi^{-\alpha}}{\sqrt{D_O}}\right)$$

normal pulse
voltammetry

[1620] For further information on normal and differential pulse voltammetries see Z. Stojek in: F. Scholz (Ed.) *Electrochemical Methods*, 2nd edn, Springer, 2010, pages 107–119.

[1621] unless the reaction behaves irreversibly at E_0. Occasionally pulse voltammetries are carried out with rotating electrodes which further encourages recovery.

Notice that the only item that changes, from one pulse to the next, is the ξ parameter, given by

16:23
$$\xi = \exp\left\{ \frac{F}{RT}\left[E_0 + \Delta E_{\text{pulse}} - E^{\text{o}\prime} \right] \right\}$$

and reflecting the increasing pulse heights ΔE_{pulse}, which might typically augment by 25 mV between consecutive pulses. Accordingly, a **normal pulse voltammogram** is a point-by-point plot of the measured currents versus the corresponding electrode potential $E = E_0 + \Delta E_{\text{pulse}}$. The shapes predicted[1622] for three such voltammograms are shown in Figure 16-7, in which the reversibility index, defined as

16:24
$$k^{\text{o}\prime}\left(\sqrt{D_R} + \sqrt{D_O}\right)\sqrt{\frac{\Delta t_{\text{pulse}}}{D_R D_O}} = \lambda \qquad \begin{array}{l}\text{reversibility index for} \\ \text{normal pulse voltammetry}\end{array}$$

has values of 10, 1, and 0.1, corresponding to reversible, quasireversible, and irreversible behavior.

When reaction 16:21 behaves reversibly, equation 16:22 simplifies to a concise

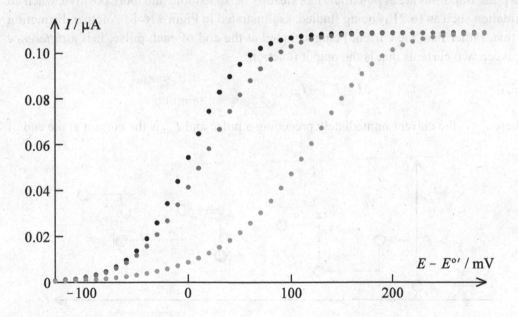

Figure 16-7 Normal pulse voltammograms for a **reversible**, quasireversible, and irreversible electrode reactions. The graph is based on equations 16:22 and 16:23, with the following values: $T = T^{\text{o}}$, $A = 1.0 \times 10^{-5}$ m^2, $c_R^b = 0.0010$ mM, $= 1.0 \times 10^{-3}$ mol m^{-3}, $D_R = D_O = 1.0 \times 10^{-9}$ m^2 s^{-1}, $\alpha = 0.5$, $\Delta t_{\text{pulse}} = 25$ ms, $k^{\text{o}\prime} = \mathbf{1.0 \times 10^{-3}}$, 1.0×10^{-4}, and 1.0×10^{-5} m s^{-1}, $E_0 = E^{\text{o}\prime} - 130$ mV, $\Delta E_{\text{pulse}} = 10, 20, 30, ..., 420$ mV.

[1622] To help in understanding the principles involved, pick one of the points in Figure 16-7 and hand-calculate the current. See Web#1622.

expression that can be rewritten more informatively by referencing the potential to the nernstian half-wave potential:

$$16\!:\!25 \quad I = \frac{FAc_R^b / \sqrt{\pi \Delta t_{\text{pulse}}}}{\dfrac{1}{\sqrt{D_R}} + \dfrac{1}{\xi \sqrt{D_O}}} = \frac{FAc_R^b \sqrt{D_R}}{2\sqrt{\pi \Delta t_{\text{pulse}}}} \left[1 + \tanh\left\{ \frac{E_0 + \Delta E_{\text{pulse}} - E^h}{RT/F} \right\} \right]$$

reversible
normal
pulse
voltammetry

The latter expression exactly matches the **standard reversible wave** discussed on pages 252–253. In the absence of reversibility, the wave is more elongated and its shape begins to be influenced somewhat by the value of the transfer coefficient α.

Because the wave height is proportional to the concentration c_R^b, normal pulse voltammetry is useful in chemical analysis and, because the method offers good discrimination against capacitive interference, it can analyze down to very low concentrations. Even better[1623] in this respect is **differential pulse voltammetry**[1620]. The principles of this method are similar to those of normal pulse voltammetry in that there is a succession of brief pulses each followed by a long waiting time. However, both the waiting periods and the pulse durations are at potentials that steadily become more and more positive, when an oxidation such as 16:21 is being studied, as illustrated in Figure 16-8. Another distinction is that, rather than the current being sampled at the end of each pulse, it is a *difference* between two currents that is the output function.

$$16\!:\!26 \quad \Delta I = I_{\text{end}} - I_{\text{pre}}$$

output in differential
pulse voltammetry

where I_{pre} is the current immediately preceding a pulse and I_{end} is the current at the end of

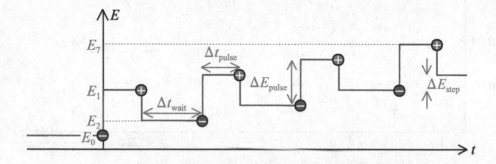

Figure 16-8 The potential-time profile early in differential pulse voltammetry. The diagram is not to scale because, in practice, the waiting time Δt_{wait} exceeds the pulse duration Δt_{pulse} by a factor typically of 50. The dots indicate the instants, before (⊖) and at the end of (⊕) each pulse, when currents are sampled. The output is the difference between these samples.

[1623] Analyses in the subnanomolar range are reported in A.M. Bond, *Modern Polarographic Methods in Analytical Chemistry*, Dekker, 1980.

that pulse, just before the potential returns for the next waiting period. The waveform is characterized by the waiting period Δt_{wait}, the pulse duration Δt_{pulse}, the pulse height ΔE_{pulse}, and the step height ΔE_{step}, all of which are positive.

As in the normal variant, the long interpulse waiting time in differential pulse voltammetry effectively insulates each pulse from the effects of its predecessors. Making plausible assumptions, mathematical modeling shows[1624] that, in the case of a reversible reaction, the voltammogram resembles Figure 16-9 and obeys the equation

16:27 $\quad \Delta I_N = \dfrac{F^2 A c_R^b \Delta E_{\text{pulse}}}{4RT} \sqrt{\dfrac{D_R}{\pi \Delta t_{\text{pulse}}}} \text{sech}^2 \left\{ \dfrac{F[E_N - E^h]}{2RT} \right\}$ nernstian differential pulse voltammogram

It therefore has the **standard peak shape** discussed on pages 254-255, with a peak height of $(F^2 A c_R^b \Delta E_{\text{pulse}} / 4RT)\sqrt{D_R / \pi \Delta t_{\text{pulse}}}$. However, if the pulse height exceeds about 25 mV, the peak shape becomes degraded[1625] and its height is better described by the formula $(F A c_R^b / RT)\sqrt{D_R / \pi \Delta t_{\text{pulse}}} \tanh\{F\Delta E_{\text{pulse}} / 4RT\}$.

The "square" in the name **square-wave voltammetry** indicates that the "waiting time" and "pulse duration" are equal, though these terms are no longer really apposite. It is one

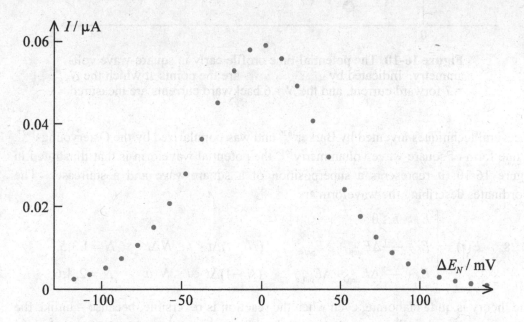

Figure 16-9 Reversible differential pulse voltammogram. The points are based on equation 16:27, with the following values: $T = T^{\circ}$, $A = 1.0 \times 10^{-5}$ m^2, $c_R^b = 0.0010$ mM, $D_R = 1.0 \times 10^{-9}$ m^2 s^{-1}, $\Delta t_{\text{wait}} = 250$ ms, $\Delta t_{\text{pulse}} = 5$ ms, $E_N = E_0 + \frac{1}{2}\Delta E_{\text{pulse}} + \frac{N-1}{2}E_{\text{step}}$, where $E_0 = E^h - 130$ mV, $\Delta E_{\text{pulse}} = 25$ mV, $\Delta E_{\text{step}} = 10$ mV, and $N = 1, 3, 5, ..., 53$.

[1624] See Web#1619 and the second example therein.

[1625] It then obeys equation (27) in Web#1619. The potential to associate with each ΔI_N measurement is ambiguous, and this is important if ΔE_{pulse} is large.

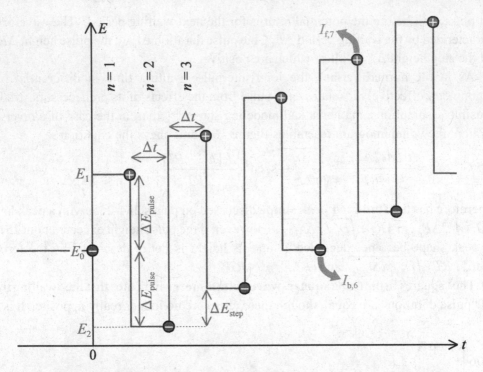

Figure 16-10 The potential-time profile early in square-wave volt-ammetry. Indicated by green arrows are the points at which the N = 7 forward current, and the N = 6 backward currents are measured.

of several techniques invented by Barker[1626] and was popularized by the Osteryoungs[1627]. In one form of square-wave voltammetry[1628] the potential waveform is that illustrated in Figure 16-10; it represents a superposition of a square-wave and a staircase. The coordinates describing the waveform are

$$16{:}28 \quad E(t) = \begin{cases} E_0 & t \le 0 \\ E_0 + \frac{N-1}{2}\Delta E_{\text{step}} + \Delta E_{\text{pulse}} & (N-1)\Delta t < t \le N\Delta t \quad N = 1,3,5,\cdots \\ E_0 + \frac{N-2}{2}\Delta E_{\text{step}} - \Delta E_{\text{pulse}} & (N-1)\Delta t < t \le N\Delta t \quad N = 2,4,6,\cdots \end{cases}$$

The theory is quite elaborate, even when the reaction is reversible, because – unlike the pulse voltammetries discussed previously in this section – there is now no prolonged waiting time during which the solution adjacent to the electrode is renewed; instead, the current during any one pulse contains contributions arising from all previous pulses.

[1626] Geoffrey Cecil Barker, 1915–2000, English electrochemist who introduced electronics into polarography.

[1627] Robert Allen Osteryoung, 1927–2004, and his wife Janet G. Osteryoung, U.S. electroanalytical chemists.

[1628] For a review and references, see M. Lovrić in: F. Scholz (Ed.), *Electrochemical Methods*, 2nd edn, Springer, 2010, pages 121–145.

A further complication, though one that confers additional power to square-wave voltammetry, is that there are three alternative readouts. In the "forward mode", the current is sampled at times Δt, $3\Delta t$, $5\Delta t$, ..., that is, at the points indicated by a "⊕" in Figure 16-10. The measured current will be denoted $I_{f,N}$, where N is odd. Sampling for the "backward mode" current $I_{b,N}$ (usually, but not always, negative) occurs at $N\Delta t$ with N even; that is, at the points marked "⊖". In the third mode, sometimes called the "net mode", the current readout is the difference $\Delta I_N = I_{f,N} - I_{b,N-1}$. The formula[1629]

$$16{:}29 \qquad \left. \begin{matrix} I_{f,N} = \\ I_{b,N} = \end{matrix} \right\} \frac{FAc_R^b}{2}\sqrt{\frac{D_R}{\pi\Delta t}}\left[\frac{1}{\sqrt{N}}+\Upsilon_N+\sum_{n=1}^{N-1}\left(\frac{1}{\sqrt{N-n+1}}-\frac{1}{\sqrt{N-n}}\right)\Upsilon_n\right] \quad \left\{ \begin{matrix} N=1,3,5,\cdots \\ N=2,4,6,\cdots \end{matrix} \right.$$

applies irrespective of the parity of N, but the values of the parameter Υ do depend on whether N (or n) is odd or even. The formula

$$16{:}30 \qquad \Upsilon_N = \tanh\left\{\frac{E_0+\frac{2N-3}{4}\Delta E_{step}+\left[\frac{2N-1}{4}-\text{Int}\left\{\frac{N}{2}\right\}\right]\left[4\Delta E_{pulse}+\Delta E_{step}\right]-E^h}{2RT/F}\right\}$$

caters to both possibilities. Examples of the voltammograms generated by a reversible reaction are shown in Figure 16-11.

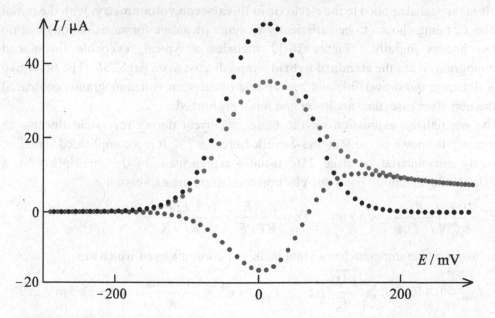

Figure 16-11 The forward, backward, and net square-wave voltammograms for a reversible electrode reaction. Data used: $T = T°$, $A = 1.0\times10^{-5}$ m^2, $c_R^b = 1.0$ mM, $D_R = 1.0\times10^{-9}$ m^2 s^{-1}, $E^h = 0$ mV, $E(0) = -300$ mV, $\Delta E_{pulse} = 50$ mV, $\Delta E_{step} = 10$ mV, and $\Delta t = 10$ ms.

[1629] The derivation is reported in Web#1629.

Square-wave voltammetry is a powerful method, both for electrochemical analysis and for the investigation of electrode reaction mechanisms. The published theory of the method goes far beyond the simple nernstian case addressed here to include quasireversible reactions, spherical diffusion, adsorbed reactant, and so on. Theory/experiment comparisons have been employed in all such milieu to investigate mechanisms and measure parameters. In a related technique bearing the name "**cyclic square-wave voltammetry**", the pulse heights at first increase, then diminish in size. Square-wave voltammetry is a common "modulation" employed in the measurement stage of stripping analysis, as described on pages 177–180. At the present time, square-wave voltammetry is one of the more popular voltammetric techniques, perhaps second only to the technique to which our attention now turns.

Ramped Potentials: linear-scan voltammetry and cyclic voltammetry

The words "scan", "sweep" and "ramp", with or without the adjective "linear", are all used to describe a potential that increases (or decreases) steadily with time:

16:31 $E(t) = E_0 \pm vt$ linear scan for $^{\text{oxidation}}_{\text{reduction}}$

and this is the signal applied to the electrode in **linear-scan voltammetry**, with the initial potential E_0 being chosen to be sufficiently extreme (negative for an oxidation) that no reaction occurs initially. Figure 16-12 includes a typical reversible linear-scan voltammogram; it has the **standard hybrid shape** discussed on page 256. The other two curves illustrate quasireversible and irreversible linear-scan voltammograms; compared with the nernstian case, they are lower and more prolonged.

The normalized expression for the faradaic current during reversible linear-scan voltammetry is known as the **Randles-Ševčik function**[1630]. It is a complicated function, previously encountered on page 256, usually represented, by $\sqrt{\pi}$ multiplied by a dimensionless χ function. For a one-electron oxidation, the expression is:

16:32 $\dfrac{I(t)}{Ac_R^b}\sqrt{\dfrac{RT}{F^3 D_R v}} = \sqrt{\pi}\,\chi\{\varepsilon\}$ $\varepsilon = \dfrac{E - E^h}{RT/F} = \dfrac{vt + E_0 - E^h}{RT/F}$ Randles-Ševčik function

The nernstian voltammogram has a blunt peak, the coordinates of which are

16:33 $I_{\text{peak}} = 0.44629\,Ac_R^b\sqrt{\dfrac{F^3 D_R v}{RT}}\,;$ $E_{\text{peak}} = E^h + \dfrac{1.1090RT}{F} = E^h + 28.5\,\text{mV at } T^\circ$

The equation in 16:33 that identifies the current peak is known as the **Randles-Ševčik**

[1630] Two electrochemists – the Englishman John Edward Brough Randles, 1912–1998, and the Czech Augustin Ševčik, 1926–2006 – independently calculated approximate values of the function and established its maximal value. One analytical formulation of the complete Randles-Ševčik formula is given as equation 12:50; others will be found at Web#1241.

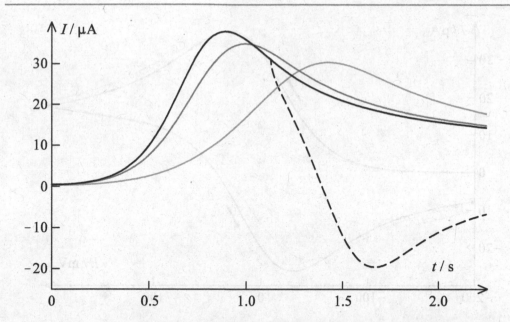

Figure 16-12 The upper curve shows a linear-scan voltammogram for a **reversible** electrode reaction. The lower curves are examples of corresponding linear-scan voltammograms for reactions that behave quasireversibly and irreversibly. The dashed line shows the effect on one of the chronoamperograms of reversing the direction of scan at a particular instant, as practiced in cyclic voltammetry. Data used: $T = T^\circ$, $A = 1.0 \times 10^{-5}$ m^2, $c_R^b = 1.0$ mM, $D_O = D_R = 1.0 \times 10^{-9}$ m^2 s^{-1}, $\alpha = 0.5$, $E^t = E_0 + 150$ mV, $v = 200$ mV s^{-1}, $t_{rev} = 1.1$ s, and $k^{o\prime} = \mathbf{10^{-2}}$, 10^{-4}, 10^{-5} m s^{-1}.

equation. Some other characteristics of the reversible hybrid will be found on page 256. As noted on that page, the semiintegration of a reversible hybrid yields a standard reversible wave. The simplicity of the formula for the semiintegrated reversible linear-scan voltammogram

16:34
$$\frac{M(t)}{FAc_R^b \sqrt{D_R}} = \frac{1}{2} + \frac{1}{2}\tanh\left\{\frac{\varepsilon}{2}\right\} = \frac{1}{1 + \exp\{-\varepsilon\}}$$

semiintegrated reversible linear-scan voltammogram

contrasts with the complexity of equation (16:32+12:50). Notice that whereas the current in linear-scan voltammetry is proportional to the scan rate v, its semiintegrated counterpart is independent of this parameter.

If the direction of the scan is reversed, at some time t_{rev} after the voltammogram's peak, then the technique is named **cyclic voltammetry** and the resulting chronoamperometric plot is as shown by the dashed line in the latter half of Figure 16-12. However, a two-branched graph[1631] of current, not versus time but electrode potential, as in Figure 16-13,

[1631] A third branch can sometimes provide useful information.

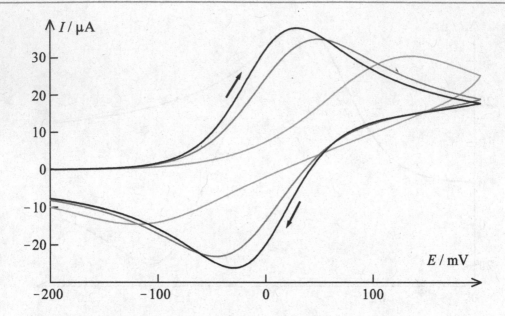

Figure 16-13 Examples of **reversible**, quasireversible and irreversible cyclic voltammograms. Data used: $T = T°$, $A = 1.0 \times 10^{-5}$ m^2, $c_R^b = 1.0$ mM, $D_O = D_R = 1.0 \times 10^{-9}$ m^2 s^{-1}, $\alpha = 0.5$, $E^h = E_0 + 200$ mV $= E_{rev} - 200$ mV, $v = 200$ mV s^{-1}, and $k^{o\prime} = \mathbf{10^{-2}}$, 10^{-4}, 10^{-5} m s^{-1}.

is customarily used to display the results of a cyclovoltammetric experiment. The instant $t = t_{rev}$ is known as the **reversal time** and the potential, E_{rev}, at which it occurs is the **reversal potential** or the **switching potential**. The "cyclic" descriptor in the name "cyclic voltammetry" reflects the fact that the *potential* returns to its initial potential E_0; the current does *not* cycle. If very many triangular potential cycles are imposed, the current does indeed eventually settle into a repetitive pattern and occasionally cyclic voltammetry is carried out in this fashion. The procedure, **ultimate cyclic voltammetry**, yields voltammograms similar to the example shown on page 274.

Whereas the pulse voltammetries discussed earlier are excellent tools for low-level chemical analysis, this is not the case for linear-scan or cyclic voltammetries, which are seldom used analytically. The reason is that capacitive interference is severe unless the concentration of the analyte is quite high. Cyclic voltammetry is, however, the preeminent method used to elucidate the mechanisms of electrode reactions. Capacitive interference is still present in these mechanistic studies, but the reactant concentration can usually be made much larger than the levels of interest to analysts, overwhelming capacitive and other background currents.

The next two sections in this chapter address mechanistic issues; attention here being restricted to cyclic voltammograms derived from the simple one-step, one-electron oxidation or reduction reactions

16:35 $$R(soln) \rightleftarrows e^- + O(soln) \qquad \text{or} \qquad O(soln) + e^- \rightleftarrows R(soln)$$

taking place under the usual voltammetric conditions summarized on pages 232–233. During cyclic voltammetry[1632], the potential-versus-time program has the shape of an isosceles triangle described by the formula

16:36 $$E(t) = E_{rev} \mp \left| -vt \pm \left(E_{rev} - E_0 \right) \right|$$ potential program in oxidative reductive cyclic voltammetry

where the $||$ symbol denotes the absolute value of its content. Of course, during its first (or "forward") branch, cyclic voltammetry is identical to linear-scan voltammetry and the coordinates of the forward peak are those given, for reversible behavior, by equations 16:33. There are no corresponding formulas for the coordinates of the second (or "backwards") peak because these depend, as does the entire location of the second branch, on the potential chosen for reversal[1633]. This arbitrariness undermines attempts to model cyclic voltammetry mathematically[1634] and has led to almost total reliance on digital simulation for theory/experiment comparisons.

The forward peak in oxidative cyclic voltammetry arises from a competition between two effects. As the potential becomes more positive, so does the rate of the $R \rightarrow O$ reaction, leading to an increasing current. However, the vicinity of the electrode becomes steadily more depleted of R, serving to decrease the current. The second effect triumphs eventually, leading to the current decline after the peak. The depletion of R is matched by a progressive enrichment of O in the vicinity of the electrode and, after an interval following reversal, this product species starts to re-reduce (unless the $R \rightarrow O$ reaction is totally irreversible), slowly at first, but faster as the potential becomes ever more negative. Once more there is a competition – this time between the increasing reduction rate of O and its decreasing availability – leading to a peak in the negative backwards current. Much of the O produced escapes diffusionally, so the area enclosed by the entire cyclic voltammogram is always positive.

Electrode reactions will behave reversibly, quasireversibly or irreversibly in cyclic voltammetry, depending on the magnitude of the quantity

16:37 $$\lambda = k^{o\prime} \sqrt{\frac{RT}{FD_R v}}$$ reversibility index for cyclic voltammetry

Figure 16-13 shows an example of each variety. As an alternative to digital simulation, the algorithm in 16:6 is an excellent vehicle for predicting the shape of linear-scan

---

[1632] Read F.Marken, A. Neudeck, and A.M. Bond in: F. Scholz (Ed.), *Electroanalytical Methods*, 2nd edn, Springer, 2010, pages 57–106, for further details of cyclic voltammetry and related issues.

[1633] For example, the separation between the two peaks, for a process behaving reversibly at 25°C, may vary between 69 mV (if reversal occurs at the forward peak) and 57 mV (if reversal is indefinitely delayed).

[1634] Nevertheless, an exact analytical formula, described in Web#1634, exists to describe both branches of reversible cyclic voltammetry, though it is too complicated for casual use.

voltammograms for any degree of reversibility. Based on the semianalytical algorithm 16:6, we provide[1635] a spreadsheet specially dedicated to cyclic voltammetry. It will accurately predict any cyclic voltammogram for a one-electron, one-step mechanism. This spreadsheet algorithm was used to construct Figures 16-13 and 16-14.

Recognize that whether or not a reaction behaves reversibly is not solely dependent on the rate constant. As equation 16:37 makes clear, the **scan rate** is also a factor. It is the ability to adjust the scan rate so as to "tune" a reaction into reversibility – thereby simplifying the voltammogram – or into quasireversibilty – thereby permitting the elucidation of $k^{o\prime}$ – that is one of the major advantages of cyclic voltammetry. There are limitations to the adjustments possible, however. With increasing scan rate, capacitive interference becomes a greater problem because the capacitive current increases proportionally to v, whereas the faradaic current increases only as $\sqrt{v}$. At the other end of the scan rate spectrum, too small a scan rate will cause the experiment to last long enough for convection to occur. Moreover, note that the reversibility index responds to the *reciprocal square-root* of the scan rate, so that a quadrupling of v is needed to halve the index.

At one time, analyses of cyclic voltammograms were largely based on the locations of the forward and backward peaks, the rate constant being estimated from the peak separation. Alternatively, simulated voltammograms can be used to generate various diagnostics[1632] that are then used to analyze experimental cyclic voltammograms. Fortunately, these crude methods are being replaced by a procedure in which the entire cyclic voltammogram is matched against a sequence of digital simulations in which the parameters in doubt are varied, until the best match with experiment is found. A thorough analysis will involve running several experimental cyclic voltammograms, employing a variety of scan rates, reactant concentrations, and reversal potentials. A single set of $k^{o\prime}$, α, E^h, and D values should provide matches to *all* these voltammograms. In this modeling exercise, extraneous factors, such as nonfaradaic currents and uncompensated resistance, may be included in the model.

Semiintegration can be a valuable adjunct to cyclic voltammetry but, notwithstanding facilities being built into many commercial instruments, it is seldom employed. Figure 16-14 depicts the result of semiintegrating the voltammograms shown in Figure 16-13. The peaks have now disappeared because semiintegration precisely compensates for the depletion effects that cause the current to peak. Instead, a plateau, named the **limiting semiintegral**, of height

16:38 $M_{\lim} = FAc_R^b \sqrt{D_R}$ plateau of semiintegrated cyclic voltammogram

is developed, irrespective of the reversibility of the electrode reaction. Of course, the

---

[1635] The spreadsheet algorithm, which provides values of $M(t)$ as well as $I(t)$, simply requires you to insert parameter values and click a button. You may, or may not, apply corrections for uncompensated resistance and/or double-layer capacitance. See Web#1635.

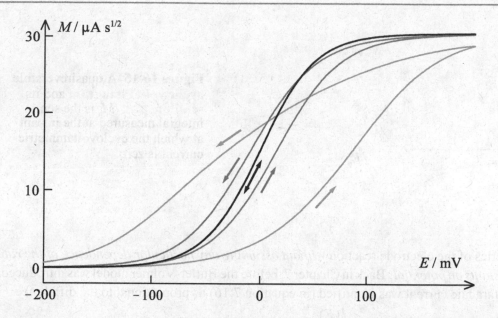

Figure 16-14 The result of semiintegrating the voltammograms from Figure 16-13.

plateau semiintegral is negative in reductive cyclic voltammetry, but hereabouts we are illustrating the oxidative variety. Notice that the backward branch of the semiintegrated cyclic voltammogram exactly retraces the forward branch, each being a **standard voltammetric wave**, when the electrode reaction behaves reversibly. When the reaction does not behave reversibly there is hysteresis, the forward and backward waves being displaced from each other by a voltage that depends on $k^{o\prime}$. In all cases the semiintegral eventually returns to zero; the semiintegrated voltammogram is truly cyclic! The similarity of these waves to the steady-state waves encountered in Chapter 12 will be evident.

One application of semiintegration is to the accurate determination of the nernstian half-wave potential from a cyclic voltammogram. Except for totally irreversible cases, all cyclic voltammograms display a null potential E_n where the current trace crosses the axis. At this instant, there is no kinetic (or ohmic) polarization and the Nernst equation holds, even though the behavior need not be reversible elsewhere. Hence the semiintegral at this instant can provide access to E^h through the formula[1636]

16:39 $$\frac{M_n}{M_{lim} - M_n} = \exp\left\{\frac{F(E_n - E^h)}{RT}\right\} \quad \text{whence} \quad E^h = E_n + \frac{RT}{F}\ln\left\{\frac{M_{lim}}{M_n} - 1\right\}$$

where M_n is identified in Figure 16-15 overleaf.

By employing the semiintegral and the current in concert, it is possible to study the

---

[1636] Derive this formula by setting the current to zero in the pan-voltammetric relation (Web#1547), or see Web#1636.

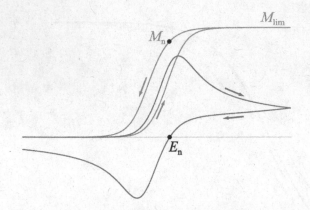

Figure 16-15 A quasireversible cyclic voltammogram and its semiintegral. M_n is the semiintegral measured at the instant at which the cyclovoltammetric current is zero.

kinetics of the electrode reaction *without assuming any particular dependence of the rate constants on potential*. Back in Chapter 7, before the Butler-Volmer model was introduced, the faradaic current was identified (in equation 7:16) as proportional to the difference

16:40
$$\frac{I(E)}{FA} = k'_{ox}(E)c_R^s(E) - k'_{rd}(E)c_O^s(E)$$

between the oxidative and reductive reaction rates. Note that there are five variables in this equation, all of which change as the potential is scanned in cyclic voltammetry. However the two rate constants are distinct in that their value depends *only* on the potential, whereas the two concentrations and the current also depend on the past history of the experiment. Thus, these latter three variables will have different values, on encountering any particular value of E during the forward scan, than they will later, on encountering the same potential during the backward scan. The notations

16:41
$$\frac{\vec{I}(E)}{FA} = k'_{ox}(E)\vec{c}_R^s(E) - k'_{rd}(E)\vec{c}_O^s(E) \qquad \text{when forward scan encounters potential } E$$

and

16:42
$$\frac{\overleftarrow{I}(E)}{FA} = k'_{ox}(E)\overleftarrow{c}_R^s(E) - k'_{rd}(E)\overleftarrow{c}_O^s(E) \qquad \text{when the backward scan encounters the same potential } E$$

use arrows to distinguish between the branches, as illustrated in Figure 16-16. Likewise, the Faraday-Fick expression 16:7 adopts different forms during the two scans

16:43
$$\vec{M}(E) = FA\sqrt{D}\left[c_R^b - \vec{c}_R^s(E)\right] = FA\sqrt{D}\,\vec{c}_O^s(E) \qquad \text{forward scan}$$

and

16:44
$$\overleftarrow{M}(E) = FA\sqrt{D}\left[c_R^b - \overleftarrow{c}_R^s(E)\right] = FA\sqrt{D}\,\overleftarrow{c}_O^s(E) \qquad \text{backward scan}$$

In the interest of brevity, we are here treating the two diffusivities as equal. The four expressions 16:41-44, as well as the formula

16:45
$$M_{lim} = FA\sqrt{D}\,c_R^b \qquad \text{limiting semiintegral}$$

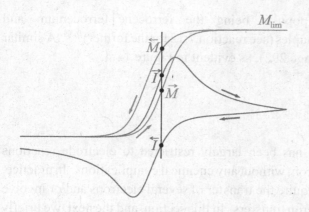

Figure 16-16 Illustration of the **global method** for measuring rate constants. Four quantities are measured, all relating to the same potential E which, in principle, may have any value.

may be solved[1637] simultaneously to produce the formulas

16:46 $\quad k'_{ox}(E) = \dfrac{\sqrt{D}\left[\bar{I}(E)\vec{M}(E) - \vec{I}(E)\bar{M}(E)\right]}{M_{lim}\left[\bar{M}(E) - \vec{M}(E)\right]}$ oxidative rate constant at potential E

and

16:47 $\quad k'_{rd}(E) = \dfrac{\sqrt{D}\left[\bar{I}(E)\left(M_{lim} - \vec{M}(E)\right) - \vec{I}(E)\left(M_{lim} - \bar{M}(E)\right)\right]}{M_{lim}\left[\bar{M}(E) - \vec{M}(E)\right]}$ reductive rate constant at E

Of course, these formulas may be applied at many potentials and, in this way, a single cyclic voltammogram can, in principle, provide values of the two rate constants over a range of potentials *without any presumption of the form of their dependence on potential*. This method of investigating electrochemical rate constants has been called the **global method**. Because formulas 16:46 and 16:47 incorporate many subtractions, they require high quality data if errors are to be avoided, A way of testing whether the data are, indeed, of sufficiently high quality to yield reliable rate constants is to check obedience to the Nernst law, by carrying out a "Nernst check"[1637].

Because of the widespread interest in unraveling the mechanisms of electrode reactions, as described in the next two sections, cyclic voltammetry has become the most popular voltammetric method in vogue today. Its popularity has reached such an extent that it is frequently used for electrochemical studies that would be better served by another technique. Mostly, the conditions stipulated on pages 232–233 are adhered to during cyclic voltammetry. A wide variety of solvents has been employed in these mechanistic studies. Two problems frequently arise when unfamiliar liquids are used as solvents in voltammetry. One is the difficulty of finding salts that are inert and sufficiently soluble to serve as supporting electrolytes. A second is that traditional reference electrodes are not always available to provide a reference scale. In this circumstance, **internal references** are

---

[1637] Solve them, but do not make the equal diffusivity approximation (that is, do *not* assume $D_R = D_O$). See Web#1637. This Web also contains details of the Nernst check.

commonly employed, the most popular being the ferrocene | ferrocenium and cobaltocene | cobaltocenium cation couples (see reaction 7:12 for the former)[1638]. A similar stratagem is used in ITIES studies (page 292), as evident in Figure 14-4.

Multiple Electron Transfers: the EE scheme

Hitherto in this book, attention has been largely restricted to electrode reactions involving the transfer of a single electron, without any chemical complications. In practice, however, many electrode reactions require the transfer of several electrons and/or involve chemical reactions in addition to electron transfers. In this section, and the next, we briefly address the way in which cyclic voltammetry is affected by some of these mechanistic complications. Of course, other voltammetric techniques are similarly affected, but here we limit discussion to cyclic voltammetry, because this is the technique that is favored by electrochemists in seeking to identify and quantify the mechanism that is operative in each case of interest.

We shall firstly discuss the sequential transfer of two electrons; the extension to three or more is straightforward. The mechanism is summarized by the scheme

$$S^b$$

16:48 $\Updownarrow$ $\Updownarrow$ $\Updownarrow$ EE scheme

$$S \underset{\longleftarrow}{\longrightarrow} I \underset{\longleftarrow}{\longrightarrow} P$$

in which an intermediate species I, produced by electron transfer to or from the substrate species S, then undergoes a second electron transfer to form the final product P. The scheme is designated "**EE**" because it involves *two* consecutive **E**lectrochemical steps. The red arrows in scheme 16:48 are the two electron transfer steps. The electrons are not shown because, in the interest of generality, we do not specify whether the transfers are oxidations or reductions. The **blue double-headed arrows** represent diffusion to and from the electrode. S^b indicates the substrate species S present in the bulk (from which I and P are absent). In what follows, we address either oxidative or reductive cyclic voltammetry[1639] indiscriminately, with the upper of any alternative signs signifying the former, the lower signs the latter. Note that it is possible for the electrochemistry to be confounded by the occurrence of the chemical **proportionation** reactions $S + P \rightleftharpoons 2I$, but

---

[1638] A minute sample of ferrocene is dissolved in the cell solution and it generates its own small cyclic voltammogram, superimposed on the voltammogram of interest. Features of the latter curve can then be referenced to the mid-peak potential of the ferrocene | ferrocenium voltammogram and reported in the style "Potential of the forward peak = −0.564 V versus ferrocene".

[1639] Of course, both oxidation *and* reduction occur in most cyclovoltammetric experiments; by *oxidative* cyclic voltammetry we mean cases in which the forward potential scan is towards more positive potentials. Notice that the adjective "forward" is applied to the first scan in cyclic voltammetry, even when, as in the reductive variety, both the potential and current may be proceeding negatively.

we discount this possibility here.

In the EE mechanism, each of the three species diffuses independently and therefore each obeys a Faraday-Fick law. These may be written

16:49 $\quad c_S^b - c_S^s(t) = \dfrac{\pm M_1(t)}{FA\sqrt{D_S}}; \quad c_I^s(t) = \dfrac{\pm M_1(t) \mp M_2(t)}{FA\sqrt{D_I}}; \quad c_P^s(t) = \dfrac{\pm M_2(t)}{FA\sqrt{D_P}} \quad$ Faraday -Fick

where $M_1(t)$ and $M_2(t)$ are respectively the semiintegrals of the current generated by the first and second electron transfers. Each electron transfer has its own nernstian half-wave potential and, if both reactions are reversible, each obeys Nernst's law in the forms

16:50 $\quad \dfrac{c_I^s(t)}{c_S^s(t)} = \exp\left\{ \dfrac{E(t) - E_1^{o\prime}}{\pm RT/F} \right\} = \xi_1; \quad \dfrac{c_P(t)}{c_I^s(t)} = \exp\left\{ \dfrac{E(t) - E_2^{o\prime}}{\pm RT/F} \right\} = \xi_2 \quad$ Nernst

If one solves[1640] the five equations in 16:49 and 16:50 simultaneously, one finds

16:51 $\quad M(t) = M_1(t) + M_2(t) = \pm FAc_S^b \sqrt{D_S} \; \dfrac{\xi_1\sqrt{D_I} + 2\xi_1\xi_2\sqrt{D_P}}{\sqrt{D_S} + \xi_1\sqrt{D_I} + \xi_1\xi_2\sqrt{D_P}} \quad$ reversible EE scheme

where the ξ abbreviations are defined in 16:50. Figure 16-17 shows examples of the

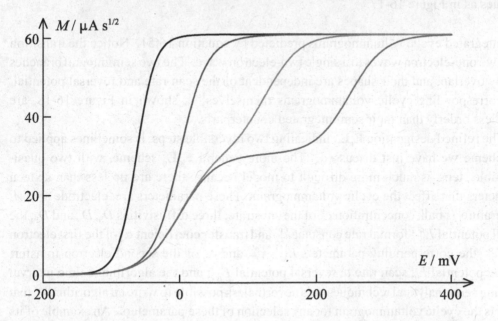

Figure 16-17 Semiintegrated oxidative cyclic voltammograms for an E_rE_r mechanism with four values of the interval separating the two formal potentials. Data used: $T = T°$, $A = 1.0 \times 10^{-5}$ m^2, $c_S^b = 1.0$ mM, $D_S = D_I = D_P = 1.0 \times 10^{-9}$ m^2 s^{-1}, $E_1^{o\prime} = 0$ mV, $E_2^{o\prime} = $ **−125**, **0**, **125**, and **250** mV, $v = 200$ mV s^{-1}.

[1640] Do so, or see Web#1640.

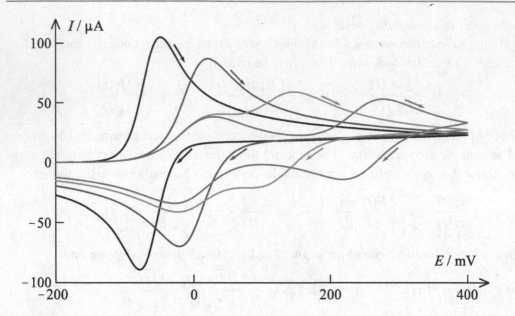

Figure 16-18 Oxidative cyclic voltammograms for an E_rE_r mechanism in which the separations between the two formal potentials, and other parameters, have the same values as in Figure 16-17.

semiintegrated cyclic voltammograms predicted by equation 16:51. Notice the transition from two one-electron waves to a single two-electron wave. The two semiintegral branches exactly overlap, and their shapes are independent of the scan rate and reversal potential. The corresponding cyclic voltammograms themselves[1641], shown in Figure 16-18, are much less orderly than their semiintegrated counterparts.

The refined designation E_rE_r, indicating two reversible steps, is sometimes applied to the scheme we have just discussed. The more general E_qE_q scheme, with two quasi-reversible steps, is much more difficult to model because there are no less than sixteen parameters that affect the cyclic voltammogram. These parameters are: electrode area A; temperature T; bulk concentration c_R^b of the substrate; three diffusivities D_S, D_I, and D_P; the formal potential $E_1^{o'}$; formal rate constant $k_1^{o'}$ and transfer coefficient α_1 of the first electron transfer; the corresponding parameters $E_2^{o'}$, $k_2^{o'}$, and α_2 of the second electron transfer; starting potential E_0; scan rate v; reversal potential E_{rev}; and the algorithmic time interval δ. Using semianalytical techniques, it is nevertheless possible to write an algorithm[1642] that predicts the cyclic voltammogram for any selection of these parameters. An example of its output is shown as Figure 16-19. Alternatively, of course, digital simulation packages may be used.

[1641] Constructed by semidifferentiating the data for Figure 16-17 using the algorithm in Web#1604. See Web#1641 for details.

[1642] See Web#1642 for the derivation and its implementation.

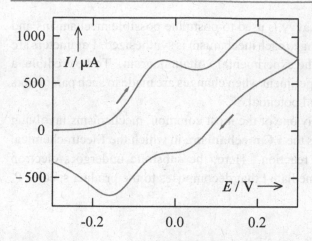

Figure 16-19 An example of an oxidative cyclic voltammogram generated by an E_qE_q mechanism. Parameters: $T = T^o$, $A = 1 \times 10^{-4}$ m^2, $c_S^b = 1$ mM, $D_S = D_I = D_P = 1 \times 10^{-9}$ m^2 s^{-1}, $E_0 = -0.3$ V, $E_{rev} = 0.3$ V, $v = 1$ V s^{-1}, $E_1^{o\prime} = -0.1$ V, $k_1^{o\prime} = 1 \times 10^{-4}$ m s^{-1}, $E_2^{o\prime} = 0.1$ V, $k_2^{o\prime} = 5 \times 10^{-5}$ m s^{-1}, $\alpha_1 = \alpha_2 = 0.5$.

If the second reaction in a consecutive electron-transfer pair occurs much more readily than the first, as in the **black** curves of Figures 16-17 and 16-18, then the effect is as if the two electrons were transferred simultaneously. All voltammetric theory then resembles the one-electron, one-step case except that the ubiquitous FA term becomes replaced by $2FA$. Similarly, the nerstian F/RT term becomes replaced by $RT/2F$. The current doubles while the potential scale becomes contracted.

Chemistry Combined with Electrochemistry: a plethora of mechanistic possibilities

Frequently, one or more homogeneous reactions play a role in cyclic voltammetry. These reactions take place in the electrolyte solution near to, rather than at, the electrode surface. Recall from Chapter 2 that the first-order chemical reaction A $\rightleftarrows$ B has a rate given by $\vec{k}'c_A - \vec{k}'c_B$, where the unit of the rate constants is s^{-1}, not the m s^{-1} that applies to electrochemical rate constants.

The chemical reactions that participate in electrochemical mechanisms occur to dissolved species in a space in which those species are also diffusing, so it becomes necessary to examine the behavior of solutes when both factors affect the concentration. The synthesizing of theoretical cyclic voltammograms, in this milieu, can be accomplished in two ways: by digital simulation, or semianalytically. Either method is satisfactory, although digital simulation[1609] is the standard tool, and the more versatile of the two. A semianalytical method is used here because the workings of this approach[1643] are in plain view, not concealed within the "black box" of a commercial product.

The strategy adopted in allocating a mechanism to a particular electrode reaction is similar to that illustrated in Figure 7-6 (page 139) with "CYCLIC VOLTAMMOGRAM"

[1643] See Web#1643 for the basic diffusion + interconversion theory underlying the method.

replacing "RATE LAW". Chemical savvy is used to postulate possible mechanisms and the cyclic voltammogram corresponding to each mechanism is synthesized. Parameters are adjusted to give the best match with the experimental voltammogram. To be credible, a mechanism must continue to match experiment when changes are made to such parameters as concentration, scan rate and reversal potential.

Perhaps the simplest, and certainly one of the most common, mechanisms involving both chemistry and electrochemistry is the **EC mechanism**, in which the **E**lectrochemical reaction is followed by a **C**hemical reaction. Here, the substrate undergoes electron transfer to produce an unstable intermediate I that decomposes to the product species P. With similarities to 16:48, the scheme

$$
\begin{array}{ccc}
 & S^b & \\
\Updownarrow & \Updownarrow \rightleftharpoons \Updownarrow & \text{EC scheme} \\
S \rightleftharpoons I & P &
\end{array}
$$

16:52

is operative. The back $P \rightarrow I$ step may, or may not, be important. The green arrows indicate the homogeneous chemical reaction occurring in concert with diffusion. These are unimolecular homogeneous reactions (pages 48–51) with rate constants $\vec{k}'$ and $\vec{k}'$.

When, as for I and P in scheme 16:52, two species are simultaneously diffusing and interconverting, it is not possible to obtain an explicit analytical solution describing the joint process unless the two species are assumed to share the same diffusivity. Species S in scheme 16:52 could be allocated a distinct diffusivity but, for uniformity, it will be assumed that all three species have the same diffusivity D. We shall often make this **equidiffusivity approximation** throughout the present section. This is not a serious retreat from exactitude because, in practice, values of the pertinent diffusivities are generally unknown and must be guessed.

We provide a spreadsheet algorithm[1644] to predict the form of cyclic voltammograms when an EC mechanism is involved. An example of its output is shown in Figure 16-20.

A variant to the EC mechanism is the **catalytic mechanism**, sometimes designated the **EC′ mechanism**. This scheme serves as an important synthetic tool in organic chemistry, where interest is in a homogeneous reaction

16:53 $A(soln) \rightleftharpoons Z(soln)$

that is thermodynamically favored, but kinetically impeded. A small quantity of an electroactive **mediator** S is present in the solution and it electrochemically generates a product P which reacts homogeneously with A to form Z and regenerate S:

$$
\begin{array}{ccc}
 & S^b & \\
Z + \Updownarrow \rightleftharpoons \Updownarrow + A & & \text{EC′ scheme} \\
S \rightleftharpoons P & &
\end{array}
$$

16:54

[1644] The algorithms that generated Figures 16-20 and 16-21 are derived and implemented in Web#1644.

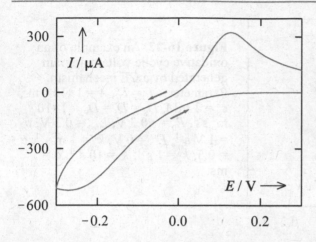

Figure 16-20 An example of a reductive cyclic voltammogram generated by an E_qC mechanism. Parameters: $T = T^\circ$, $A = 1\times10^{-4}$ m^2, $c_S^b = 1$ mM, $D_S = D_I = D_P = 1\times10^{-9}$ m^2 s^{-1}, $E_0 = 0.3$ V, $E_{rev} = -0.3$ V, $v = 1$ V s^{-1}, $E^{\circ\prime} = 0$ V, $k^{\circ\prime} = 1\times10^{-5}$ m s^{-1}, $\alpha = 0.3$, $\vec{k}' = 1$ s^{-1}, $\overleftarrow{k}' = 10$ s^{-1}, $\delta = 1$ ms.

The large excess of A serves to make the homogeneous $A + P \xrightarrow{\vec{k}} S + Z$ reaction pseudo-first order in P, with a rate constant $\vec{k}$ that incorporates the concentration of A. In many experimental manifestations of this mechanism, the rate of the reverse homogeneous reaction is negligible but, in the interest of generality, it may be assumed that the reverse reaction occurs with a rate constant $\overleftarrow{k}$ also pseudo-first order. Figure 16-21 shows an example of a catalytic cyclic voltammogram[1644]. This mechanism is often exploited by biosensors, such as the glucose sensor described on pages 175–176. Of course, species P will serve equally well as a mediator, though the designation CE$'$ might then be more appropriate.

In the **CE mechanism**[1645], the substrate is electropassive but undergoes a homogeneous reaction to produce an electroactive isomer I. The scheme is

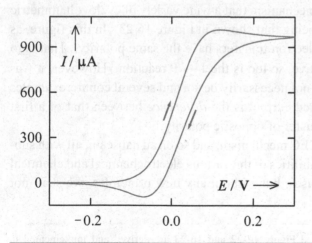

Figure 16-21 An example of the cyclic voltammogram for the catalytic scheme shown in 16:54. Parameters: $T = T^\circ$, $A = 1\times10^{-4}$ m^2, $c_S^b = 1$ mM, $c_A^b = c_Z^b = 1000$ mM, $D_S = D_P = 1\times10^{-9}$ m^2 s^{-1}, $E_0 = -0.3$ V, $E_{rev} = 0.3$ V, $v = 1$ V s^{-1}, $E^{\circ\prime} = 0$ V, $k^{\circ\prime} = 1\times10^{-4}$ m s^{-1}, $\alpha = 0.5$, $\vec{k}' = 10$ s^{-1}, $\overleftarrow{k}' = 0$ s^{-1}, $\delta = 1$ ms.

[1645] Sometimes, by analogy with the specialization of the E symbol into E_r, E_q, and E_i categories, one encounters C_r, C_q, and C_i referring to chemical reactions that are at equilibrium, bidirectional, or unidirectional. Our algorithms cater to all these categories.

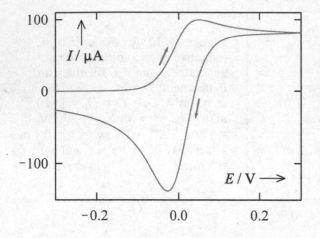

Figure 16-22 An example of an oxidative cyclic voltammogram generated by a CE mechanism. Parameters: $T = T°, A = 1\times10^{-4} \text{ m}^2$, $c_S^b = 1 \text{ mM}, D_S = D_I = D_P = 1\times10^{-9}$ $\text{m}^2 \text{ s}^{-1}, E_0 = -0.3 \text{ V}, E_{rev} = 0.3 \text{ V}, v$ $= 1 \text{ V s}^{-1}, E^{o\prime} = 0 \text{ V}, k^{o\prime} = 1 \text{ m s}^{-1}, \alpha$ $= 0.5, \underset{\rightarrow}{k^\prime} = 1 \text{ s}^{-1}, \underset{\leftarrow}{k^\prime} = 10 \text{ s}^{-1}, \delta = 1$ ms.

16:55
$$S^b$$
$$\Updownarrow \underset{\leftarrow}{\rightharpoonup} \Updownarrow \quad \Updownarrow \qquad \text{CE scheme}$$
$$S \qquad I \underset{\leftarrow}{\rightharpoonup} P$$

Figure 16-22 shows an example of a CE cyclic voltammogram[1646].

Two electron transfers occur in the **ECE mechanism** but the species I formed by the first electron transfer must isomerize to J before the second transfer can occur, the scheme being

16:56
$$S^b$$
$$\Updownarrow \quad \Updownarrow \underset{\leftarrow}{\rightharpoonup} \Updownarrow \quad \Updownarrow \qquad \text{ECE scheme}$$
$$S \underset{\leftarrow}{\rightharpoonup} I \qquad J \underset{\leftarrow}{\rightharpoonup} P$$

There are so many parameters in this mechanism that a wide variety of cyclovoltammetric shapes might result, one example[1647] being that shown in Figure 16-23. In this figure, as in most experimental cases, the two electron transfers have the same polarity. That is to say that if the S → I transfer is oxidative, so too is the J → P reaction. However, it was realized by Feldberg[1648] that this need not necessarily be so, and several counterexamples have been found in which the measured current is the *difference* between that of a first electron transfer and a subsequent transfer of opposite polarity.

There is a CEC mechanism, an ECEC mechanism, and so on ad nauseam, all with sub-classifications according to the reversibilities of the various electrochemical and chemical steps. Because the complexity increases, but without any new principles, we shall not

---

[1646] The spreadsheet algorithms that generated Figure 16-22 and 16-23 are derived and implemented in Web#1646.

[1647] The algorithm used to generate this voltammogram is derived and implemented in Web#1647.

[1648] Steven W. Feldberg, U.S. electroanalytical chemist, coinventor of Digisim®.

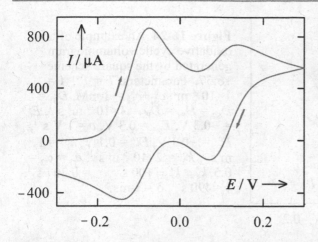

Figure 16-23 An example of an oxidative cyclic voltammogram generated by the mechanism diagrammed in scheme 16:56. Parameters: $T = T°$, $A = 1 \times 10^{-4}\,\text{m}^2$, $c_S^b = 1$ mM, $D_S = D_I = D_J = D_P = 1 \times 10^{-9}\,\text{m}^2\,\text{s}^{-1}$, $E_0 = -0.3$ V, $E_{rev} = 0.3$ V, $v = 1$ V s^{-1}, $E_1^{o'} = -0.1$ V, $E_2^{o} = 0.1$ V, $k_1^{o'} = k_2^{o'} = 1$ m s^{-1}, $\alpha_1 = \alpha_2 = 0.5$, $\vec{k} = \vec{k}' = 10$ s^{-1}, $\delta = 1$ ms.

address these mechanistic schemes. It would be remiss, however, to end this section without mention of the commonly encountered **square scheme**. Many compounds exist in solution as equilibrium mixtures. The most familiar example is provided by acids in aqueous solution, where **proton-transfer reactions**, such as

16:57 $$H_2O(\ell) + HB(aq) \rightleftarrows B^-(aq) + H_3O^+(aq)$$ equilibrium

are commonplace. If each of the two equilibrating forms is electroactive, current will arise from each electron transfer. It is likely that the two products will establish an equilibrium, too. A general scheme is

16:58
$$S_1^b \Leftrightarrow S_1 \rightleftarrows P_1 \Leftrightarrow$$
$$\uparrow\downarrow \qquad\qquad \uparrow\downarrow \qquad \text{square scheme}$$
$$S_2^b \Leftrightarrow S_2 \rightleftarrows P_2 \Leftrightarrow$$

The origin of the name "square scheme" is evident. The chemical equilibria may be fast or be subject to kinetic constraints, as may be the electrochemical reactions. Another complication that arises in some instances is a second-order chemical **cross-reaction**, such as

16:59 $$S_2 + P_1 \rightleftarrows S_1 + P_2$$ bimolecular cross-reaction

No such reactions are included in the semianalytical derivation[1649] that leads to the example illustrated in Figure 16-24 overleaf. The theory for the square scheme is particularly elaborate because there are two instances of concurrent diffusion + interconversion taking place. As many as twenty-one parameters may be needed.

[1649] See Web#1649 for details.

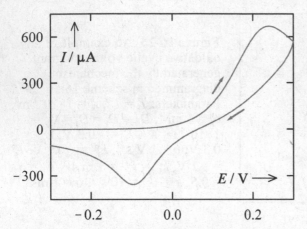

Figure 16-24 An example of an oxidative cyclic voltammogram generated by the square scheme 16:57. Parameters: $T = T°$, $A = 1 \times 10^{-4}$ m^2, $c_{S1}^b + c_{S2}^b = 1$ mM, $D_{S1} = D_{S2} = D_{P1} = D_{P2} = 1 \times 10^{-9}$ m^2 s^{-1}, $E_0 = -0.3$ V, $E_{rev} = 0.3$ V, $v = 1$ V s^{-1}, $E_1^{o\prime} = -0.1$ V, $E_2^{o\prime} = 0.1$ V, $k_1^{o\prime} = 1$ m s^{-1}, $k_2^{o\prime} = 2 \times 10^{-5}$ m s^{-1}, $\alpha_1 = \alpha_2 = 0.5$, $k_S' = k_P' = 100$ s^{-1}, $= 0.204$ s^{-1}, $k_P'=490$ s^{-1}, $\delta = 1$ ms.

Controlling Current Instead of Potential: chronopotentiometry

Voltammetry is concerned with the relationship between two time-dependent properties: the potential $E(t)$ of the working electrode, and the current $I(t)$ that flows through that electrode. Hitherto in this chapter we have described only the most common procedure, in which a potential program is imposed on the electrode, and the resulting current (or sometimes its integral or semiintegral) is recorded and analyzed. There are alternatives, however, and one of these is to impose a constant current on the cell and observe the way in which the electrode potential changes. This experiment is called **chronopotentiometry**.

Chronopotentiometry is rarely practiced nowadays, but it does have some advantages and finds application in such challenging systems as high-temperature fused-salt electrochemistry. In such conditions, uncompensated resistance is rarely a problem, though resistance associated with films on the electrode may be. The effect of such a resistance is simply to shift the measured potential by an ohmic constant proportional to the resistance and the current.

The response of the reversible electrode reaction $R \rightleftarrows e^- + O$ to a constant current can easily be found by combining the reversible pan-voltammetric relation with the formula for the semiintegral of a constant[1650], thus:

16:60
$$\frac{FAc_R^b}{\dfrac{1}{\sqrt{D_R}} + \dfrac{1}{\xi(t)\sqrt{D_O}}} = M(t) = 2I\sqrt{\frac{t}{\pi}} \qquad \xi(t) = \exp\left\{\frac{E(t) - E^{o\prime}}{RT/F}\right\}$$

[1650] See equation 16:63 for the pan-voltammetric relations and item (103) of Web#1242 for the appropriate semiderivative.

It follows that[1651]

$$16:61 \qquad E(t) = E^{h} - \frac{RT}{F} \ln\left\{ \frac{FAc_{R}^{b}}{2I} \sqrt{\frac{\pi D_{R}}{t}} - 1 \right\} \qquad \text{Karaoglanoff equation}$$

Figure 16-25 illustrates the chronopotentiometric response of an electrode behaving reversibly[1652]. The Karaoglanoff[1653] equation predicts that the potential will become infinite at a time[1654] known as the **transition time**, t_{trans}

$$16:62 \qquad t_{\text{trans}} = \frac{\pi D_{R}}{4} \left(\frac{FAc_{R}^{b}}{I} \right)^{2} \qquad \text{Sand[1655] equation}$$

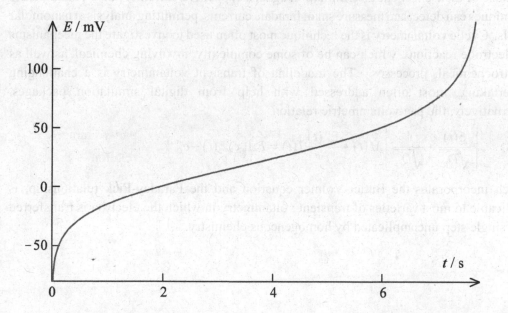

Figure 16-25 The faradaic response of an electrode (behaving reversibly) to a constant anodic current, suddenly imposed at $t = 0$. Data: $T = T^{\circ}$, $A = 1.0 \times 10^{-5}$ m^{2}, $c_{R}^{b} = 1$ mM, $c_{O}^{b} = 0$, $D_{R} = D_{O} = 1 \times 10^{-9}$ m^{2} s^{-1}, $E^{\circ\prime} = 0$, $I = 9.7$ μA.

[1651] Derive Karaoglanoff's equation and construct an expression for the time required for the potential to reach the nernstian half-wave potential. How does this time interval compare with the transition time? See Web#1651.

[1652] If, however, the reaction behaves quasireversibly, it is impossible to write an explicit formula for $E(t)$. That is, no function f can be found, such that $E = f(t)$. On the other hand, it is possible to find a function f such that $t = f(E)$, and this is almost as useful. Use the pan-voltammetric relation to find this function, or see Web#1652. What is the transition time?

[1653] Z. Karaoglanoff, 1878 – 1983, Bulgarian analytical chemist.

[1654] Use the Sand equation to predict the transition time for the data used for Figure 16-25. See Web#1654.

[1655] Henry J.S. Sand, British chemist, author of one of the first textbooks on electroanalytical chemistry.

The double-layer capacitance is one reason why, in fact, the potential does not acquire an infinite value. Even if the electrode is totally polarized faradaicly, the parallel capacitance C will cause the potential to drift at a rate of I/C.

Summary

The major applications of transient voltammetry are in chemical analysis and in the elucidation of reaction mechanisms. Almost always, the voltammetry is conducted by imposing a potential program on an electrode in the presence of a substrate of interest. Because of their excellent discrimination against capacitive currents, pulse voltammetric techniques can detect and measure small faradaic currents, permitting analysis at nanomolar levels. Cyclic voltammetry is the technique most often used to investigate the mechanisms of electrode reactions, which can be of some complexity, involving chemical, as well as electrochemical, processes. The modeling of transient voltammetry is a challenging undertaking, most often addressed with help from digital simulation packages. Alternatively, the pan-voltammetric relation

16:63
$$\left[\frac{\xi(t)}{\sqrt{D_R}} + \frac{1}{\sqrt{D_O}} \right] M(t) + \frac{\xi^\alpha(t)}{k^{o\prime}} I(t) = FA \left[c_R^b \xi(t) - c_O^b \right] \qquad \text{pan-voltammetric relation}$$

which incorporates the Butler-Volmer equation and the Faraday-Fick relationship, is applicable to most varieties of transient voltammetry in which the electron is transferred in a single step uncomplicated by homogeneous chemistry.

Appendix

Glossary: symbols, abbreviations, constants, definitions, and units

References are to the first instance of the symbol's use. When the number in the fourth column is prefixed by #, it refers to a footnote and/or a Web; otherwise the reference is to a page. Where alternative units are cited, those in green relate to homogeneous reactions, whereas those in red relate to heterogeneous reactions. Symbols shown in bold type (such as **t** or **x**) are undimensioned counterparts of the standard symbols (such as t or x). They are defined locally, not in this glossary.

Symbol	Name	Unit, definition or value	See
A	ampere, *SI* unit of electric current	base *SI* unit	14
A,B,...	reactants		46
A	area (often of WE)	m^2	14
A_a	area of anode	m^2	217
A_c	area of cathode	m^2	219
a.c.	alternating current		24
^{an} (suffix)	anodic		204
arccos$\{y\}$	inverse cosine function of y	arccos$\{\cos\{y\}\} = y$	#1225
arccot$\{y\}$	inverse cotangent function of y	arccot$\{\cot\{y\}\} = y$	26
arcoth$\{y\}$	inverse hyperbolic cotangent function of y	$\frac{1}{2}\ln\{(y+1)/(y-1)\}$	#1317

Symbol	Name	Unit, definition or value	See
$\arcsin\{y\}$	inverse sine function of y	$\arcsin\{\sin\{y\}\} = y$	#1612
$\arctan\{y\}$	inverse tangent function of y	$\arctan\{\tan\{y\}\} = y$	26
$\operatorname{arsinh}\{y\}$	inverse hyperbolic sine function of y	$\ln\left\{y + \sqrt{1+y^2}\right\}$	201
^{av} (suffix)	average value		319
a_i	activity of species i	(no unit)	33
(ads)	adsorbed on a surface		#205
($amal$)	dissolved in mercury		#205
(aq)	in aqueous solution		#205
B	adsorption coefficient	$m^3\ mol^{-1}$	268
b	characteristic volume	m^3	164
b	Tafel slope	V	220
C	coulomb, SI unit of electric charge	$=$ A s	1
C	a chemical step		143
C_r, C_q, C_i	reversible, quasireversible, irreversible chemical step		#1645
CE	counter electrode		209
C	capacitance	F	12
^{cath} (suffix)	cathodic		204
^{cell} (suffix)	of the cell		197
^{conv} (suffix)	due to convection		161
$\cos\{\theta\}$	cosine function of θ	$\sin\left\{\frac{\pi}{2} + \theta\right\}$	6
$\cosh\{y\}$	hyperbolic cosine function of y	$\frac{1}{2}\exp\{y\} + \frac{1}{2}\exp\{-y\}$	263

Symbol	Name	Unit, definition or value	See
$\coth\{y\}$	hyperbolic cotangent function of y	$\dfrac{\exp\{y\}+\exp\{-y\}}{\exp\{y\}-\exp\{-y\}}$	#1317
$\csc\{\theta\}$	cosecant function of θ	$1/\sin\{\theta\}$	#1225
$\operatorname{csch}\{y\}$	hyperbolic cosecant function of y	$1/\sinh\{y\}$	#1317
c_A, c_C	concentration of the anion, cation	$\mathrm{mol\ m^{-3}} = \mathrm{mM}$	263
c	charge-carrier concentration	$\mathrm{mol\ m^{-3}} = \mathrm{mM}$	19
c_i	concentration of solute i	$\mathrm{mol\ m^{-3}} = \mathrm{mM}$	20
c_i^s	concentration of i at (electrode) surface	$\mathrm{mol\ m^{-3}} = \mathrm{mM}$	51
$\lvert c_i^s \rvert$	amplitude of the periodically varying concentration of i at electrode surface	$\mathrm{mol\ m^{-3}} = \mathrm{mM}$	306
c_i^b	concentration of i in the bulk	$\mathrm{mol\ m^{-3}} = \mathrm{mM}$	156
c°	standard concentration	$1.0000\times10^3\ \mathrm{mol\ m^{-3}}$	34
D	debye, a non-SI unit of dipole moment	$3.3356\ \mathrm{C\ m}$	#102
D_i	diffusivity (diffusion coefficient) of i	$\mathrm{m^2\ s^{-1}}$	154
d (prefix)	infinitesimal change in		6
$d^{1/2}/dt^{1/2}$	semidifferentiation operator	$\mathrm{s^{-1/2}}$	#1242
$d^{-1/2}/dt^{-1/2}$	semiintegration operator	$\mathrm{s^{1/2}}$	#1242
d.c.	direct current		23
dif (suffix)	due to diffusion		159
dmFc	decamethylferrocene	$\mathrm{Fe[C_5(CH_3)_5]_2}$	294
d	superficial diameter	m	234
∂ (prefix)	partial differential operator		162
E	an electrochemical step		143
E_r, E_q, E_i	reversible, quasireversible, irreversible electrochemical step		356

Symbol	Name	Unit, definition or value	See
E	potential, electrode potential	V	12
$\lvert E \rvert$	voltage amplitude	V	24
E_{app}	applied potential	V	#1635
E_{cor}	corrosion potential	V	217
E^{h}	nernstian half-wave potential	V	240
E_{int}	interfacial potential	V	#1635
E_{n}	null potential, open-circuit potential	V	109
E_{rev}	reversal (switching) potential	V	348
E_{zc}	potential of zero charge	V	261
$E_{1/2}$	half-wave potential	V	#1216
E°	standard electrode potential	V	110
$E^{\circ\prime}$	formal (conditional) electrode potential	V	113
e^{-}	an electron		1
$\mathrm{erf}\{y\}$	error function of y	$\frac{2}{\sqrt{\pi}}\int_{0}^{y}\exp\{-x^2\}\,\mathrm{d}x$	156
$\mathrm{erfc}\{y\}$	error function complement of y	$1-\mathrm{erf}\{y\}$	#828
$\mathrm{eerfc}\{y\}$	"experfc" function of y	$=\exp\{y^2\}\mathrm{erfc}\{y\}$	#723
$_{equil}$ (suffix)	at equilibrium		37
eV	electron volt	1.6022×10^{-19} J	#124
$\exp\{y\}$	exponential function of y	$=\mathrm{e}^{y}$	22
e	base of natural logarithms	2.7183	#164
F	farad, SI unit of capacitance	$=\mathrm{C}\,\mathrm{V}^{-1}$	13
FAD	flavin adenosine dinucleotide		175
Fc, Fcd	ferrocene, a derivative of ferrocene	$\mathrm{Fc}=(\mathrm{C}_5\mathrm{H}_5)_2\mathrm{Fe}$	109
FFT	fast Fourier transform		#1545

Symbol	Name	Unit, definition or value	See
F_n	nth member of a time series	as appropriate	#1545
F	Faraday's constant	96485 C mol^{-1}	19
f (prefix)	femto	10^{-15}	
f	force	N	2
f	repetition frequency	Hz	#1545
(fus)	in the fused (molten) state		#205
Glox	glucose oxidase		175
$_{GC}$ (suffix)	according to the Gouy-Chapman model		263
G_i	(molar) Gibbs energy of species i	J mol^{-1}	31
G_i	Goldman permeability to ion i	m s^{-1}	189
G_i°	standard Gibbs energy of species i	J mol^{-1}	31
g	gram	$=10^{-3}$ kg	
gd$\{y\}$	gudermannian function of y	arctan$\{\sinh\{y\}\}$	#1225
g	gravitational acceleration	9.8066 m s^{-2}	#813
g	gain of an amplifier	(no unit)	#1029
(g)	in the gaseous state		#205
H	henry, *SI* unit of inductance	$=$ V s A^{-1}	313
HUP	hydrogen uranyl phosphate tetrahydrate		69
Hz	hertz, *SI* unit of frequency	$=$ s^{-1}	24
^{H} (suffix)	of the hydrogen evolution reaction		219
$_H$ (suffix)	according to the Helmholtz model		259
H	(molar) enthalpy	J mol^{-1}	30
h	hour	$=$ 3600 s	86
h$^+$	a hole		288

Symbol	Name	Unit, definition or value	See
h	thickness of a monomolecular layer	m	282
h	Planck's constant	6.6261×10^{-34} J s	290
I, J, ...	intermediates		258
I_n	nth "imaginary" transform	as appropriate	#1545
Int$\{y\}$	integer-value function of y	largest integer $\leq y$	#177
ISE	ion-selective electrode		119
ITIES	interface between two immiscible electrolyte solutions		291
IUPAC	International Union of Pure and Applied Chemistry		xi
$^{I, II}$ (suffix)	in phase I, in phase II		35
I	electric current	A	13
$\vec{I}, \overleftarrow{I}$	current during the forward, backward branch of cyclic voltammetry	A	352
$|I|$	amplitude of sinusoidal current	A	25
I_{cor}	corrosion current	A	220
I_{far}	faradaic current	A	#1635
I_{pre}, I_{end}	current before, at the end of, a pulse	A	342
I_{meas}	measured current	A	#1635
I_{mod}	modeled current	A	#1635
i	an arbitrary species		19
ierfc$\{y\}$	integral of the error function complement of y	$\int_y^\infty \text{erfc}\{x\}\mathrm{d}x$	#828
^{in} (suffix)	inside the (biological) cell		189
i	current density	A m^{-2}	14
i_{ox}, i_{rd}	current densities of opposing reactions	A m^{-2}	137

Symbol	Name	Unit, definition or value	See
i_n	exchange current density	$A\ m^{-2}$	134
J	joule, *SI* unit of energy and work	$= kg\ m^2\ s^{-2}$	5
j	imaginary operator	$\sqrt{-1}$	313
J	total flux	$mol\ s^{-1}$	#802
j_i	flux density of species i	$mol\ m^{-2}\ s^{-1}$	145
K	kelvin, *SI* unit of temperature	base *SI* unit	82
K	activity-based equilibrium constant	(no unit)	32
K'	concentration-based equilibrium constant	$(mol\ m^{-3})^{\bar{\Omega}-\bar{\Omega}}$	#818
K_w	activity-based ionic product for water	1.005×10^{-14}	39
kg	kilogram, *SI* unit of mass	base *SI* unit	#225
$_{kin}$ (suffix)	kinetic, kinetically controlled		200
k	frequency counter	(no unit)	#1545
k_B	Boltzmann's constant	$1.3807\times10^{21}\ J\ K^{-1}$	#124
$\vec{k}, \bar{k}$	activity-based rate constants	$mol\ m^{-3\ or\ -2}\ s^{-1}$	47
$\vec{k}', \bar{k}'$	forward, backward experimental rate constants	$m^{3\Omega-3\ or\ 3\Omega-2}\ mol^{\Omega-1}s^{-1}$	50
$k'_\#, k'_{-\#}$	forward, backward rate constants of mechanistic step #	$m^{3\Omega-3\ or\ 3\Omega-2}\ mol^{\Omega-1}\ s^{-1}$	50
$k_{ox}(E), k_{rd}(E)$	individual activity-based rate constants of oxidative, reductive electrode reaction	$mol\ m^{-2}\ s^{-1}$	131
$k'_{ox}(E), k'_{rd}(E)$	individual rate constants of electrode reaction moieties	$m\ s^{-1}$	131
$k^{o'}$	formal rate constant	$m\ s^{-1}$	132

Symbol	Name	Unit, definition or value	See
k'_{ox}, k'_{rd}	composite oxidative, reductive rate constants	$m^{3\Omega-2}\,mol^{\Omega-1}\,s^{-1}$	141
L	liter (litre)	$= 10^{-3}\,m^3$	38
$^{L,\,R}$ (suffix)	on the left, on the right		37
$^{L,\,U}$ (suffix)	lower, upper		153
L	length	m	4
L	inductance	H	313
$_{lim}$ (suffix)	limiting		204
$\log_{10}\{y\}$	decadic logarithm of y	$\ln\{y\}/2.3026$	39
$\ln\{y\}$	(natural) logarithm of y		32
ℓ	length coordinate	m	5
(ℓ)	in the liquid state		#205
M	molar	$mol_{solute}\,L^{-1}_{solution}$	38
M (prefix)	mega	10^6	
^{M} (suffix)	of the metal dissolution reaction		219
$_{M}$ (suffix)	of the metal		283
M	semiintegral of the current	$A\,s^{1/2}$	#162
$\vec{M}, \overset{\leftarrow}{M}$	current semiintegral during the forward, backward branch of cyclic voltammetry	$A\,s^{1/2}$	352
M_n	semiintegral at the null potential	$A\,s^{1/2}$	351
M_i	molar mass of species i	$kg\,mol^{-1}$	282
m	meter (metre), SI unit of length	base SI unit	22
m (prefix)	milli	10^{-3}	
$_{meas}$ (suffix)	measured		271

Symbol	Name	Unit, definition or value	See
^{mig} (suffix)	due to migration		159
mol	mole, *SI* unit of chemical amount	base *SI* unit	19
m	harmonic number	(no unit)	322
m_i	transport coefficient of species i	$m\ s^{-1}$	169
N	newton, *SI* unit of force	$= kg\ m\ s^{-2}$	2
NASA	National Aeronautical and Space Administration (of the United States)		100
NHE	normal hydrogen electrode		#605
$N\ell$	naphthyl radical	$C_{10}H_7$	142
N	collection efficiency	(no unit), $0 < N < 1$	252
N_A	Avogadro's constant	$6.0221 \times 10^{23}\ mol^{-1}$	19
N_i	number of channels for the ion i	(no unit)	#932
n (prefix)	nano	10^{-9}	
$_{net}$ (suffix)	of the net reaction	$y_{net} = y_{ox} - y_{rd}$	131
$_{nf}$ (suffix)	nonfaradaic		271
$_{nuc}$ (suffix)	of the nucleus or nucleation		283
n	number of electrons in electrode reaction	(no unit)	59
n	time counter	(no unit)	#1545
n_i	amount of species i	mol	46
$_n$ (suffix)	of the nth member		283
O	oxidized member of redox pair		130
$_{ohm}$ (suffix)	ohmic		197
$_{osm}$ (suffix)	electroosmotic		300
$_{ox,\ rd}$ (suffix)	of the oxidation, reduction		131

Symbol	Name	Unit, definition or value	See
° (suffix)	standard value		31
^{out} (suffix)	outside the (biological) cell		189
(*org*)	dissolved in an organic liquid		292
P	a product species		358
Pa	pascal, *SI* uit of pressure	$= N\,m^{-2}$	33
PolyLog{,}	polylogarithm (Mathematica® symbol)	(no unit)	#1241
P_i	permeability towards species i	$m^2\,s^{-1}$	80
P	period	s	24
p (prefix)	pico	10^{-12}	
per(*t*)	periodic function of time	(no unit)	322
ppb	parts per billion	per 10^9	#924
ppm	parts per million	per 10^6	171
p	pressure	Pa	33
$p°$	standard pressure	$10^5\,Pa = 1\,bar$	33
p_i	partial pressure of gas i	Pa	33
pH	basicity index	$-\log_{10}\left\{a_{H_3O^+}\right\}$	39
Q	electric charge, quantity of electricity	C	2
Q_i	charge on species i	C	19
Q_0	elementary charge	$1.6022 \times 10^{-19}\,C$	1
Q_{test}	test charge (vanishingly small magnitude)	C	3
q	(superficial) charge density	$C\,m^{-2}$	5
R	reduced member of redox pair		130
R	an organic group		76

Symbol	Name	Unit, definition or value	See
RE	reference electrode		105
R_n	nth "real" transform	as appropriate	#1545
R	distance, radius	m	2
R	resistance	Ω	15
R	gas constant	$8.3145 \, \mathrm{J\,K^{-1}\,mol^{-1}}$	32
R_c	radius of central ion in the Debye-Hückel model	m	44
R_{ct}	charge-transfer resistance	Ω	202
R_{pol}	polarization resistance	Ω	221
R_u	uncompensated resistance	Ω	210
$RT°$		$2.4790 \, \mathrm{kJ\,mol^{-1}}$	32
$RT°/F$		25.693 mV	
$\dfrac{RT°}{F}\ln\{10\}$		59.159 mV	
$^R, ^L$ (suffix)	on the right, on the left		37
rad	radian	primary *SI* unit	162
$_{rem}$ (suffix)	removal controlled		241
$_{rms}$ (suffix)	root-mean-square		24
rpm	revolutions per minute		#1229
r	distance along radial coordinate		4
r_{hemi}	radius of a hemispherical electrode	m	198
S	siemens, *SI* unit of conductance	$= \mathrm{A\,V^{-1}} = \Omega^{-1}$	15
S	a substrate species		354
SCE	saturated calomel electrode		#604
SHE	standard hydrogen electrode		106

Symbol	Name	Unit, definition or value	See
$_s$ (suffix)	according to the Stern model		264
S	(molar) entropy	$J\,K^{-1}\,mol^{-1}$	30
S	sheared area	m^2	#842
SI	le Système International d'unités		ix
s	second, SI unit of time	base SI unit	14
$\text{sech}\{y\}$	hyperbolic secant function of y	$2/[\exp\{y\}+\exp\{-y\}]$	254
$\sin\{y\}$	sine function of y	$y - y^3/3! + y^5/5! - ...$	24
$\sinh\{y\}$	hyperbolic sine function of y	$\frac{1}{2}\exp\{y\} - \frac{1}{2}\exp\{-y\}$	201
s	Laplace variable	Hz	#162
(s)	in the solid state		#205
(sc)	in the semiconductor		288
(sep)	in the separator		99
$(soln)$	in solution (not necessarily aqueous)		51
soc	state of charge	%	#547
T	temperature	K	30
$T°$	standard temperature	$298.15\,K = 25.00°C$	32
$\tan\{\theta\}$	tangent function of θ	$\sin\{\theta\}/\cos\{\theta\}$	#1225
$\tanh\{y\}$	hyperbolic tangent function of y	$\dfrac{\exp\{y\} - \exp\{-y\}}{\exp\{y\} + \exp\{-y\}}$	252
$_{trans}$ (suffix)	transport		203
t	time	s	14
t_{trans}	transition time	s	363
t_∞	time to reach completion	s	126
$U(t_0)$	Heaviside (or unit-step) function	0 if $t < t_0$; 1 if $t > t_0$	#1242

Symbol	Name	Unit, definition or value	See
u	mobility of charge carrier	$m^2\,V^{-1}\,s^{-1}$	19
u_i	mobility of species i	$m^2\,V^{-1}\,s^{-1}$	20
u_{osm}	electroosmotic mobility	$m^2\,V^{-1}\,s^{-1}$	#1419
V	volt, *SI* unit of electrical potential	$J\,C^{-1}$	6
Vm	valinomycin	$(C_{18}H_{30}N_2O_6)_3$	295
V	volume	m^3	47
$\dot{V}$	flow rate	$m^3\,s^{-1}$	126
v	velocity	$m\,s^{-1}$	#842
v	scan (or sweep or ramp) rate	$V\,s^{-1}$	271
v	rate of reaction, net reaction rate	$mol\,m^{-3\ or\ -2}\,s^{-1}$	47
$\vec{v}, \overleftarrow{v}$	forward, backward reaction rates	$mol\,m^{-3\ or\ -2}\,s^{-1}$	47
v_{ox}, v_{rd}	rates of the electrode reaction moieties	$mol\,m^{-2}\,s^{-1}$	131
$\overline{v}$	average velocity	$m\,s^{-1}$	19
v_x, v_r, v_θ	velocity in the axial, radial, angular directions	$m\,s^{-1}$	163
W	watt, *SI* unit of power	$J\,s$	#159
WE	working electrode		105
W	molar work	$J\,mol^{-1}$	32
W	magnitude of Warburg element	$\Omega\,s^{-1/2}$	309
w	work	J	5
$w_i^{x \to \infty}$	work to carry one i ion from x to infinity	J	263
w_n	nth weight	(no unit)	332
X	electric field strength	$V\,m^{-1}$	3
x	integration variable		#828

Symbol	Name	Unit, definition or value	See
x, y	distances along cartesian coordinates	m	2
x_H	thickness of the Helmholtz layer	m	259
x_i	mole fraction of solution component i	(no unit)	35
Y	admittance	S	#170
y	an arbitrary variable		
$\bar{y}(s)$	the Laplace transform of $y(t)$	s × (unit of y)	#162
Z, Y, ...	products		46
Z	impedance	Ω	25
z_i	charge number of species i	(no unit)	20
α (alpha)	proportionality constant	m^2	188
$\alpha, 1-\alpha$	reductive, oxidative transfer coefficient	(no unit), $0 < \alpha < 1$	133
α_{ox}, α_{rd}	composite transfer coefficients	(no unit)	138
β (beta)	Debye length	m	42
Γ_i (Gamma)	surface concentration of i	$mol\ m^{-2}$	268
$\Gamma\{y\}$	(complete) gamma function of y	$\int_0^\infty x^{y-1} \exp\{-x\}dx$	164
γ_i (gamma)	activity coefficient of solute i	(no unit)	34
$\gamma_\pm$	mean ionic activity coefficient	(no unit)	44
$\gamma\{\frac{1}{3}, y\}$	incomplete gamma function of y of one-third order	$\int_0^y x^{-2/3} \exp\{-x\}dx$	164
Δ (prefix) (Delta)	change in, difference	$\Delta y = y_{new} - y_{old}$	12
ΔE_n	null (or open-circuit) cell voltage	V	62
ΔE_{gap}	band gap	V	#124

Symbol	Name	Unit, definition or value	See
$\Delta\phi^{mem}$	transmembrane potential difference	V	188
ΔE_{step}, Δt_{step}	step height, step width	V, s	276
δ (delta)	Nernst diffusion layer thickness	m	#857
δ	small time interval	s	332
δ (prefix)	small change in		5
ε (epsilon)	permittivity	F m^{-1}	6
ε	abbreviation for $F(E-E^h)/RT$	(no unit)	256
ε_{CB}, ε_{VB}	energy of conduction, valence band	J	290
ε_H	permittivity of the Helmholtz layer	F m^{-1}	260
ε_0	electric constant, permittivity of space	8.8542×10^{-12} F m^{-1}	2
ζ (zeta)	zeta, or electroosmotic, potential	V	300
$\zeta(y)$	zeta number of y	$1^{-y}+2^{-y}+3^{-y}+4^{-y}+\cdots$	#1240
κ (kappa)	conductivity	S m^{-1}	15
Λ (Lambda)	abbreviation for $\lambda/(RT)$	(no unit)	#723
λ (lambda)	parameter in double-layer theory	(no unit)	#1317
λ	(molar) reorganization energy	J mol^{-1}	#723
λ	reversibility index	(no unit)	242
λ	wavelength	m	290
$\lambda\{y\}$	lambda function of y	$1^{-y}+3^{-y}+5^{-y}+7^{-y}+\cdots$	#1240
μ (mu)	ionic strength	mol m^{-3} = mM	41
μ	dipole moment	C m	#102
μ (prefix)	micro	10^{-6}	
μ_{DH}	Debye-Hückel ionic-strength constant	mol m^{-3} = mM	43
μ_i	chemical potential of species i	J mol^{-1}	#836

Symbol	Name	Unit, definition or value	See
$\tilde{\mu}_i$	electrochemical potential of species i	$J\,mol^{-1}$	#836
v (nu)	frequency of radiation	Hz	290
v	order of differintegration	(no unit)	#1245
v_i	stoichiometric coefficient of species i	(no unit)	45
v_C	Cochran number	0.6159	#844
v_K	von Kármán number	0.51023	163
v_L	Levich number	0.62046	163
ξ (xi)	abbreviation for $\exp\{F(E-E^{\circ\prime})/RT\}$	(no unit)	#1547
ξ^h	abbreviation for $\exp\{F(E-E^h)/RT\}$	(no unit)	#1603
ξ, ψ (xi, psi)	oblate spheroidal coordinates	(no units)	#1225
Ψ (Psi)	abbreviation defined in equation (17) of the cited Web	(no unit)	#1619
π (pi)	Archimedes' constant	3.1416	2
η (eta)	viscosity (dynamic viscosity coefficient)	$kg\,m^{-1}\,s^{-1}$	148
η	overvoltage	V	194
$\eta(y)$	eta function of y	$1^{-y}-2^{-y}+3^{-y}-4^{-y}+\cdots$	#1630
θ (theta)	angle	radian or degree	7
θ_i	fractional coverage by adsorbate i	(no unit), $0 < \theta_i < 1$	35
ρ (rho)	(volumetric) charge density	$C\,m^{-3}$	7
ρ	density	$kg\,m^{-3}$	#914
σ (sigma)	abbreviation for $\exp\{F(E_{1/2}-E^h)/RT\}$	(no unit)	#1231
σ	surface tension	$N\,m^{-1}$	261
Υ (Upsilon)	Onsager coefficient	$m^{7/2}\,s^{-1}\,V^{-1}\,mol^{-1/2}$	150
Υ	abbreviation for $\tanh\{F(E-E^h)/2RT\}$	(no unit), $-1 < \Upsilon < 1$	345

Symbol	Name	Unit, definition or value	See
ϕ (phi)	electrical potential	V	6
φ (phi)	phase angle	radian or degree	25
$\chi\{y\}$ (chi)	$\sqrt{\pi}\,\chi\{y\}$ is the Randles-Ševčik function of y	(no unit)	256
$\chi'\{y\}$	a function relevant to cyclic voltammetry	(no unit)	#1634
Ω (Omega)	ohm, *SI* unit of resistance	$= kg\ m^2\ s^{-3}\ A^{-2}$	15
Ω	total reaction order	(no unit)	139
Ω_i	order with respect to species i	(no unit)	139
$\Omega_{i,ox}$, $\Omega_{i,rd}$	order of species i in oxidative, reductive step	(no unit)	138
ω (omega)	angular frequency	$rad\ s^{-1}$	24
ω	angular velocity	$rad\ s^{-1}$	162
$^\circ$ (suffix)	standard		31
$^\circ$ (suffix)	at infinite dilution		#149
$'$ (suffix)	based on concentration not activity		51
$^{\circ\prime}$ (suffix)	formal, conditional		113
‡ (suffix)	of activation		#232
! (suffix)	factorial function	$n! = (1)(2)(3)\ldots(n)$	#1604
$\lvert y \rvert$	absolute value of y	y if $y \leq 0$; $-y$ if $y < 0$	349
$\lvert E \rvert$, $\lvert I \rvert$	amplitude of a periodic potential, current	V, A	24
$(\hat{n})$	nth step is rate determining		49
$\binom{y}{n}$	binomial coefficient	$\dfrac{(y{-}n{+}1)(y{-}n{+}2)\cdots(y)}{(1)(2)\cdots(n)}$	#1242

Absolute and Relative Permittivities: also some dipole moments

The second column lists permittivities, usually measured at 20°C or 25°C. The third column lists the corresponding relative permeability, often cited as the dielectric constant. The final column gives the dipole moment of the gaseous molecule, in the debye unit[102].

Material	ε / pF m^{-1}	ε / ε°	μ / D
free space	8.8542	1	
nitrogen, $N_2(g)$	8.8590	1.005	0
teflon®, $(CF_2)_\infty(s)$	18	2.0	
1,4 dioxane, $C_4H_8O_2(\ell)$	19	2.2	0.45
tetrachloromethane, $CCl_4(\ell)$	19.7	2.22	0
polyethene, $(CH_2)_\infty(s)$	20	2.3	
mylar, $(CH_2 \cdot OOC \cdot C_6H_4 \cdot COO \cdot CH_2)_\infty(s)$	28	3.2	
silica, $SiO_2(s)$	38.3	4.3	
a typical glass	44	5.0	
chlorobenzene, $C_6H_5Cl(\ell)$	49.8	5.62	1.69
neoprene, $(CH_2{:}CCl \cdot CH{:}CH_2)_\infty(s)$	58	6.6	
tetrahydrofuran, $C_4H_4O(\ell)$	65	7.6	1.75
dichloromethane, $CH_2Cl_2(\ell)$	76	8.9	1.6
1,2 dichloroethane $Cl(CH_2)_2Cl(\ell)$	91.7	10.4	
methanol, $CH_3OH(\ell)$	288	32.6	1.7
nitrobenzene, $C_6H_5NO_2(\ell)$	308.3	34.82	4.22
acetonitrile, $CH_3CN(\ell)$	332	37.5	3.93
dimethyl sulfoxide, $(CH_3)_2SO(\ell)$	400	47	3.96
water, $H_2O(\ell)$	695.4	78.54	1.86
formamide, $HCO \cdot NH_2(\ell)$	970	109.5	3.73

Material	$\varepsilon \, / \, \text{pF m}^{-1}$	$\varepsilon \, / \, \varepsilon^\circ$	$\mu \, / \, \text{D}$
titanium dioxide $TiO_2(s)$	1500	170	
N-methylformamide, $HCO \cdot NH \cdot CH_3$ (ℓ)	1615	182.4	3.8

Properties of Liquid Water: SI values at T° and p°

Molar mass, $M = 0.0180152 \text{ kg mol}^{-1}$

Density, $\rho = 997.07 \text{ kg m}^{-3}$

Molecular volume, $M/(N_A\rho) = 3.0003 \times 10^{-29} \text{ m}^3 = (310.73 \text{ pm})^3$

Thermal expansivity $= 2.572 \times 10^{-4} \text{ K}^{-1}$

Thermal conductivity $= 0.6069 \text{ W K}^{-1} \text{ m}^{-1}$

Heat capacity $= 75.48 \text{ J K}^{-1} \text{ mol}^{-1}$

Vapor pressure $= 3167.2 \text{ Pa}$

Surface tension, $\sigma = 0.07198 \text{ N m}^{-1}$

Viscosity, $\eta = 8.937 \times 10^{-4} \text{ kg m}^{-1} \text{ s}^{-1}$

Kinematic viscosity, $\eta/\rho = 8.932 \times 10^{-7} \text{ m}^2 \text{ s}^{-1}$

Self diffusion coefficient, $D = 2.44 \times 10^{-9} \text{ m}^2 \text{ s}^{-1}$

Permittivity, $\varepsilon = 6.954 \times 10^{-10} \text{ F m}^{-1}$

Relative permittivity, $\varepsilon/\varepsilon^\circ = 78.54$

Debye length, $\beta = 9.534 \times 10^{-7} \text{ m}$

Conductivity, $\kappa = 5.696 \times 10^{-6} \text{ S m}^{-1}$

Ionic strength, $\mu = 1.003 \times 10^{-4} \text{ mol m}^{-3}$

Ionic product, $K_w = 1.005 \times 10^{-14}$

$p\text{H} = 6.998$

Conductivities and Resistivities: assorted charge carriers

Mostly at 25°C or thereabouts. Both electronic and ionic conductors are included in this listing.

Material	κ / S m^{-1}	ρ / Ω m	Charge carrier(s)
vacuum and most gases	0	∞	none
teflon®, $(CF_2)_\infty$	10^{-15}	10^{15}	impurities?
typical glass	3×10^{-9}	3×10^{-8}	univalent cations
water, $H_2O(\ell)$	5.7×10^{-6}	1.75×10^5	$H_3O^+(aq)$ and $OH^-(aq)$
$[(CH_3)_2NH]_2CO_2(\ell)$, an ionic liquid	2.8×10^{-3}	360	$(CH_3)_2NH_2^+$ and $(CH_3)_2NCO_2^-$
silicon, $Si(s)$	0.072	14	electrons and holes
$Zr_{18}Y_2O_{39}(s)$, yttria-stabilized zirconia ceramic (at 1000 K)	0.3	3	$O^{2-}(s)$
100 mM aqueous KCl solution	1.3	0.77	$K^+(aq)$ and $Cl^-(aq)$
germanium, $Ge(s)$	2.2	0.45	electrons and holes
0.500 M aqueous $CuSO_4$ solution	4.2	0.24	$Cu^{2+}(aq)$ and $SO_4^{2-}(aq)$
seawater	5.2	0.19	cations, anions, and ion pairs
1.00 M aqueous KCl solution	10.2	9.5×10^{-2}	$K^+(aq)$ and $Cl^-(aq)$
$Na_6Al_{32}O_{51}$, β'' alumina (at 350°C)	20	5×10^{-2}	$Na^+(s)$
$RbAg_4I_5(s)$	25	4×10^{-2}	$Ag^+(s)$
1.000 M aqueous HCl solution	33.2	3.01×10^{-2}	$H_3O^+(aq)$ and $Cl^-(aq)$
5.2 M aqueous H_2SO_4 solution, "battery acid"	82	1.2×10^{-2}	$H_3O^+(aq)$ and $HSO_4^-(aq)$

Material	κ / S m^{-1}	ρ / Ω m	Charge carrier(s)
molten KCl (at 1043°C)	217	4.61×10^{-3}	K$^+$(*fus*) and Cl$^-$(*fus*)
doped polypyrrole	6×10^3	1.7×10^{-4}	pi electrons
graphite, C	4×10^4	2.5×10^{-5}	pi electrons
bismuth, Bi(s)	8.2×10^5	1.22×10^{-6}	electrons
mercury, Hg(ℓ)	1.040×10^6	9.62×10^{-7}	electrons
iron, Fe(s)	1.0×10^7	1.0×10^{-7}	electrons
copper, Cu(s)	5.69×10^7	1.758×10^{-8}	electrons
silver, Ag(s)	6.17×10^7	1.62×10^{-8}	electrons
superconductors (low temperature)	∞	0	electron pairs

Elements with Major Importance in Electrochemistry: properties

The atomic mass gives the mass of one mole of the atom *in grams*; this is not an *SI* unit; to obtain the *SI* unit, kg mol^{-1}, divide by 1000. Densities, at $p°$ and $T°$, are listed in the common unit of kg L^{-1} or g cm^{-3}; to obtain the *SI* unit, kg m^{-3}, multiply by 1000. All uncombined elements exist in oxidation state 0; this oxidation number has been omitted from the listing. An orange color in the fifth column indicates that this oxidation state is rare or unstable; unusual oxidation state are often discovered electrochemically.

Atomic number	Symbol & Name Nonmetals in red	Atomic mass	Standard state and density / kg L^{-1}	Oxidation numbers Rare values in orange
1	H, hydrogen	1.0079	$H_2(g)$ 0.00008132	−1, +1
3	Li, lithium	6.941	Li(s) 0.534	+1
6	C, carbon	12.011	C(*graphite*) 2.250	−4, +2, +4
7	N, nitrogen	14.0067	$N_2(g)$ 0.0011300	−3, −2, −1, +1, +2, +3, +4, +5
8	O, oxygen	15.9994	$O_2(g)$ 0.0012909	−2
9	F, fluorine	18.998403	$F_2(g)$ 0.0015328	−1
11	Na, sodium	22.98977	Na(s) 0.971	+1
12	Mg, magnesium	24.305	Mg(s) 1.740	+2
13	Al, aluminum	26.98154	Al(s) 2.702	+1, +3
14	Si, silicon	28.0855	Si(s) 2.329	−4, +4
15	P, phosphorus	30.97376	$P_4(s)$ 1.82	−3, +3, +5
16	S, sulfur	32.06	$S_8(s)$ 2.070	−2, +4, +6
17	Cl, chlorine	35.453	$Cl_2(g)$ 0.0028604	−1, +1, +3, +5, +7
19	K, potassium	39.0983	K(s) 0.862	+1
20	Ca, calcium	40.078	Ca(s) 1.54	+2
22	Ti, titanium	47.88	Ti(s) 5.54	+2, +3, +4
23	V, vanadium	50.9415	V(s) 6.11	+1, +2, +3, +4, +5
24	Cr, chromium	51.996	Cr(s) 7.19	−2, +1, +2, +3, +6
25	Mn, manganese	54.9380	Mn(s) 7.30	+1, +2, +3, +4, +5, +6, +7
26	Fe, iron	55.847	Fe(s) 7.87	−2, −1, +1, +2, +3, +4

Atomic number	Symbol & Name Nonmetals in red	Atomic mass	Standard state and density / kg L^{-1}		Oxidation numbers Rare values in orange
27	Co, cobalt	58.9332	Co(s)	8.90	+1, +2, +3, +4
28	Ni, nickel	58.69	Ni(s)	8.902	−2, −1, +2, +3, +4
29	Cu, copper	63.546	Cu(s)	8.96	+1, +2, +3
30	Zn, zinc	65.38	Zn(s)	7.134	+2
33	As, arsenic	74.9216	As(*grey*)	5.73	−3, +3, +5
35	Br, bromine	79.9904	Br$_2$(ℓ)	3.12	−1, +3, +5, +7
42	Mo, molybdenum	95.94	Mo(s)	10.2	−1,+1,+2,+3,+4,+5,+6
46	Pd, palladium	106.4	Pd(s)	12.02	+2, +4
47	Ag, silver	107.8682	Ag(s)	10.49	+1
48	Cd, cadmium	112.41	Cd(s)	8.65	+2
49	In, indium	114.76	In(s)	7.31	+1, +2, +3
50	Sn, tin	118.69	Sn(s)	7.31	+2, +3, +4
53	I, iodine	126.9045	I$_2$(s)	4.93	−1, +1, +3, +5, +7
58	Ce, cerium	140.12	Ce(s)	6.770	+3, +4
74	W, tungsten	183.85	W(s)	19.3	−1,+1,+2,+3,+4,+5,+6
78	Pt, platinum	195.08	Pt(s)	21.45	+1, +2, +3, +4
79	Au, gold	196.9665	Au(s)	19.32	+1, +2, +3
80	Hg, mercury	200.59	Hg(ℓ)	13.534	+1, +2
81	Tl, thallium	204.383	Tl(s)	11.85	+1, +3
82	Pb, lead	207.2	Pb(s)	11.35	+2, +4
92	U, uranium	238.0289	U(s)	18.95	+3, +4, +5, +6

Transport Properties: mostly of ions in water

Values of conductivities, mobilities and diffusivities are listed here for many ions "at infinite dilution" in water at 25°C. The three properties are mutually proportional, being connected by the relationships $RTu_i^\circ = z_i FD_i^\circ$ and $\lambda_i^\circ = z_i Fu_i^\circ$. Mostly, these relationships were used to populate the third and fourth columns from values of the ionic conductivities, measured from the extrapolation to zero concentration of the conductances of very dilute aqueous solutions. The final column contains a few diffusivity values, measured voltammetrically in the stated media. Note that these differ significantly from the infinite dilution values and depend on the identity of the supporting electrolyte and its concentration.

species i	$\dfrac{10^3 \times \lambda_i^\circ}{\text{S m}^2\,\text{mol}^{-1}}$	$\dfrac{10^9 \times u_i^\circ}{\text{m}^2\,\text{V}^{-1}\,\text{s}^{-1}}$	$\dfrac{10^9 \times D_i^\circ}{\text{m}^2\,\text{s}^{-1}}$	$\dfrac{10^9 \times D_i}{\text{m}^2\,\text{s}^{-1}}$
$H_3O^+(aq)$	34.96	362.3	9.31	(Grothuss mechanism)
$OH^-(aq)$	19.8	-205.2	5.27	(Grothuss mechanism)
$H_2O\,(\ell)$	0	0	2.44	(self diffusion)
$O_2(aq)$	0	0	2.26	
$Br^-(aq)$	7.81	-80.9	2.080	
$Rb^+(aq)$	7.78	80.6	2.072	
$Cs^+(aq)$	7.72	80.0	2.056	
$I^-(aq)$	7.68	-79.6	2.045	
$Cl^-(aq)$	7.631	-79.09	2.032	
$Tl^+(aq)$	7.47	77.4	1.989	
$NH_4^+(aq)$	7.35	76.2	1.96	
$K^+(aq)$	7.348	76.16	1.957	
$NO_3^-(aq)$	7.142	-74.02	1.902	
$Zn(amal)$	0	0	1.89	(Zn metal in mercury)
$ClO_4^-(aq)$	6.73	-69.8	1.792	
$Cd(amal)$	0	0	1.66	(Cd metal in mercury)
$Ag^+(aq)$	6.19	64.16	1.648	
$F^-(aq)$	5.54	-57.4	1.475	
$Na^+(aq)$	5.008	51.90	1.344	

species i	$\dfrac{10^3 \times \lambda_i^\circ}{S\ m^2\ mol^{-1}}$	$\dfrac{10^9 \times u_i^\circ}{m^2\ V^{-1}\ s^{-1}}$	$\dfrac{10^9 \times D_i^\circ}{m^2\ s^{-1}}$	$\dfrac{10^9 \times D_i}{m^2\ s^{-1}}$
$(CH_3)_4N^+(aq)$	4.49	46.5	1.196	
$CH_3COO^-(aq)$	4.09	-42.4	1.089	
$IO_3^-(aq)$	4.05	-42.0	1.078	1.015 in 100mM KCl
$SO_4^{2-}(aq)$	16.00	-82.91	1.065	
$Li^+(aq)$	3.866	40.07	1.029	
$Fe(CN)_6^{3-}(aq)$	32.07	-110.8	0.949	
$Pb^{2+}(aq)$	14.2	73.6	0.945	0.828 in 100 mM KNO_3 0.715 in 100 mM KCl 0.681 in 1000 mM KCl
$C_6H_5COO^-(aq)$	3.24	-33.6	0.863	
$Ba^{2+}(aq)$	12.72	65.92	0.847	
$Ca^{2+}(aq)$	11.89	61.62	0.792	
$Fe(CN)_6^{4-}(aq)$	44.16	-114.4	0.735	0.650 in 100 mM KCl
$Fe^{2+}(aq)$	10.8	56.0	0.719	
$Cd^{2+}(aq)$	10.8	56.0	0.719	0.715 in 100 mM KCl
$Cu^{2+}(aq)$	10.72	55.55	0.714	
$Mg^{2+}(aq)$	10.60	54.9	0.706	
$Zn^{2+}(aq)$	10.56	54.72	0.703	0.654 in 100 mM NaOH 0.638 in 100 mM KNO_3 0.620 in 1000 mM KNO_3
$PO_4^{3-}(aq)$	20.7	-71.5	0.613	
$Fe^{3+}(aq)$	20.4	70.5	0.604	
$(C_4H_9)_4N^+(aq)$	1.95	20.2	0.519	

Standard Gibbs Energies: key to calculating ΔE^o and E^o values

Listed here are the standard Gibbs energies of assorted neutral species and aqueous ionic species. More exactly, each G^o value is the molar Gibbs energy of formation of the cited species, in its standard state, from its elements, also in their standard states, all at the standard temperature $T^o = 298.15$K and standard pressure $p^o = 1$ bar $= 100$ kPa. For aqueous ions, the convention is adopted that the hydronium ion $H_3O^+(aq)$ and the water molecule $H_2O(\ell)$ share the same standard Gibbs energy. Because all elements in their standard states have zero values of G^o they are omitted from the table. The table is mostly compiled from the voluminous (but atrociously arranged) articles "Standard Thermo-dynamic Properties of Chemical Substances" and "Thermodynamic Properties of Aqueous Ions" in D.R. Lide (Ed.), *CRC Handbook of Chemistry and Physics, 89th Edition (Internet Version 2009)*, Taylor & Francis. All manner of equilibrium physicochemical properties (equilibrium constants, solubilities, vapor pressures, reaction feasibilities, equilibrium compositions, etc), as well as electrochemical data, are calculable from G^o values. Recall that a G^o value applies to a species only in its standard state. Whenever its state is nonstandard, the operative molar Gibbs energy is $G = G^o + RT\ln\{a\}$.

The change in standard Gibbs energy accompanying the reaction

A:1
$$v_A A + v_B B + \cdots \rightarrow v_Z Z + v_Y Y + \cdots$$

is

A:2
$$\Delta G^o = v_Z G_Z^o + v_Y G_Y^o + \cdots - v_A G_A^o - v_B G_B^o - \cdots$$

The standard null voltage ΔE^o of a cell at which reaction A:1 could occur is $-\Delta G^o/nF$, where the positive integer n is determined by the procedure described on pages 58–59.

To find the standard potential of an electrode operating in aqueous (*only!*) solution, write the electrode reaction as

A:3
$$v_A A + v_B B + \cdots + ne^- \rightarrow v_Z Z + v_Y Y + \cdots$$

where n is positive, then calculate $E^o = -\Delta G^o/nF$, where ΔG^o is given by equation A:2, ignoring the electrons. This will not always generate exactly the same value as from tables of electrode potentials; discrepancies on the order of one millivolt are common. Divergence arises from experimental error and from compromises made in correlating thermodynamic data from diverse sources.

Neutrals		Cations		Anions	
i	$\dfrac{G_i^\circ}{\text{kJ mol}^{-1}}$	i	$\dfrac{G_i^\circ}{\text{kJ mol}^{-1}}$	i	$\dfrac{G_i^\circ}{\text{kJ mol}^{-1}}$
$AgBr(s)$	-95.92	$Ag^+(aq)$	77.08	$Br^-(aq)$	-102.76
$AgCl(s)$	-109.59	$Al^{3+}(aq)$	-485.0	$Cl^-(aq)$	-131.2
$Ag_2O(s)$	-11.2	$Ca^{2+}(aq)$	-533.6	$ClO_3^-(aq)$	-8.0
$Ag_2S(s)$	-40.7	$CaOH^+(aq)$	-718.4	$ClO_4^-(aq)$	-8.5
$Al(OH)_3(s)$	-1140.7	$Cu^{2+}(aq)$	65.6	$CO_3^{2-}(aq)$	-527.82
$Al_2O_3(s)$	-1582.3	$Fe^{2+}(aq)$	-78.9	$Fe(CN)_6^{3-}(aq)$	729.4
$C(diamond)$	2.9	$Fe^{3+}(aq)$	-4.5	$Fe(CN)_6^{4-}(aq)$	695.1
$CO_2(g)$	-394.38	$FeOH^+(aq)$	-277.4	$HCO_3^-(aq)$	-586.8
$CO_2(aq)$	-186.2	$FeOH^{2+}(aq)$	-229.4	$HSO_4^-(aq)$	-755.9
$Fe_2O_3(s)$	-742.2	$Fe(OH)_2^+(aq)$	-437.0	$I^-(aq)$	-51.65
$HCl(g)$	-95.3	$H_3O^+(aq)$	-237.14	$I_3^-(aq)$	-51.48
$H_2O(\ell)$	-237.14	$In^{3+}(aq)$	-97.95	$IO_3^{2-}(aq)$	-128.0
$H_2O(g)$	-228.58	$Mg^{2+}(aq)$	-454.8	$MnO_4^-(aq)$	-447.2
$Hg_2Br_2(s)$	-181.1	$Mn^{2+}(aq)$	-228.1	$MnO_4^{2-}(aq)$	-500.7
$Hg_2Cl_2(s)$	-210.7	$NH_4^+(aq)$	-79.3	$OH^-(aq)$	-157.2
$HgCl_2(s)$	-178.6	$Ni^{2+}(aq)$	-46.4	$PtCl_4^{2-}(aq)$	-384.5
$NH_3(g)$	-16.4	$Pb^{2+}(aq)$	-24.3	$S^{2-}(aq)$	$+85.8$
$PbCl_2(s)$	-313.94	$Tl^+(aq)$	-32.4	$SO_4^{2-}(aq)$	-744.5
$PbO_2(s)$	-217.3	$Tl^{3+}(aq)$	214.6	$VO_3^-(aq)$	-783.6
$PbSO_4(s)$	-813.0	$Zn^{2+}(aq)$	-147.2	$VO_4^{3-}(aq)$	-899.0
$ZnO(s)$	-320.5	$ZnOH^+(aq)$	-330.1	$ZnO_2^{2-}(aq)$	-384.4

Standard Electrode Potentials: some examples

The writing of an electrode reaction and its standard potential is a way of summarizing thermodynamic information about the reaction. It does not imply that the reaction will actually occur. For example, the first and third of the reactions below have never been observed at any electrode. The standard electrode potential does not change when the reaction is written in the reverse direction, nor is it affected if all the stoichiometric coefficients are doubled or multiplied by any other number.

$$MnO_4^-(aq) + 8H_3O^+(aq) + 5e^- \rightleftarrows Mn^{2+}(aq) + 12H_2O(\ell) \qquad E^\circ = +1.512 \text{ V}$$

$$Cl_2(g) + 2e^- \rightleftarrows 2Cl^-(aq) \qquad E^\circ = +1.3578$$

$$Cr_2O_7^{2-}(aq) + 14H_3O^+(aq) + 6e^- \rightleftarrows 2Cr^{3+}(aq) + 21H_2O(\ell) \qquad E^\circ = +1.33 \text{ V}$$

$$O_2(g) + 4H_3O^+(aq) + 4e^- \rightleftarrows 6H_2O(\ell) \qquad E^\circ = +1.2288 \text{ V}$$

$$Ag^+(aq) + e^- \rightleftarrows Ag(s) \qquad E^\circ = +0.7989 \text{ V}$$

$$Hg_2^{2+}(aq) + 2e^- \rightleftarrows 2Hg(\ell) \qquad E^\circ = +0.7958 \text{ V}$$

$$Fe^{3+}(aq) + e^- \rightleftarrows Fe^{2+}(aq) \qquad E^\circ = +0.771 \text{ V}$$

$$O_2(g) + 2H_3O^+(aq) + 2e^- \rightleftarrows H_2O_2(aq) + 2H_2O(\ell) \qquad E^\circ = +0.6946 \text{ V}$$

$$I_3^-(aq) + 2e^- \rightleftarrows 3I^-(aq) \qquad E^\circ = +0.5362 \text{ V}$$

$$Fe(CN)_6^{3-}(aq) + e^- \rightleftarrows Fe(CN)_6^{4-}(aq) \qquad E^\circ = +0.355 \text{ V}$$

$$Ag_2O(s) + H_2O(\ell) + 2e^- \rightleftarrows 2Ag(s) + 2OH^-(aq) \qquad E^\circ = +0.3428 \text{ V}$$

$$Cu^{2+}(aq) + 2e^- \rightleftarrows Cu(s) \qquad E^\circ = +0.340 \text{ V}$$

$$Hg_2Cl_2(s) + 2e^- \rightleftarrows 2Hg(\ell) + 2Cl^-(aq) \qquad E^\circ = +0.2680 \text{ V}$$

$$AgCl(s) + 2e^- \rightleftarrows Ag(s) + Cl^-(aq) \qquad E^\circ = +0.22216 \text{ V}$$

$$2H_3O^+(aq) + 2e^- \rightleftarrows H_2(g) + 2H_2O(\ell) \qquad E^\circ = \;\; 0 \;\; \text{definition}$$

$$Pb^{2+}(aq) + 2e^- \rightleftarrows Pb(s) \qquad E^\circ = -0.1207 \text{ V}$$

$$V^{3+}(aq) + e^- \rightleftarrows V^{2+}(aq) \qquad E^\circ = -0.255 \text{ V}$$

$$Zn^{2+}(aq) + 2e^- \rightleftarrows Zn(s) \qquad E^\circ = -0.7628 \text{ V}$$

$$2H_2O(\ell) + 2e^- \rightleftarrows H_2(g) + 2OH^-(aq) \qquad E^\circ = -0.8280 \text{ V}$$

Index

An entry prefixed by # refers to a footnote and/or a Web. Other entries reference pages.